Tools of American Mathematics Teaching, 1800–2000

Johns Hopkins Studies in the History of Mathematics
Ronald Calinger, Series Editor

Tools of American Mathematics Teaching, 1800–2000

PEGGY ALDRICH KIDWELL
AMY ACKERBERG-HASTINGS
DAVID LINDSAY ROBERTS

Smithsonian Institution
WASHINGTON, D.C.
The Johns Hopkins University Press
BALTIMORE

Printed in the United States of America on acid-free paper
2 4 6 8 9 7 5 3 1

The Johns Hopkins University Press
2715 North Charles Street
Baltimore, Maryland 21218-4363
www.press.jhu.edu

Library of Congress Cataloging-in-Publication Data

Kidwell, Peggy Aldrich.
Tools of American mathematics teaching, 1800–2000 /
Peggy Aldrich Kidwell, Amy Ackerberg-Hastings, David Lindsay Roberts.
p. cm. — (Johns Hopkins studies in the history of mathematics)
Includes bibliographical references and index.
ISBN-13: 978-0-8018-8814-4 (hardcover : alk. paper)
ISBN-10: 0-8018-8814-X (hardcover : alk. paper)
1. Mathematical instruments—United States—History.
2. Mathematics—Study and teaching—United States—History.
I. Ackerberg-Hastings, Amy.
II. Roberts, David Lindsay. III. Title.
QA71.K46 2008
510.71′073—dc22 2007040065

A catalog record for this book is available from the British Library.

Special discounts are available for bulk purchases of this book.
For more information, please contact Special Sales at 410-516-6936
or specialsales@press.jhu.edu.

The Johns Hopkins University Press uses environmentally
friendly book materials, including recycled text paper that is composed
of at least 30 percent post-consumer waste, whenever possible. All of
our book papers are acid-free, and our jackets and covers are printed
on paper with recycled content.

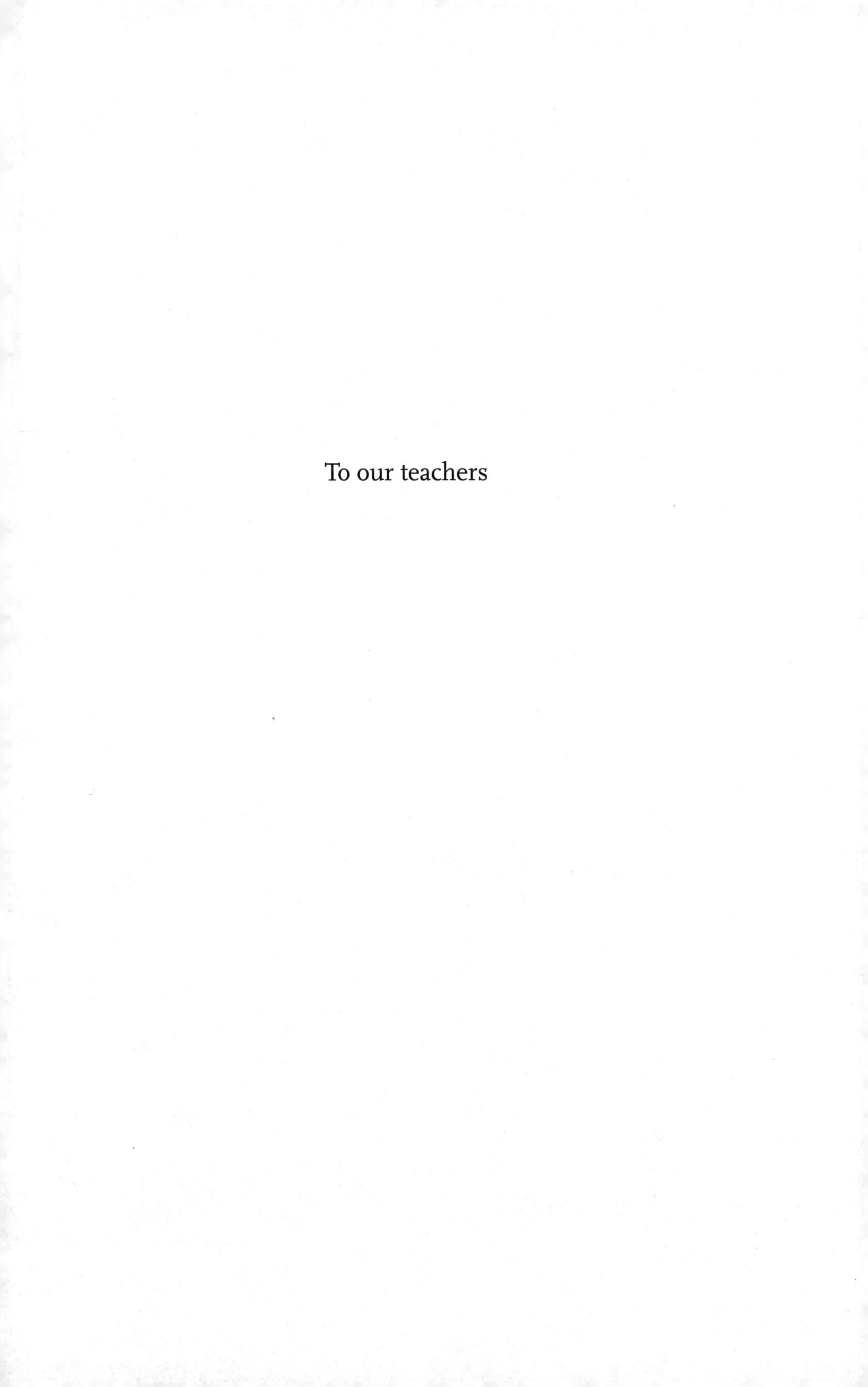

To our teachers

Contents

Acknowledgments

This book is grounded in the examination of objects and the perusal of texts: trade literature and archival collections, journals, newspapers, textbooks, monographs, and reference works. In locating and using these sources, we have been ably assisted by a host of curators, librarians, archivists, and other helpful people. It is a privilege to acknowledge their aid.

We have relied especially heavily on materials at the Smithsonian Institution's National Museum of American History (NMAH). We thank present and former colleagues at the museum, members of the staff of the Smithsonian Institution Libraries, and Smithsonian photographers. Numerous donors have done much to preserve diverse objects representing the nation's heritage in mathematics and education. A few of them are acknowledged individually in photograph captions; our debt includes many more. We also have learned much about objects from visits to and correspondence with such organizations as the Archives of American Psychology at the University of Akron; Dartmouth College; the Harvard Collection of Scientific Instruments; the Library of Congress; the Rockport Public Library in Rockport, Indiana; Stanford University Archives; Union College; the U.S. Military Academy at West Point; the University of Maryland College Park; the Westwood Historical Society in Westwood, Massachusetts; and the Working Men's Institute Library in New Harmony, Indiana.

For access to archival material, books, journals, trade literature, and photographs, we are grateful to numerous institutions—and have the copy cards to prove it. Particular thanks to American University, the American Philosophical Society, Boston College, the Boston Public Library, the Charles Babbage Institute, Columbia University (including Teachers College), the Connecticut Historical Society, Dartmouth College, the archives of Hewlett-Packard Corporation, Lafayette College, Lehigh University, the Library of

Congress, 3M Corporation, the Massachusetts Institute of Technology (MIT) Archives, the MIT Museum, Monroe Systems for Business, the Stephens Institute of Technology, the University of Chicago, the University of Illinois, the University of Iowa, the University of Texas, the U.S. Naval Academy, and Yale University.

This book is an outgrowth of long-standing research, including a temporary exhibit at the National Museum of American History. Some chapters have been presented in earlier form at meetings of the American Mathematical Society, the Mathematical Association of America, the History of Science Society, the Philadelphia Area Seminar on the History of Mathematics, and the Scientific Instrument Commission. They also have been given to groups at the museum, at American University, and at West Point. A few chapters are revised from earlier publications in *Historia Mathematica* and in *Rittenhouse*. The authors thank all of these audiences for their insightful questions and comments.

We also are indebted to numerous individuals who have shared their knowledge of and insight on specific topics. Many of them are associates of the repositories and organizations mentioned here or donors to the museum. We also thank Ira Chinoy, Jane Hess, Christopher Wells, Ron Calinger, Karen Hunger Parshall, an anonymous reviewer, and the able staff of the Johns Hopkins University Press.

Peggy Kidwell thanks Lorenzo Traldi and his colleagues in the Department of Mathematics at Lafayette College for providing space, access to classrooms, and library resources during a year-long sabbatical there and her Smithsonian colleagues, who approved this break from routine duties.

Finally, we acknowledge the patient assistance and unfailing goodwill of our families. Particular thanks go to our spouses, Mark Kidwell, Nelson Hastings, and Jenny Scott.

Introduction

Learning mathematics has been important to life in the United States from the earliest years of the republic. As one enthusiast for numbers put it in 1836, the early citizens of the United States were a "calculating people."[1] They knew and used the basic arithmetic needed for trade and commerce. Indeed, the application of mathematics was written into the U.S. Constitution, which requires a decennial census, regular counting of votes, and standard units of weight and measure. Knowledge of the elements of arithmetic, geometry, and trigonometry also was essential to navigation, surveying, engineering, and gunnery. Students preparing for the learned professions of theology, medicine, and law were expected to master the rudiments of Euclidean geometry, both to promote general mental discipline and as training in deductive thinking.

In the two centuries from 1800 to 2000 formal instruction in mathematics expanded in several ways. The population of the United States increased from about 5 million to over 280 million.[2] A much larger fraction of these citizens began to attend school, for considerably longer times. Urbanization and improved transportation made it easier for students to assemble. New institutions, complementing existing colleges, academies, and town schools, provided formal education at many levels. Particularly in the 1820s and 1830s, locally funded common schools, designed to provide free basic education for students of all economic classes, were established throughout most of the Northeast. In time they spread to the Mid-Atlantic states and, after the Civil War, throughout the country. Also beginning in the 1820s, experimental schools were established for young children of the poor in cities such as Philadelphia and Boston. Although these "infant schools" were not widespread, their apparatus was adapted for primary education. Infant schools also were forerunners of the kindergartens that were established after the Civil War as well as of later schools run on principles developed by the Italian educator Maria Montessori. At a more advanced level academies, high schools, and seminaries were founded to train teachers and prepare students for college. In the late nineteenth and especially the twentieth centuries such schools became far more popular. They provided not only general instruction but also vocational preparation for office workers and those entering various trades.

The mission of advanced educational institutions—and demand for mathematics teaching—also expanded. Beginning with the founding of the U.S. Military Academy at West Point, New York, in 1802, special technical institutes, modeled on the École Polytechnique in Paris, were established for training engineers. These schools proved particularly influential in establishing general curricula in mathematics. Formal training in technical subjects soon extended to private universities and, especially after passage of the Morrill Act of 1862, to state universities. The late nineteenth century saw the founding of the first graduate schools in the United States, encouraging both students and faculty to make contributions to mathematical research. Related programs in the physical, biological, and social sciences, combined with developments within mathematics, created a demand for instruction in mathematical subjects ranging from statistics to mathematical physics to real analysis. Here German ideals were particularly important.

The growth of mathematics education was by no means the only factor that encouraged the development of diverse forms of teaching apparatus. Many schools acquired objects to attract faculty, new pupils, and favorable publicity. Other influences came from the larger society. The changing economic infrastructure and growing technical expertise of the nation shaped the tools of mathematics teaching. A small, largely agricultural country became wealthier and more urban and began to manufacture uniform, inexpensive consumer goods, often using new materials. Cheap paper and pencil replaced slates and slate pencils, for example, and made possible both the widespread use of graph paper and standardized paper-and-pencil tests. Brass protractors gave way to instruments made from stainless steel, sheet metal, paper, and plastic. Plastic also came to be widely used in the manufacture of slide rules. New manufacturing techniques also enabled production of inexpensive wooden rulers and other demonstration apparatus, and better transportation permitted widespread sale of these educational products. In addition to these economic and technical changes, pedagogical theories encouraged the use of specific devices.

Patronage for education expanded with time. Churches and other charitable institutions established both colleges and schools for impoverished children, with suitable equipment. Local communities and states not only funded schools but encouraged the use of specific textbooks and apparatus. Wealthy entrepreneurs endowed universities, colleges, and technical institutes. From the late 1950s the federal government provided funds for the purchase of classroom teaching apparatus.

Professional organizations of mathematicians, mathematics teachers, and educators also encouraged schools to improve their facilities. The American Mathematical Society (AMS), which dates its origins to 1888, has focused much of its attention on problems of mathematical research. Yet its publications and its members have also made substantial contributions to establishing and changing classroom practice. In 1915 the Mathematical Association of America (MAA) formed as an organization separate from the AMS, particularly concerned with mathematics teaching. Its meetings, committees, and publications provided a rich forum for discussion of objects. By 1920 separate organizations for college and secondary school teachers seemed desirable. The National Council of Teachers of Mathematics (NCTM), an organization for the latter group, was founded that year. Its work plays a particularly prominent role in our narrative. Societies for schoolteachers and administrators such as the National Education Association and its departments also took part in the stories we present. Other organizations served the educational psychologists who developed standardized tests, those developing specific techniques such as overhead projection, and those using electronic technologies. Advocates of specific reforms, such as the adoption of the metric system, have also played a role. Others have described many of these institutions.[3] Our goal is to suggest how the efforts of these diverse organizations were translated into the common things of mathematical learning.

This complex interplay between ideas, individuals, institutions, and techniques is best understood by considering the stories of specific teaching tools. Because mathematics education is part of the past of millions of Americans, these narratives are written for a wide range of readers. To teachers, parents, and former students the accounts give a glimpse into the rich past that lies behind tools that they might take for granted. To historians of mathematics they suggest how broad developments in mathematics and the mathematical community were and were not reflected in material objects. To historians of education they show how pedagogical theories, individual innovations, and government policies played out within one discipline. To those portraying historic classrooms, they suggest what might—and what might not—be appropriate to include in period settings. To historians of mathematical instruments and computing, who have often focused on the invention of new devices, these tales warn that one must distinguish between the existence of a technology and its importance within education and society more generally. For those concerned about what they consider to be the slow pace of classroom adoption of recent electronic technologies, these essays offer historical perspective. At a time of international turmoil they also attest to rich and productive interchanges between the United States and other nations.

This volume has four broad topical parts. Each chapter within these parts discusses the history of a specific type of object, with the chapters arranged roughly chronologically according to the date devices were introduced. Our focus is on the introduction of new practices, with brief mention of tools that disappeared. While the enduring classroom role of some instruments is noteworthy, we rarely attempt to explore this persistence in detail.

The first part concerns general tools of pedagogy as they have been used to convey mathematical ideas. All of these tools have been used at several levels of mathematics teaching, from the elementary school to the college. The first two chapters consider the two most important objects to become common in antebellum American mathematics teaching, the inexpensive, uniform textbook and the blackboard. Both had European roots, and both became commercial products distributed throughout the United States. The third chapter discusses the introduction of another teaching instrument, the paper-and-pencil standardized test. In the early twentieth century experts in the new discipline of experimental psychology, many of whom were American, developed these examinations to evaluate school systems and then to measure student aptitude and achievement. Many early tests covered arithmetic, with a smaller number devoted to algebra, geometry, and more advanced mathematical topics. Tests would be widely used to sort students into different classes, to evaluate applicants to college and graduate schools, and to judge the performance of schools. The fourth chapter considers the history of the overhead projector. First used by a few lecturers in the nineteenth century, overhead projectors became commonplace as a result of military interest. During World War II the United States armed forces used a variety of teaching aids, including movies, filmstrips, and transparencies, to teach large numbers of trainees specific skills. Funding for such audiovisual aids continued after the war, and schools of education developed special departments devoted to instruction in their use. Military patronage encouraged improvements in the production of overhead projectors to the point that they became sturdy and inexpensive enough to be considered for ordinary classroom use. Federal funds made available to schools in the 1960s offered further incentives. As we show in chapter 5, military needs also prompted experts in educational technology to take a new look at teaching machines. Psychologists had proposed such machines as early as the 1920s. The suggestion that teachers might be aided—if not supplanted—by machines received serious attention in the late 1950s and 1960s. Both federal funding and the booming school population encouraged such thinking. Teaching machines, and related programmed textbooks, were prepared for instruction in everything from arithmetic to advanced algebra. They would greatly influence the early educational use of computers.

The second part of the book recounts the history of an array of instruments used especially to teach—or to carry out—arithmetic. Some of these instruments had a long history before they found their way into the classroom. The teaching abacus, for example, discussed in chapter 6, had its roots in a computing device that dates to antiquity. In the early nineteenth century special forms of the abacus were introduced in both France and England and from there spread to the United States. The "numeral frame," as it was called, was first used in schools for very young children and then spread to primary and grammar schools. Other objects were meant not so much to demonstrate principles of arithmetic as to substitute manual dexterity for mental effort. One such instrument was the slide rule, which is the subject of chapter 7. Americans first learned about slide rules outside the classroom, either from printed manuals or from other users. From the 1880s use of the instrument became popular among engineers, and it was introduced in a variety of courses in engineering schools. By the early twentieth century slide rule manipulation increasingly was taught as part of mathematics not only at the college level but in high schools and even junior highs.

Attempts to teach arithmetic to young children encouraged the development of another piece of demonstration apparatus, the cube root block. The use—and abandonment—of this device is the subject of chapter 8. Devices designed for direct use by children came to be popular after the Civil War. Several forms of blocks intended, at least in part, for teaching counting and arithmetic are considered in chapter 9. Some of these materials were associated with the nineteenth-century kindergarten movement, others with the educational program of Maria Montessori. More recently, authorities such as Catherine Stern, Georges Cuisenaire, and Charles Tacey have proposed systems of blocks designed exclusively for teaching arithmetic and counting.

In the third part of the book we consider devices used for measuring and representing mathematical structures. Some of these instruments were part of everyday work long before they entered the classroom. The protractor, for instance, was a tool of surveyors, draftsmen, and navigators invented in the sixteenth century. It became commonplace in American schools only with changes in geometry teaching that occurred in the late nineteenth century and which are described in chapter 10. Weights and measures had an even longer history in ordinary life and were often discussed in texts on commercial arithmetic. Several mid-nineteenth-century authors suggested that students would benefit from actually handling common measures, as discussed in chapter 11. It was only in the 1870s, however, that American advocates of the metric system began to design and sell inexpensive demonstration weights and measures. Even when most American schools chose not to adopt metric units, inexpensive rulers and other demonstration apparatus remained in the classroom.

Chapter 12 describes three-dimensional models of plane figures, solids, surfaces, and other structures that mathematicians, mathematics teachers, and students designed and built. In the early nineteenth century, just as the abacus was used to introduce arithmetic and counting to young children, models of simple forms were prepared to introduce basic geometry. Later in the century more complex models were purchased or built to teach engineers descriptive geometry and to introduce formulas relating to the area and volumes of solids. Universities beginning graduate programs in mathematics often imported more complex models from Europe, especially Germany. A few Americans also began making models. The mathematical content of these models intrigues some scholars to this day. Their beauty has attracted, and continues to attract, the attention of artists and hobbyists. It is more difficult to judge how they have functioned as pedagogical tools. Examination of the extensive collections of models made by Richard P. Baker of the University of Iowa and by A. Harry Wheeler of Massachusetts suggests the complexity of this judgment.

Other ways of representing mathematical concepts are less widely discussed than models but more generally used. For most of the twentieth century, for example, graph paper was widely used in courses on algebra and more advanced mathematics. As chapter 13 demonstrates, this practice was closely associated with the pedagogical ideas of the University of Chicago mathematician E. H. Moore and had wide influence not only in mathematics but in engineering and the sciences. The use of graphical methods also was tied to instruction in another computational tool, the nomogram. Also of interest to both engineers and mathematicians, though not as generally known, were kinematic models of linkages. Chapter 14 examines their introduction and use.

The final part of the book concerns the use of electronic technologies in mathematics teaching. Because of the numerous possible uses of programmable devices, this is a large subject, which we tie to earlier developments. Chapter 15 traces the pedagogical application of calculators from the mechanical adding machine to the handheld electronic calculator. From the 1890s American scientists and businessmen increasingly had used machines to do routine arithmetic. Such machines were expensive and not nearly as easy to transport as the slide rule. Nonetheless, from the 1950s a few American educators experimented with using adding machines as aids to teaching arithmetic. In the 1970s inexpensive handheld electronic calculators displaced the slide rule as tools of technical practice. How much these instruments should be used for arithmetic teaching was and remained a matter of debate.

We then turn to the use of computers in mathematics teaching. Early computers were bulky, temperamental, and extremely expensive. Nevertheless, experiments with using them for mathematics instruction began in the early 1960s, with the advent of minicomputers and computer time-sharing. Some built on the idea of the teaching machine, using computers as tutors and drillmasters. Others believed that students would gain understanding, particularly of disciplines such as mathematics, if they learned to program computers themselves. Others considered the possibility of using computers more like tools and developed programs for graphing and symbolic manipulation. The first educational computer games also date from this period. These efforts are the subject of chapter 16.

With the advent of microcomputers in the late 1970s, computers became accessible to millions of Americans. Chapter 17 describes early applications of microcomputers in mathematics education, particularly as they are represented in the collections of the National Museum of American History of the Smithsonian Institution. Microcomputers, like minicomputers, were used for drill, for programming, for graphing, for symbolic manipulation, and for games. An enormous range of users tried out the new tool. Some wrote computer software for their own purposes, some shared programs at cost, and others developed successful commercial products. The late 1980s and 1990s saw the introduction of both sophisticated mathematical software systems and graphing calculators. The final chapter considers the early history of a few of these tools. The ready availability and mathematical prowess of these instruments may fundamentally change assumptions about what should be taught in the mathematics curriculum. This topic, however, will be grist for the mill of future historians.

Our stories amply demonstrate the rich heritage represented by the material culture of mathematics teaching. At the same time, they are somewhat apart from usual approaches to the history of mathematics. More conventional attitudes are suggested by a 1939 interchange in a distinguished journal of the history of science, *Isis*. That year the founding editor of the publication, Harvard University professor George Sarton, received an inquiry from a reader concerning the origins of graph paper. The reader asked when logarithmic paper was first used and who introduced it. Sarton himself provided one reply, which read in part: "The question you are interested in, the history of log paper, does not really concern the history of mathematics. No new principle is involved, it is simply a matter of convenience. Hence it is not surprising that even the most elaborate histories of mathematics . . . do not refer to it."[4] He went on to cite several early sources that he had found relating to the topic as well as recent information obtained from a vendor of the paper. This book is about precisely the "matters of convenience" that Sarton thought were not a concern of our discipline. Moreover, we focus not on the first uses of various objects by scientists and mathematicians but on their adoption into the American mathematics classroom. "Matters of convenience," like graph paper and blackboards, tell us much about the ways mathematical ideas have diffused into a wider culture. We hope readers will not only enjoy learning more about this part of the American past but will join in the effort to preserve and to understand it.

PART ONE

Tools of Presentation and General Pedagogy

CHAPTER ONE

Textbooks
Creating a National Standard

At the beginning of this [nineteenth] century the great want of this country in the department of pure mathematics was adequate text-books.

He has incorporated in his text-books from time to time the advantages of every improvement in methods of teaching, and every advance in science. During all the years in which he has been laboring, he constantly submitted his own theories and those of others to the practical test of the class-room . . . Davies' System is the acknowledged National Standard for the United States.

Even in the age of the Internet, most American mathematics classes are organized around a printed textbook. Teachers and students alike assume that each member of a class will have the same book. Moreover, this book is written for a specific age group and academic level, meeting criteria set by state adoption committees and professional organizations. The ubiquity of uniform, graded mathematics textbooks obscures their relatively brief classroom history. It was only in the mid-nineteenth century that the availability of inexpensive printed books, the conscious efforts by authors to prepare materials on specific subjects for specific kinds of schools, and the marketing of publishers combined forces. The result was the textbook series—a group of books that covered topics ranging from arithmetic through the calculus, designed for classrooms from primary school through college.

Historians, mathematicians, and educators have written about textbooks at length. Since at least the 1860s they have compiled catalogs of these publications: recording titles, authors, and changes in editions.[1] They also have explored patterns in the content and transmission of texts. Florian Cajori, in his pioneering work *The Teaching and History of Mathematics in the United*

States (1890), argued, for example, that professors in American colleges first adopted British, synthetic textbooks before turning to French, analytical approaches around 1820. Scholars have not only revisited this model but have also traced the influence of later ideas from Germany and elsewhere.[2] Others have examined such topics as the cultural and mathematical content of word problems in arithmetic and algebra textbooks for children.[3] Still others have considered textbooks as specific instructors and institutions used them.[4]

This substantial literature reveals that certain volumes were far more popular than others. At least forty-two geometry textbooks were published in the United States between 1800 and 1850, for example, but just thirteen of them were printed at least five times, indicating long-term and widespread use. Certain names gained recognition that persists today, as readers may find familiar authors such as Nathan Daboll, Charles Davies, Jeremiah Day, William Granville, Benjamin Peirce, Joseph Ray, and George Wentworth or publishers such as John Wiley and Sons, Harper & Brothers, or A. S. Barnes & Company. Textbooks produced by these men successfully treated standard subjects in a manner acceptable to the intended audience. They also achieved staying power because their creators saw textbooks as classroom commodities that would boost their personal fortunes and academic reputations.

This chapter examines how Americans consciously shaped textbooks into mathematical teaching apparatus. It first suggests when and where it became necessary for each mathematics student to possess a textbook. These books were initially imported or reprinted European textbooks that offered a compendium of useful mathematics. Gradually, materials came to be compiled and created in the United States, while compendia were replaced by subject-specific textbooks within series. Both the expansion of education at all levels that occurred in the nineteenth century and the development of mass publication techniques shaped this process. This transformation of the textbook into a common tool and lucrative product is particularly well illustrated by the work of the partnership between the author Charles Davies and the publisher A. S. Barnes. Combining their interest in graded series and considerable marketing skills, Davies and Barnes prepared a course of mathematics books comprehensive both in subject and age range. In a country where educational decisions were made at disparate local levels, this series set a de facto national standard for mathematics textbooks.

Teaching and Textbooks before 1815

In the seventeenth and eighteenth centuries European mathematical treatises that were accessible to learners evolved into textbooks.[5] Like Euclid's *Elements of Geometry,* upon which they were modeled in structure and style, most works used in classrooms initially were written for practitioners and brought into the classroom unchanged. Even as authors such as Alexis Clairaut or John Bonnycastle began to simplify their language for student audiences, they continued to write in the Euclidean format. Because books were relatively expensive, only the professor, tutor, or teacher generally owned the textbook used for a class. Students would write out specified sections in cipher- or copybooks (Figure 1.1).[6] Besides those men employed in education, only mechanics and others pursuing self-improvement purchased personal copies of manuals and textbooks for individual study.

After about 1750, American college presidents required each student to purchase a textbook covering a range of mathematical subjects. Such compendia were commonly used in Great Britain. Thanks to a thriving Atlantic trade, the cost of such books had declined, making them affordable to students. Meanwhile, increasing yearly class sizes meant that presidents needed accessible learning materials so that the tutors, recent graduates who led mathematics classes, could manage their students, restless middle- and upper-class adolescent boys. Moreover, college studies had expanded from three years to four. Mathematics was taught throughout the uniform course of study, instead of only in the final year.

Those teaching mathematics emphasized that its study had two distinct benefits. First, its practical applications in commerce, surveying, navigation, and architecture made it a cornerstone of learning. Second, American educators believed that mathematics promoted general mental discipline. Learning and memorizing mathematics, they argued, developed the worthwhile habits of fixed attention, abstraction, and proper reasoning. To help train the mind, tutors at colonial colleges began to employ the recitation method of teaching. They regularly assigned sections of the text for students to memorize and recite aloud in class.

Recitation proved to be an efficient way to teach large, diverse, and sometimes rowdy classes. After independence, the method diffused into grammar schools, academies, and the increasing number of common (essentially public elementary) schools. These institutions were funded by individuals,

Figure 1.1. Page from Jesse Harmon Alexander's Mathematical Exercise Book, 1825. National Museum of American History collections, gift of David Challinor. Smithsonian Negative no. 97-252.

by local towns or through tuition. From early in the nineteenth century students' oral presentations took place at a blackboard, where they reproduced the diagram for a proof or worked the steps of a problem, as we will see in chapter 2. Nevertheless, Americans were only beginning to realize the potential that textbooks held as teaching tools for bringing structure and uniformity to classrooms. Especially in the common schools, pupils brought the arithmetics that their families already had at home, so that in one room there might be as many different textbooks as children.

First Steps: From Compendia to Series

A nascent American mathematics textbook industry both responded to and created these demands for its products. The first American authors to achieve wide success were Nicholas Pike (*The New and Complete System of Arithmetic,* 1788) and Nathan Daboll (*The Schoolmaster's Assistant,* 1799)

(Figure 1.2). Their books were compendia, covering not only simple arithmetic but also fractions, units of measurement, and simple volumes. They were notable for promoting the American system of currency and for paving the way for a flood of arithmetics by American authors in the nineteenth century. On the college level Samuel Webber of Harvard assembled a compendium (*Mathematics, Compiled from the Best Authors,* 2 vols., 1801), which included algebra, geometry, trigonometry and its applications to heights and distances and navigation, mensuration of surfaces and solids, conic sec-

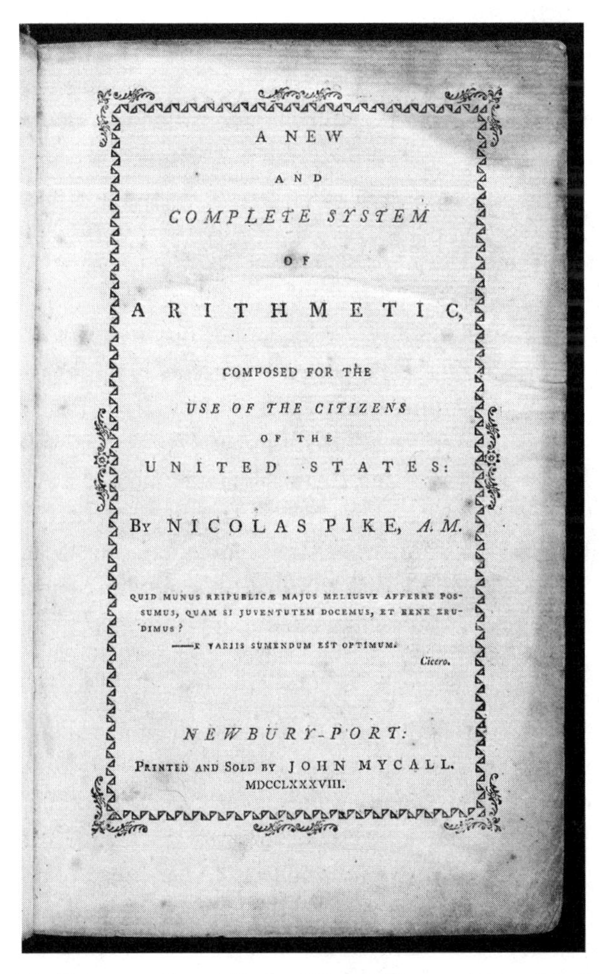

A NEW
AND
COMPLETE SYSTEM
OF
ARITHMETIC,
COMPOSED FOR THE
USE OF THE CITIZENS
OF THE
UNITED STATES:
BY NICOLAS PIKE, *A. M.*

QUID MUNUS REIPUBLICÆ MAJUS MELIUSVE AFFERRE POSSUMUS, QUAM SI JUVENTUTEM DOCEMUS, ET BENE ERUDIMUS?
——E VARIIS SUMENDUM EST OPTIMUM.
Cicero.

NEWBURY-PORT:
PRINTED AND SOLD BY JOHN MYCALL.
MDCCLXXXVIII.

Figure 1.2. Title page from Nicolas Pike, *A New and Complete System of Arithmetic,* 1788. Smithsonian Negative no. 2006-0205.

tions, and spherical geometry. Like Pike's and Daboll's books, *Mathematics* was published by a local printer. Imported copies of Euclid's *Elements of Geometry* were already being used at Harvard and Yale alongside the compendia when American reprints of the classic text began to appear in Philadelphia in 1803 and 1806. Thus, geometry was the first subject to be taught with a separate textbook.

These American editions of Euclid were pirated from translations prepared by Scottish mathematicians Robert Simson and John Playfair. By the end of the War of 1812 some professors had grown dissatisfied with their reliance on such reprints as well as on books purchased overseas. Blockades had disrupted both imports from abroad and international intellectual exchanges. At the same time, the United States was entering a second round of nation building. Both the overall population and the proportional size of the middle class of the United States grew, creating a greater demand for education. Literacy rates of around 85 percent for adult white men and 50 percent for white women were typical. The diversity of nineteenth-century locales for instruction was maintained throughout expansions in demographics and national identity. Colleges, for example, offered a uniform course of study from geometry and algebra through astronomy and sometimes fluxions for elite young men. Grammar schools taught arithmetic and algebra, along with the Greek and Latin classics, for college preparation. Common schools offered "reading, 'riting, and 'rithmetic" to middle- and working-class boys and girls. Finally, adult craftsmen and mechanics often engaged in self-study. All of these types of institutions were concentrated most heavily in New England, where laws requiring towns to fund common schools dated to the seventeenth century. Colleges and schools were also rapidly founded through land grants in the Old Northwest in the early nineteenth century.[7]

Near the center of these transformations, Jeremiah Day of Yale College was the first to try to modernize the uniform, mandatory college curriculum by preparing a series of separate mathematics textbooks for each course. Between 1814 and 1817 he produced four volumes on algebra, trigonometry, the mensuration of surfaces and solids, and navigation and surveying. Day drew materials from a variety of European and American mathematicians, including John Bonnycastle, Nathaniel Bowditch, Leonhard Euler, Silvestre-François Lacroix, and Robert Woodhouse. As was standard for mathematics textbooks, Day's series presented the process for solving each type of situation but did not include large sets of problems that students could use for practice. Before these textbooks were adopted, Yale catalogs listed only the subjects in which students were to be instructed. Afterward the Yale course was publicized in terms of the books studied: "Day's Algebra," "Playfair's Euclid," "Day's Mathematics, Parts II and III" (the volumes on trigonometry and mensuration), and the like.[8] This practice was adopted at other colleges. As these institutions also gradually began to require entrance examinations in arithmetic, algebra, and geometry, prospective students were informed of the books they needed to master in preparing for the tests.

Day made much of his personal fortune from these textbooks, as New Haven printer Oliver Steele continued to issue them until 1858. Yet Day took little interest in which institutions adopted his series. In 1817 he was elevated from mathematics professor to president at Yale. His expanded duties, ranging from student discipline to restoring the financial solvency of the college after its bank lost the school's investments in Ohio, left Day no time to revise the textbooks. Indeed, dozens of academies and grammar schools adopted *An Introduction to Algebra,* even though it was presented as a college textbook. Younger students thus purchased a significant portion of the nearly ninety thousand copies made of this work.

Although a market across educational levels had been shown to exist, American mathematics textbook authors who followed Day tended to focus either on college audiences or on beginning learners. Between 1818 and 1831, for example, John Farrar of Harvard translated thirteen textbooks, mainly by Lacroix and Jean-Baptiste Biot, which were sold as the Cambridge Series of Mathematics and Natural Philosophy. Although the content of the series no longer represented the most advanced knowledge, the material was accessible only to college students with a solid base of instruction. These books were published by independent bookseller and former Harvard College printer William Hilliard. Along with his second wife's etiquette manuals and stories for children, the Cambridge Series provided financial security to Farrar. Yet even after decades of incapacitating illness, Farrar defined himself only as a college professor—never as an author. Similarly, Warren Colburn, who helped introduce Pestalozzian ideals into American mathematics teaching through his 1821 *Intellectual Arithmetic,* spent most of his working life as a factory supervisor in Waltham and Lowell, Massachusetts. He also wrote for young boys and girls, instead of for adolescent men in college.[9]

Becoming a Mathematics Textbook Author

By the 1820s some local printers were building their firms into the first full-fledged American publishing houses. These men saw lucrative possibilities in taking a more systematic interest in schoolbooks and college tomes. John Wiley and James and John Harper, both of New York City, concentrated, for example, on technical and scientific textbooks when developing their book-lists.[10] At least one mathematics professor stood ready to oblige them.

Charles Davies had completed an abbreviated course of study at the United States Military Academy in West Point, New York, during the War of 1812.[11] He returned in late 1816 to serve as assistant professor of mathematics and then, still only nineteen years old, embraced the educational reforms introduced by Sylvanus Thayer, who became superintendent of the academy in 1817. Thayer regularized the academic calendar and examinations, eliminating the previous ability of cadets to enter and graduate at will; encouraged the use of French-language textbooks, which he saw as the cutting edge in engineering education; and instituted a numerical daily grading scale for classroom recitations.[12] (A similar assessment system was already in use at Yale.) In 1823, after the cadets had struggled to master French and mathematics simultaneously, West Point professors turned to Farrar's Cambridge Series. These books, however, were regarded as being out-of-date. They also did not cover the additional subjects in West Point's curriculum: descriptive geometry, analytical geometry, and perspective drawing.

This situation presented an opportunity to Davies, whose career was already on the rise. He succeeded David Douglass as the lead professor of mathematics in 1823, and he was courting the daughter of natural and experimental philosophy professor Jared Mansfield. Yet Davies hungered for more. He had supplemented his assistant professor's salary, ten dollars per month in addition to a room and board stipend, with a second job at a woolen manufacturer a few miles north of West Point. This led his future mother-in-law to describe him as a "man of business" who was characterized by "exertion and an obliging disposition."[13] Davies also studied law and was admitted to the New York bar in 1828, although he apparently only litigated one case. Keith Hoskin has argued persuasively that Davies's West Point textbooks show how he followed Thayer's philosophy of adapting mental discipline to serve technical education and to prepare students for the daily routines of civil engineers and businessmen.[14] It is clear, however,

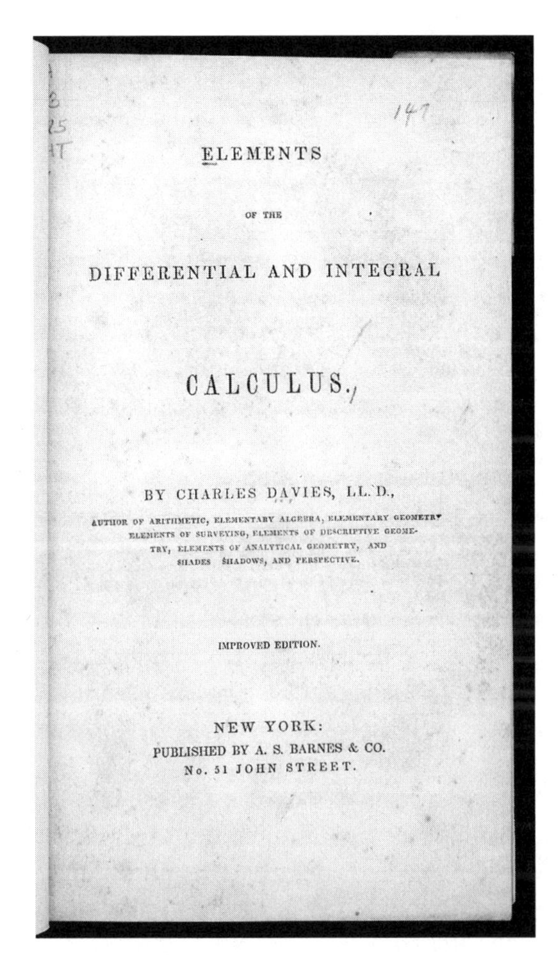

ELEMENTS

OF THE

DIFFERENTIAL AND INTEGRAL

CALCULUS,

BY CHARLES DAVIES, LL.D.,

AUTHOR OF ARITHMETIC, ELEMENTARY ALGEBRA, ELEMENTARY GEOMETRY ELEMENTS OF SURVEYING, ELEMENTS OF DESCRIPTIVE GEOMETRY, ELEMENTS OF ANALYTICAL GEOMETRY, AND SHADES SHADOWS, AND PERSPECTIVE.

IMPROVED EDITION.

NEW YORK:
PUBLISHED BY A. S. BARNES & CO.
No. 51 JOHN STREET.

Figure 1.3. Title page from Charles Davies, *Elements of the Differential and Integral Calculus,* 1836. Smithsonian Negative no. 2006-0206.

that personal and financial ambition were also powerful motivators for Davies's unprecedented burst of activity.

What, then, did Davies do? Between 1826 and 1876 he compiled nearly fifty separate mathematics textbooks, ranging from *Mental and Practical Arithmetic* (1838) for very young children to *Elements of the Differential and Integral Calculus* (1836) for college seniors and *The Logic and Utility of Mathematics* (1850) for secondary school mathematics teachers (Figure 1.3).[15] Davies was both exceptionally prolific and popular as a writer. By 1866 more than five million copies of his books had been printed and sold in the United States.[16] Although other authors, such as Joseph Ray, sold larger numbers of a single textbook, no one prepared a range of mathematics textbooks for common schools, academies, and colleges or marketed to a national audience earlier than Davies.[17] Unlike any other previous or contemporary author, Davies also identified himself openly and primarily as a writer of mathematics textbooks.

Davies began by replacing the textbooks used at West Point. In 1826 he completed *Elements of Descriptive Geometry,* which Mathew Carey of Philadelphia, the leading publisher in the United States, produced and sold.[18] In 1828 Davies took his second textbook, a co-optation of Thomas Carlyle's 1822 translation of the 1817 ninth edition of Adrien-Marie Legendre's *Elements of Geometry and Trigonometry,* to James Ryan of New York City. In 1830 Davies took his next textbook, *Elements of Surveying,* to New York's Harper brothers, whose publishing house soon outranked Carey's and who were known for their quality cloth bindings and pioneering use of stereotyping, which allowed books to be printed and reprinted much more rapidly.[19] Davies also turned over his next work, *Treatise on Shades and Shadows* (1832), as well as reprint rights for *Elements of Descriptive Geometry* and *Elements of Geometry and Trigonometry,* to the Harpers. His next three textbooks, *Elements of Algebra, Elements of Analytical Geometry,* and *Elements of the Differential and Integral Calculus,* however, went to Wiley of New York in 1835 and 1836.

Such "shopping around" was unusual. It indicates Davies was dissatisfied with the terms of each contract he signed. In his dealings with the Harpers, for instance, he retained rights to the figures. They offered their usual royalties of one-half the books' profits after deducting expenses and an overhead fee of one-third of the gross profits.[20] Compared to other firms, this agreement appears generous. William McGuffey signed a contract in 1833 with Truman & Smith for only 10 percent of the profits from his readers and allowed his earnings to be capped at $1,000.[21] There is evidence, however, that Davies wanted his own piece of the publishing business as well as sizable royalties. In 1833 he began to assemble textbooks for younger students, the first of which was *Common School Arithmetic.* This book was published in Hartford, Connecticut, likely by a printing and bookselling business that Davies operated in the 1830s with a relative of his future son-in-law, William Guy Peck.[22]

Graded Textbooks as Teaching Tools

The partner Davies was searching for turned out to be Alfred Smith Barnes. Twenty-two years younger than Davies, Barnes learned business from bookseller D. F. Robinson in New York City and returned to Hartford in 1838 to establish his own firm. Local resident Hiram F. Sumner then provided a letter of introduction to Davies, who had retired from West Point the previous year with a throat ailment that no longer troubled him. Davies and Barnes started to publish in a 12-foot by 9-foot room in Hartford. They initially agreed that Barnes would earn six hundred dollars per year and Davies would receive sales royalties, but they later decided to split their firm's profits evenly. Davies also retained copyright to his textbooks, although no cases are known in which any of his works were republished elsewhere once A. S. Barnes & Company was established. The publishing relationship endured for forty years.

For their starting list the pair used Davies's seven West Point textbooks. Although they began with no capital, the two men keenly understood the value of a West Point association. Because the academy was considered the national leader in scientific and engineering education in the 1830s and 1840s, its graduates were frequently hired as professors at other technical institutes and liberal arts colleges. They chose the textbooks they had studied, Davies's series, to teach to their own students. *Elements of Geometry and Trigonometry* was in the West Point curriculum from 1828 to 1902, was used at the University of Alabama from 1843 to 1872, and was studied at Dartmouth College from 1839 to 1870. Additionally, other West Point authors recommended their former colleague. The 1837 first edition of Dennis Hart Mahan's *Elementary Course of Civil Engineering* instructed the reader to learn mathematics before beginning to study civil engineering. When Mahan continued to specific recommendations, he wrote: "Without wishing to prejudice the works of others, the author would call attention to the very complete course of Mathematics of *Professor* Davies, late Professor of Mathematics in the Military Academy, as the best that has fallen under his observation in the English language."[23]

Barnes utilized this reputation for quality in more direct marketing efforts. He traveled to colleges and academies around New England for two years, thus bypassing the usual schoolbook agents. First, Barnes would befriend the students, often playing an early version of baseball with them.[24]

Next, he turned his "frank and winning manner" to the instructors.[25] He inquired into the difficulties raised by the texts they were using. Finally, he demonstrated a close familiarity with Davies's West Point series that convinced the instructors that these works were superior in content to any other American mathematics textbook.

In a similar fashion Barnes visited common schools. Between 1838 and 1842 his partner prepared eight volumes on arithmetic, algebra, and geometry for children. Davies was teaching his own children at home, so perhaps the material was tested on this captive audience.[26] Davies and Barnes benefited additionally from starting their business at an opportune time and place. Political and educational leaders in Connecticut had led the way in the process of ensuring that local common schools were founded everywhere they were required. They encouraged local school boards to select and provide uniform textbooks rather than to ask teachers to work with all the different books children brought from home.[27] By the 1839–40 school year, for instance, Barnes's efforts resulted in the adoption of *Common School Arithmetic* (1833) in 53 common schools in Connecticut, trailing only textbooks by Roswell C. Smith (167 schools), Nathan Daboll (133 schools), and Warren Colburn (67 schools).[28] Davies ultimately prepared as many as fifteen separate arithmetic textbooks for common and secondary school students that appeared in a total of 122 printings between 1833 and Davies's death in 1876. Various incarnations of his *School Arithmetic, Analytical and Practical* (1852) alone sold 1.25 million copies.[29]

Indeed, by 1850 Davies and Barnes were reaping substantial financial benefits from their unusual approach to textbooks. A. S. Barnes & Company found larger quarters in Philadelphia in 1840 and even more space in New York City in 1845. In New York Barnes began to add authors in other subjects to the company "stable," including Emma Willard, head of the Troy Female Seminary, writer of history textbooks, and teacher of Davies's wife, Mary Ann.[30] As the company's original "talent," Charles Davies had never concerned himself with the daily affairs of the business. He sold his interest to Edmund Dwight in 1848; Dwight in turn sold his share of the partnership to Henry L. Burr in 1849. The firm operated as "A. S. Barnes & Burr" until Burr died in 1865, when the original name was restored.

Although Davies resumed teaching as professor of mathematics for Trinity College in Hartford from 1839 to 1841, he increasingly saw writing as his full-time occupation. He was recommissioned into the U.S. Army in 1841 in order to work as West Point's treasurer, but he retired again in 1845. Davies then moved his family into a country home on the Hudson River, emerging in 1848 and 1849 to substitute for Elias Loomis at the University of the City of New York (New York University today); to teach briefly at the normal school in Albany, New York, in the 1850s; and to be the professor of higher mathematics at Columbia College in New York City from 1857 to 1865. Finally, Davies and Barnes became involved in the effort to professionalize education. Their Connecticut connections included leading educators such as Henry Barnard, while Davies was elected president of the New York Teachers' Association in 1853 and 1854.[31]

Expanding the Market, Horizontally and Vertically

Although Davies and Barnes had achieved positions of high respectability, they did not rest on their laurels. Rather, they continued to adjust Davies's series to suit new audiences that were opening up with the further evolution of the American educational system. Private colleges, academies, and normal schools proliferated around 1850, but the instruction and curriculum they offered was generally and necessarily at a relatively low level. Meanwhile, in 1838 the city of Philadelphia established the first truly public high school in the United States, and soon cities across the Northeast and Midwest were adopting this model.[32] Male students in these various types of secondary schools learned arithmetic and algebra to meet entrance requirements of more established colleges, such as Harvard or the University of Virginia, or to find positions as bank or shop clerks.[33] Coeducational secondary schools did not ordinarily offer mathematics to girls in the middle of the nineteenth century. When courses were given, the usual purpose was to develop the girls into pleasant intellectual and social companions who were also prepared to teach their own children.[34] After the Civil War the Freedmen's Bureau founded schools for emancipated slaves in the South, while settlers streamed West to establish towns and schools.

The first step Davies and Barnes took in marketing their wares in this changing environment was to organize Davies's course into subseries. Advertisements were sporadic in early editions of Davies's textbooks, when he moved from printer to printer. When Wiley issued *Elements of Analytical Geometry* in 1836, however, four of Davies's five earlier textbooks were listed in the front matter.[35] By 1842 Barnes was including the list—now up to fourteen works—and titling it "Davies's Course of Mathematics." The books were arranged in the order that a student ought to encounter them, from *First Lessons in Arithmetic* to *Differential and Integral Calculus.*[36] By 1852 Barnes made the graduated nature of the series more explicit by dividing it into three levels: the arithmetical course "for schools," the academic course, and the collegiate course.[37] Barnes usually included Davies's entire course when he organized the list this way, but Davies's *Metric System, Explained and Adapted to the Systems of Instruction in the United States* (1867) promoted only the arithmetical series. Teachers were to understand that four of the arithmetics Davies had prepared, along with *Practical Mathematics* (1852), were "designed as a full Course of Arithmetical Instruction necessary for the practical duties of business life; and also to prepare the Student for the more advanced Series of Mathematics by the same Author."[38]

A slightly different, 1860 version of the advertisement in *New University Arithmetic* (1856) simply divided Davies's series into "elementary" and "advanced" texts (Figure 1.4). Unlike the announcements we have described so far, this advertisement appeared in the back of the textbook. It was part of a multipage presentation of the schoolbooks A. S. Barnes & Burr offered for all subjects, including geography, English grammar, and reading. Two particular words, however, conveyed the significance of Davies's course: *complete* and *national.* In the range of his textbooks Davies differed from every other mathematical author; as the advertisement put it, his "methods, harmonizing as the work of one mind, carry the student onward . . . and are calculated to impart a comprehensive knowledge of the science."[39]

As Barnes recruited prominent writers to his firm, he envisioned securing the greatest possible educational market. This objective certainly reflected two mutually reinforcing trends in New York in the mid-nineteenth century: the city was becoming the center for American publishing houses, while New York publishers aspired to reach all regions of the United States. Thus, Barnes presented his entire catalog as the "national series of standard school-books." Further, he argued that Davies's course set the national standard for mathematics instruction: "Many authors and editors in this department have started into public notice, and by borrowing ideas and processes original with Dr. Davies, have enjoyed a brief popularity, but are now almost unknown. Many of the series of to-day, built upon a similar basis, and described as 'modern books,' are destined to a similar fate; while the most far-seeing eye will find it difficult to fix the time, on the basis of any data afforded by their past history, when these books will cease to in-

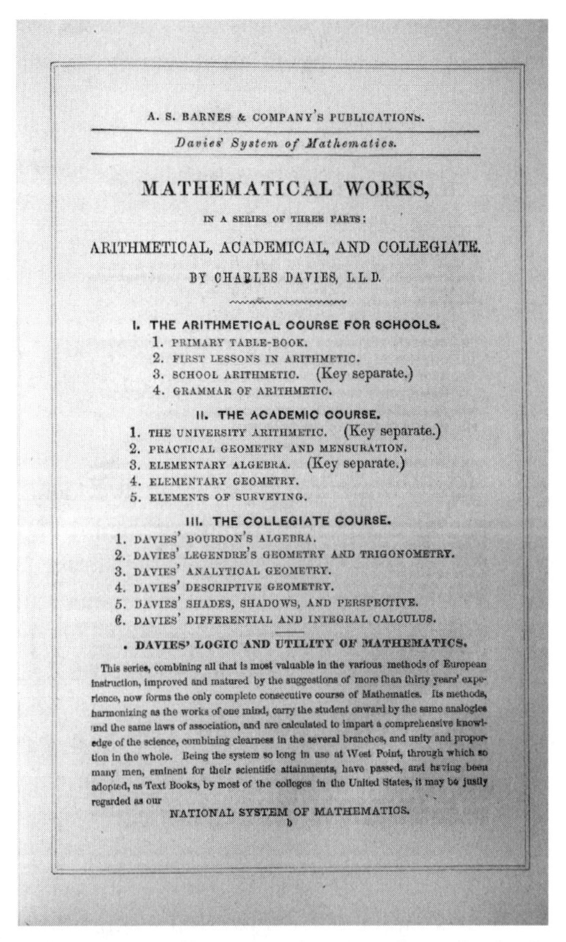

A. S. BARNES & COMPANY'S PUBLICATIONS.

Davies' System of Mathematics.

MATHEMATICAL WORKS,

IN A SERIES OF THREE PARTS:

ARITHMETICAL, ACADEMICAL, AND COLLEGIATE.

BY CHARLES DAVIES, L.L.D.

I. THE ARITHMETICAL COURSE FOR SCHOOLS.

1. PRIMARY TABLE-BOOK.
2. FIRST LESSONS IN ARITHMETIC.
3. SCHOOL ARITHMETIC. (Key separate.)
4. GRAMMAR OF ARITHMETIC.

II. THE ACADEMIC COURSE.

1. THE UNIVERSITY ARITHMETIC. (Key separate.)
2. PRACTICAL GEOMETRY AND MENSURATION.
3. ELEMENTARY ALGEBRA. (Key separate.)
4. ELEMENTARY GEOMETRY.
5. ELEMENTS OF SURVEYING.

III. THE COLLEGIATE COURSE.

1. DAVIES' BOURDON'S ALGEBRA.
2. DAVIES' LEGENDRE'S GEOMETRY AND TRIGONOMETRY.
3. DAVIES' ANALYTICAL GEOMETRY.
4. DAVIES' DESCRIPTIVE GEOMETRY.
5. DAVIES' SHADES, SHADOWS, AND PERSPECTIVE.
6. DAVIES' DIFFERENTIAL AND INTEGRAL CALCULUS.

. DAVIES' LOGIC AND UTILITY OF MATHEMATICS.

This series, combining all that is most valuable in the various methods of European instruction, improved and matured by the suggestions of more than thirty years' experience, now forms the only complete consecutive course of Mathematics. Its methods, harmonizing as the works of one mind, carry the student onward by the same analogies and the same laws of association, and are calculated to impart a comprehensive knowledge of the science, combining clearness in the several branches, and unity and proportion in the whole. Being the system so long in use at West Point, through which so many men, eminent for their scientific attainments, have passed, and having been adopted, as Text Books, by most of the colleges in the United States, it may be justly regarded as our

NATIONAL SYSTEM OF MATHEMATICS.

b

Figure 1.4. Advertisement for National Series of Standard School-Books: Davies' Complete Course of Mathematics, 1860. This is from the end matter of Charles Davies, *New University Arithmetic,* 1860. Courtesy of Smithsonian Institution Libraries.

crease and prosper, and fix a still firmer hold on the affection of every educated American."[40] Barnes provided seven justifications for this claim to national status:

> 1st. It is the basis of instruction in the great national schools at West Point and Annapolis.
> 2d. It has received the *quasi* endorsement of the National Congress.

3d. It is exclusively used in the public schools of the National Capital.
4th. The officials of the Government use it as authority in all cases involving mathematical questions.
5th. Our great soldiers and sailors commanding the national armies and navies were educated in this system. So have been a majority of eminent scientists in this country. All these refer to "Davies" as authority.
6th. A larger number of American citizens have received their education from this than from any other series.
7th. The series has a larger circulation throughout the whole country than any other, being *extensively used in every State in the Union.*[41]

Whether the technical content of Davies's textbooks lived up to Barnes's claims of originality and perfection is arguable.[42] It is evident, however, that Davies consistently emphasized a joint pedagogical approach of mental discipline and practical applications that appealed to instructors at all educational levels. In the prefaces of his textbooks Davies repeatedly pointed out that this system, which he helped develop at West Point, was so excellent because it first trained the immature mind to contemplate clear, general, exact mathematical truths and to reason by logical rules.[43] Additionally, in a revision of *Elementary Algebra* that first appeared in 1859, Davies explained that "each lesson is so arranged as to introduce a single principle, not known before, and the whole is so combined as to prepare the pupil, by a thorough system of mental training, for those processes of reasoning which are peculiar to the algebraic analysis."[44] Because the arithmetical, academic, and collegiate series were all in the format of treatises organized by topic into numbered paragraphs, the books were designed for classrooms conducted as blackboard recitations.[45] Davies's course reinforced this educational practice for all students—and perhaps enabled it to survive longer than it should have—but it also permitted school and college administrators to function with larger class sizes as the expansion of education resulted in shortages of knowledgeable teachers and of schools.

Indeed, by the time the scale of Barnes's operation necessitated this shift from personal marketing to extensive print advertising and the use of regional sales agents, Americans were accustomed to the idea that relatively inexpensive, uniform, graduated textbooks were the cornerstone of mathematics instruction at all levels of education. Teachers looked to textbooks for the content of their courses, and students expected to study a book when they entered a classroom. Daily sessions were conducted as recitations from the first years of primary school through college. Educators who selected textbooks chose particular authors and then particular publishers, largely by reputation.

As an author-publisher pair who founded a company specifically to produce mathematics textbooks, Davies and Barnes went beyond their predecessors in envisioning textbooks as commercial teaching tools. They set a standard for financial success and range of coverage to which contemporaries and successors aspired. The 7 works in Davies's collegiate or West Point series were printed over 200 times, while the 5 academic-level textbooks appeared in 116 printings. Barnes reprinted Davies's more than 35 arithmetic textbooks and resources for teachers at least 160 times.[46] Of the 800,000 books sold by A. S. Barnes & Burr in 1853, over 200,000 bore Davies's name.[47] The next generations of authors followed his example by affiliating with one publishing house throughout their careers, although none addressed the needs of common schools, secondary education, and colleges as extensively as Davies. Elias Loomis, for example, wrote 14 textbooks for Harper & Brothers from 1851 to 1868, with *Elements of Plane and Spherical Trigonometry* (1848) alone appearing in 76 printings. Edward Olney prepared 18 textbooks and 8 answer keys that were printed a total of 79 times by Sheldon & Company of New York between 1854 and 1887, while Webster Wells wrote 29 textbooks and 19 answer keys for D. C. Heath and Company of Boston and by 1916 had seen them printed a total of 138 times.[48] Nearly all of these volumes were only for high school and college students, and none laid claim to a national reputation.

Dreams of establishing national standards for mathematics teaching would reappear in efforts to eliminate recitation and to reform instruction in the late nineteenth century. When the first such initiative, the National Educational Association's 1893 Committee of Ten, evaluated the content of school mathematics courses, it referred to textbooks that were markedly different in appearance from those of the mid-nineteenth century. Long lists of student exercises had been added, and mathematical diagrams had been supplemented or replaced by drawings and photographs of mathematics applied to daily life. The thirty-three textbooks and twenty-nine answer keys authored or coauthored by George Wentworth illustrate the introduction and immediate popularity of typographical flourishes, such as boldfaced headings and italicized key terms. Ginn & Company of Boston produced at least 350 printings of these books between 1878 and 1906. Intense competition between publishers led to the merger of five firms' textbook divisions into the American Book Company in 1884.[49] The industry continued in a pattern of corporate development and consolidation in the twentieth century. These modernized publishers exerted influence beyond that of individual authors; increasingly, textbooks were prepared by pairs or teams of instructors. Nonetheless, the assumption that mathematics teaching at all levels should be structured around the textbook endured.

CHAPTER TWO

The Blackboard
An Indispensable Necessity

> Now, it is safe to say that no mechanical invention ever effected greater improvements in machinery, no discovery of new agents more signal revolutions in all the departments of science, than the blackboard has effected in the schools.

Precisely where and how the blackboard entered the American classroom is a mystery. Records of libraries, universities, and school systems tell us far more about how other forms of apparatus as well as textbooks became part of daily classroom routine. From the mid-1820s, for example, the Boston School Committee strove to supply its crowded grammar schools with improved books and equipment. Records of the committee reveal that the compendium of Nathan Daboll gave way to textbooks by more modern local authors such as Warren Colburn and Frederick Emerson.[1] They show that in 1823 these schools were equipped with globes and maps.[2] By 1831 the school system was also willing to provide all eight of these schools with blackboards as well as clocks, geometrical models, and a numeral frame, or abacus. When the school committee surveyed schoolmasters, they found that four of the eight grammar schools already had at least two blackboards, and two of them had one. Hence, by 1831 six of the eight Boston grammar schools already had blackboards.[3] They apparently had been quietly made or purchased by teachers who, at least in one case, were reimbursed by the city.[4] For other apparatus the situation was quite different; none of the schools had geometric models, and only one had an abacus. Thus, while one can date with some precision the time of the introduction of the latter objects into the Boston schools, dating exactly when the blackboard became commonplace is more difficult.

While few objects entered the classroom with less fanfare than the blackboard, few were as generally accepted. Erasable surfaces have now been part of American education, particularly mathematics education, for two centuries. Although they have been a central tool of teaching and research, they have usually been taken for granted by educators, mathematicians, and historians.

There are some parallels between the spread of the blackboard and of textbooks. Both were introduced following English and French models, and neither was used exclusively for mathematics teaching, although education in the subject played an important role in their diffusion. Many early blackboards, like cipher books, were formed by hand, and commercial versions of the tool became common only gradually. Enthusiasm for the new devices came earlier in the North than the South.[5] The histories of the objects also differ, however, in important respects. Blackboards were humble classroom furnishings, lacking both the intellectual claims—and the fascination for later bibliophiles—of textbooks. Their story, such as it survives, comes primarily through the words of pedagogical reformers and school committees, the mute testimony of objects, and the advertisements of manufacturers.

The early history of the blackboard in the United States may be divided into two broad periods. Up until about 1850 it was a pedagogical novelty introduced to supplement the handheld slate. Blackboards were first used by a few mathematics teachers in the northeastern and central Atlantic states and then widely adopted by both schools and colleges. Educators from this period gave instructions about how teachers could make their own blackboards and sometimes suggested that the devices offered an inexpensive alternative to textbooks. A second era began in the latter half of the nineteenth century, when widespread school building, improved transportation, the rhetoric of educators, and more ample school budgets combined to produce regional and national markets for school supplies. Dealers published catalogs of what Deborah J. Warner has called “commodities for the classroom.”[6] These catalogs included not only blackboards but liquid slating, slated paper and cloth, and slated globes. Blackboards came to occupy not just a few square feet of one wall of a classroom but extended around the entire room. Increasingly, they were made of natural slate.

The Blackboard as Pedagogical Novelty

Small wax tablets were used for drawings from at least the time of the ancient Greeks—an example survives in the collections of the British Museum.[7] Tablets made of slate were known and used in Europe from the Middle Ages and in colonial America. Early European printed books from the fifteenth through the seventeenth centuries include drawings of classrooms with somewhat larger slates hanging on the wall.[8]

Such blackboards were not unknown in colonial America. In July 1779, for example, John Taylor, a tutor at Queens College (forerunner of Rutgers University) in New Brunswick, New Jersey, wrote to recent graduate John Bogart, asking Bogart to take over his classes while Taylor was away on military duties. Bogart also wrote, "I have spoken to Mr. Brinson to make a blackboard, and have procured lamp black, you will hurry him on, and get Col. D [*sic*] Vroom to paint it,- and keep an account of the expense."[9] In September 1801 the English-born George Baron began his brief tenure as the civilian professor of mathematics, teaching the U.S. Army cadets stationed at West Point (Baron's teaching took place before the official founding of the U.S. Military Academy in 1802). According to a two-volume work produced to honor the centennial of West Point, use of the blackboard was "a favorite method" of Baron.[10]

Baron reportedly used a "standing slate" on which he wrote with a piece of chalk.[11] As the quotation from Taylor suggests, blackboards also could simply be boards painted black. Such instruments were of great use to instructors trying to teach large numbers of students simultaneously, as became increasingly common in the early nineteenth century. The earliest published description of such an instrument as it was used in the United States that we have found was in the fourth edition of Emmor Kimber's textbook *Arithmetic Made Easy for Children,* a volume that appeared in Philadelphia in 1809. Kimber explained in a footnote to the preface of his book that "the Black Board may be usefully applied to many purposes in teaching; it should be about 3 feet square, painted or stained with ink, and hung against the wall in a convenient place for a class to assemble around it."[12]

Kimber specifically associated the blackboard with the monitorial system of instruction developed by the English Quaker Joseph Lancaster. Lancaster advocated educating poor children inexpensively by having them taught by monitors who had just learned the material themselves. One device used in Lancasterian schools was a board painted black, on which the letters of the alphabet were painted in white. As best we can tell, Lancaster did not advocate blackboards that could be erased. His ideas did find disciples in the United States. In December 1808, for example, the Philadelphia Association of Friends for the Instruction of the Poor was formed to start a Lancasterian school. Its thirty charter members, including Emmor Kimber, were soon running a school with over two hundred students.[13]

The preface to Kimber's arithmetic specifically suggested that monitors could read out numbers to be added, which students then wrote on their slates. Alternately, the monitor could write the problem on the blackboard with chalk and have the students gather around the board. In other words, Kimber, unlike Lancaster, suggested an erasable blackboard, setting a precedent that many would follow. The children (Kimber assumed they would all be boys) worked columns of the problem in turn, with the teacher writing down results. Kimber had not mentioned the blackboard in the second and third editions of his text, published in Philadelphia in 1805 and 1808, respectively. He republished his 1809 preface unchanged in editions of his arithmetic book published in 1812, 1816, and 1819. The monitorial school operated in Philadelphia until 1822, when a more general system of primary schooling was introduced. Other publications on the Lancasterian method do not mention the blackboard.[14]

French émigrés also brought blackboards to the United States. As far as we know, the first reported use of a blackboard by a French teacher in the United States occurred in 1814. Boston merchant Samuel J. May later wrote that in that year he attended a mathematical school kept by a refugee French Catholic priest, Francis Xavier Brosius. According to May's recollection: "On entering his room, we were struck at the appearance of an ample *Blackboard* suspended on the wall, with lumps of chalk on a ledge below, and cloths hanging at either side. I had never heard of such a thing before." May would go on to become a reformer in the Boston schools and introduced the blackboard generally into the primary schools of Massachusetts.[15]

The first general use of the blackboard in collegiate mathematics and science teaching has been credited to Sylvanus Thayer and the faculty he hired at West Point. In 1817 Thayer recruited Claude Crozet, a graduate of the École Polytechnique in Paris, to teach the cadets. Crozet quickly discovered that his students could not read the French textbooks from which they were to learn descriptive geometry and turned to the blackboard. Apparently, Crozet did not use a slate but relied on a painted blackboard constructed by a local carpenter and painter.[16] Use of the blackboard soon was an established ritual at West Point. Professors gave lectures at the board, with students taking notes. They then were expected to study the material and present it the next day at recitations held at the blackboard. A professor's questioning could, in the words of one student, keep a class "up to the highest pitch of interest all the time."[17] This innovation soon spread to other parts of the curriculum. Joseph Henry, a teacher at the Albany Academy who would become one of the country's most eminent scientists, visited West Point in June 1826. He commented in his journal: "One article very necessary in teaching chemestrys is found in this room viz a black board on which the student is taught the attomic theory and all algebraical formula in chemestry. Indeed it appears to be one of the principles of teaching in this institution that every thing as far as practical should be demonstrated on the black board. The student is even required to draw all articles of chemical aparatus and explain them in this way."[18]

Blackboards were also adopted at traditional colleges and universities. By the mid-1820s they were regularly used in Harvard College geometry classes. Plates for Harvard professor John Farrar's *Elements of Geometry* were even published separately so that students could consult them when presenting proofs at the board.[19] Similar recitations also took place elsewhere in New England. In 1825 William Smyth, a tutor at Bowdoin College who had previously focused his attention on Greek, decided to offer sophomores a course in algebra taught on the blackboard. It proved to be a great success, to the extent that "a considerable portion of the Juniors requested the privilege of reviewing the algebra under the new method at an extra hour." Alpheus Spring Packard commented wryly in his history of Bowdoin College that such a request was "a wonder in college experience."[20] Smyth's success had its rewards—he soon was promoted to assistant professor of mathematics.

Students were not always as enthusiastic. By 1830 Yale students, who had been allowed to consult figures in their textbooks during recitations, were required to recite using only the figure drawn on the blackboard. That summer the order of topics covered in the mathematics curriculum was altered. Spherical trigonometry was taught before, not after, conic sections. Disturbed by what they took as decreased time for preparation, the students requested that in reciting the demonstrations on conic sections, "they might be permitted to recite from the book, and not the figure."[21] The faculty demurred. At the next class, as was the custom, names were drawn to indicate who should recite the assignment. Nine of those selected declined to recite. When the Yale faculty met to decide what to do about the recalcitrant students, it received another petition, signed by forty-two of the ninety-six members of the class. These students asked to share any punishment meted out to the reluctant nine. In response to the "Conic Sections Rebellion" the faculty dismissed the dissenting students—almost half the class—and refused to recommend them for admission to other colleges.[22]

Meanwhile, blackboards were becoming increasingly common in more elementary education. Teachers envisioned several different roles for them. When Elizabeth Peabody was running a private school for a few children of the Boston elite in the 1820s, for example, she tested her charges' command of arithmetic by using the blackboard. According to Peabody, each day she required her students "to sit round in a class and tell me how they do the sums—& following their directions with a piece of chalk upon a black board, so that every error can be made manifest to the whole class, who have the privilege of correcting each other."[23] Peabody found the method very effective, although she admitted that not all students were as attentive as they needed to be. Others followed the West Point model, having both teachers and students work out problems at the blackboard. Blackboards also played a role in more public presentations. In 1829 the *Galaxy*, a Mississippi periodical, described open examinations at the Elizabeth Academy for girls in the town of Washington in that state, reportedly the first detailed account of such an exhibition in the state that survives:

> When we heard the class in mathematics we were astonished; and certainly it is a matter of astonishment to witness little girls of 12 years of age treat the most abstruse problems of Euclid as mere playthings. Nor were they dependent on memory alone; and we will give our reasons for so thinking. During one of the solutions upon the blackboard (we forget which it was) it was suggested that the young lady was in error. "No, ma'am," replied the pupil, with great promptitude and self-possession; "I am correct. The bases of a parallelogram must be equal." The principle is indeed a simple one, but the readiness with which it was adduced in argument, and that, too, under embarrassing circumstances, was to us the most conclusive evidence of an extraordinary discipline of mind.[24]

The governess of the Elizabeth Academy at the time was the poet and author Caroline Matilda Warren Thayer, a Massachusetts native who had headed the female department of the Wesleyan Seminary in New York City before a religious dispute led to her dismissal and move south. We have found no evidence that Thayer's late husband, the physician James Thayer, was related to Sylvanus Thayer of West Point. It does seem possible that

Matilda Thayer was responsible for the introduction of the blackboard at the Elizabeth Academy.

During the 1830s and 1840s several states, particularly in New England, established free public elementary schools, called "common schools," that were intended to provide well-educated citizens for the young republic. Books and journals published at the time discussed at length the proper construction, ventilation, furnishing, and apparatus for schoolrooms. Reformers agreed that classrooms should have blackboards.[25] Joseph Felt reported in the *Annals of Salem* that by 1820 "blackboards were used in our Common Schools for arithmetical calculations."[26] In 1829 Samuel R. Hall, who had established one of the first schools for teachers in the United States, published his *Lectures in School-Keeping,* in which he affirmed the importance of teaching apparatus generally and asserted that one large or two smaller blackboards "ought to be in every school-room in the country."[27] Similarly, in an 1830 lecture on the construction and furnishing of school rooms, William J. Adams listed as essential apparatus a timepiece; maps and globes; the abacus, or numeral frame; and the blackboard.[28]

Equipping a classroom with a blackboard did not mean that a teacher would use it. In 1839 Henry Barnard, the secretary to the newly established Board of Commissioners of Common Schools of the state of Connecticut, published his first report in the newly founded *Connecticut Common School Journal.* Here he commented that "black-boards are not uncommon, but are little resorted to by the teacher."[29] During the 1840s educators continued to comment on the advantages of blackboards. They also offered specific suggestions about how teachers should use them, particularly in conjunction with slates. The statutes of the state of New York relating to common schools, as revised in 1841, read, for example, "Large black boards in frames are indispensible [*sic*] to a well conducted school. The operations in arithmetic performed on them, enable the teacher to ascertain the degree of the pupils' acquirements, better than any result exhibited on slates. He sees the various steps taken by the scholar and can require him to give the reason for each. It is in fact an exercise for the entire class and the whole school by this public process insensibly acquires a knowledge of the rules and operations in this branch of study."[30] That same year Josiah F. Bumstead published in Boston a whole book on the use of the blackboard in primary schools.[31] In 1842 four issues of the *Connecticut Common School Journal* were devoted to extracts from William A. Alcott's forthcoming *Slate and Blackboard Exercises for Common Schools.* Alcott, a Massachusetts teacher, editor, and distant relative of the writer Louisa May Alcott, firmly believed that "a blackboard, in every school house, is as indispensably necessary as a stove or fireplace." He gave specific examples of how the slate and blackboard might be used in combination to teach mapmaking, spelling, composition, reading, arithmetic, geography, drawing, and moral precepts and stressed that students should be usefully occupied throughout the day. Students who copied drawings, made measurements, and worked arithmetic problems following instructions on the blackboard would learn more, Alcott claimed, than those who simply recited by rote. Moreover, schools might even avoid the high cost of textbooks.[32]

Alcott, Henry Barnard, and other nineteenth-century educators emphasized that blackboards need not be fancy. Some were simply wooden planks, matched, leveled, set in a frame, and painted black (Figure 2.1). A bit of grit

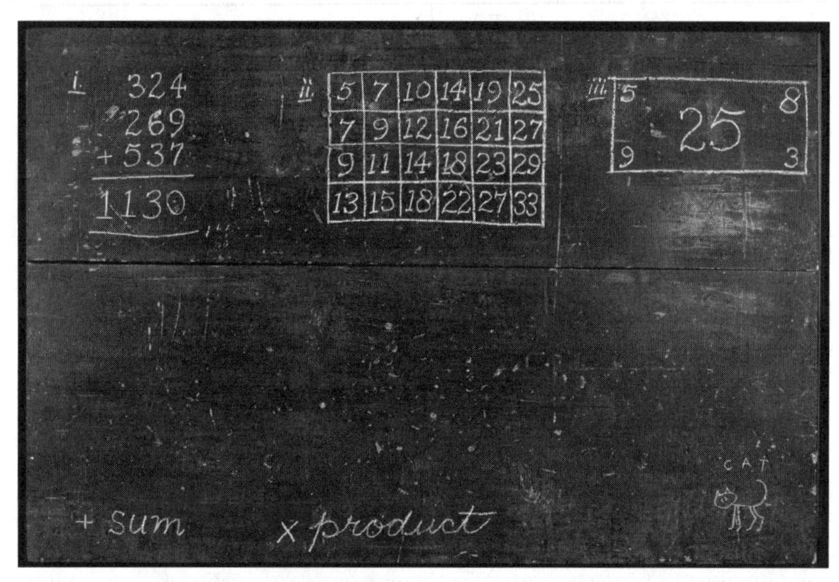

Figure 2.1. Blackboard from the Glebe Schoolhouse in Westmoreland, N.H., n.d. The object consists of two wide horizontal wooden boards, painted black. Three thin vertical boards are attached to the back. The exercises drawn on the blackboard are taken from Josiah Bumstead's book *The Black Board in the Primary School: A Manual for Teachers* (Boston: Perkins & Marvin, 1841), iv, 6–7. National Museum of American History collections, gift of Dr. and Mrs. Arthur M. Greenwood. Smithsonian Negative no. 2001-12006.

could be added to the paint to make a rougher surface that would bite into the chalk. Calvin S. Keep, a Rhode Island teacher, made such a blackboard for the school at which he taught in Burrillville, Rhode Island in 1842–43. He reportedly was severely reprimanded "for spending so much time in figuring and marking on this board when there were several slates owned by the older scholars." Keep soon moved on to another school, which took more kindly to both his introduction of a blackboard and his telegraphic experiments.[33] Barnard's *School Architecture* (1848) also suggested that one cover a surface with paper and coat it with a mixture of coloring, grit, and adhesive. The book recommended lampblack dissolved in alcohol, powder of emery, and distilled varnish. One also could use a somewhat different mixture—or, to use Barnard's phrase, "composition"—and apply it directly to the walls. Blackboards of this sort survive at the Fisher School in Westwood, Massachusetts.[34] Alternately, Barnard reported that large pieces of slate could be "substituted for the blackboard."[35]

Commodities for the Classroom

The choices Henry Barnard offered for blackboards—coated wood, slated paper or cloth, specially treated walls, and natural slate—would remain available to teachers and school planners for at least the next half-century. Increasingly, however, teachers and school boards left the task of making coatings and constructing blackboards to commercial firms. In the mid-nineteenth century several Americans patented compositions that were sold as "liquid slating," for use on boards, walls, paper, and cloth. An 1873 author saw the sale of liquid slating as an example of the remarkable expansion of American education, until it ranked as a national industry. He wrote: "'What is the price of a quart of blackboard?' Heretofore we might as well have inquired for a yard of oil, or a pound of conscience. But it is no joke at all; a material is regularly manufactured and extensively used, which is neither more nor less than liquid blackboard. It is bottled or canned for carriage and keeping; may be spread like paint on board, paper or wall, and becomes a blackboard."[36]

According to later advertisements, the first liquid slating sold commercially in this country was invented by the New Haven scientific instrument maker George Munger in 1857.[37] That same year Forrest Shepherd, also of New Haven, took out patent no. 18,931 for "apparatus for illustrating conic sections and the lines of the globe." Commercial versions of Shepherd's

Figure 2.2. Slated globe on stand, n.d. Both the globe and the stand have been refinished. National Museum of American History collections. Photograph by George Norton. Courtesy of Smithsonian Institution.

globes were coated with the "Eureka" slating invented by Munger, and it seems likely that the two men knew of one another's work from the beginning. As Deborah J. Warner has shown, globes covered with a layer of slating that could be marked with a slate pencil or, later, chalk were just becoming popular teaching devices in the United States at this time. A slated globe even graced the desk of eminent mathematician and scientist Josiah Willard Gibbs.[38] Globes were sold not only for teaching spherical trigonometry but for instruction in elementary geography and for teaching high school astronomy (Figure 2.2).[39]

Although Munger may have had slated globes in mind when he invented his liquid slating, he quickly moved to sell the product for blackboards. An

1864 advertisement in the *American Educational Monthly* reported that Munger manufactured "Eureka Slated Goods of All Kinds."[40] An undated broadside published by A. H. Andrews Company of Chicago boasted that Eureka liquid slating had "been constantly manufactured and improved by the inventor. It has been in continuous use in many schools without repairs since 1860."[41] Munger's name for liquid slating, Eureka, suggested discovery. Other makers sought to emphasize quality with brand names such as Black Diamond, Excelsior, and Acme, while another evoked patriotism, dubbing its product "American." Dealers also sold slated paper and cloth as well as wooden blackboards coated with liquid slating and framed. Liquid slating apparently was used to produce some of the first chalkboards sold in colors other than black. A. H. Andrews & Company, for example, sold slating in green, blue, and gray as well as black by 1881, charging somewhat more for the unusual colors. Similarly, in its 1884 catalog the firm of J. C. Brooks of Cincinnati offered Excelsior liquid slating in quantities ranging from a pint to a gallon and noted that green slating was available for a 25 percent surcharge.[42]

In the 1890s liquid slating began to be applied to new materials. The 1897–98 catalog of Edward E. Babb & Company of Boston showed blackboards made from "Hyloplate." The catalog explained that "Hyloplate is a product of wood pulp with cement. It therefore partakes of the desirable qualities of both paper and wood. Its manufacture resembles that of the paper car wheels that have proved so nearly indestructible."[43] Hyloplate and a material made from pressed wood that sold as "Duroplate" could be made in any size and, at least according to the distributors, were hard, lightweight, not subject to breakage on shipping, and cheap. Blackboards with a Hyloplate backing sold at least as late as the 1920s (Figure 2.3). Although blackboards made from pressed wood pulp (with or without cement) and coated with liquid slating sold widely around 1900, another material was becoming even more popular. Natural slate from quarries in Vermont, New York, and especially Pennsylvania was shipped throughout the country. To give only one example, the minutes of the faculty of Centre College in Danville, Kentucky, record that on December 3, 1886, "The Professor of Mathematics requested and obtained permission to have put up in his recitation room a blackboard of real slating, made by L. B. McClees & Co., Philadelphia, Pa."[44]

Slate was more expensive than other materials used in blackboards. Increasingly aware of the need to apply new layers of liquid slating regularly, however, school administrators came to agree with slate manufacturers that

Old Reliable Hyloplate Blackboard

Black

Green

The Blackboard of Utmost Satisfaction

Durable and permanent.
A uniformly smooth writing surface.
Will not crack, chip nor break.
Does not reflect light.
Erases easily.
Insured by a most liberal guarantee.

Economical

Low first cost and long service.
Low transportation cost.
No breakage in transit.
Uses less crayon.
Furnished in long lengths.
Easy to install.

Old Reliable Hyloplate has been in use for more than 38 years. During this time its manufacturers have made every effort to improve and to keep it to the highest standard of perfection. The best materials for the purpose, the best way of producing, has been found and adopted, no matter what the cost. Hyloplate has a worldwide reputation for high quality. Besides being used more extensively than any other composition blackboard in the schools of the United States, it is sold in more than twenty foreign countries.

Hyloplate is made only of the best quality long stock live spruce wood pulp, compressed and cemented up flat under a pressure of forty tons to the square inch. It does not contain any "chip," "solid news" or any of the other cheaper grades of wood pulp. It would not be Old Reliable Hyloplate if cheaper materials were used.

It is carefully and thoroughly dried and seasoned in kilns designed especially for this purpose. Two waterproofing coats are applied to the back of the board.

The writing surface is faced with a hard-calendered, sulphite-faced ply of an extra grade of pulp. This special, hard, smooth surface facing is coated with five coats of Old Reliable Hyloplate Liquid Slating. (The green board is given seven coats.) Each coat is carefully sanded and the last coat is rubbed to a velvety smooth finish. It is easy to write on, easy to erase and wears indefinitely.

"the best is always the cheapest."[45] Even as the slates given to individual pupils were replaced by paper and pencil, much larger sheets of slate came to cover one or more walls of the typical American classroom. When the first edition of the *Thomas Register of American Manufacturers* was published in 1905, thirteen of the eighteen makers of blackboards listed there sold slate blackboards, while only three sold boards coated with a chemical mixture.

In the first half of the nineteenth century advocates of the blackboard such as Samuel R. Hall and William A. Alcott had envisioned at most one large or two small blackboards per classroom.[46] By 1848 Henry Barnard described a few schools in which the classrooms were lined with blackboards on all four walls.[47] Blackboard vendors were happy to encourage this suggestion. In 1859, for example, James Johonnot recommended in his book *Country Schoolhouses* that schoolrooms have a large blackboard on one wall, with additional boards if this was necessary to accommodate all the students in a class.[48] In 1871 a new edition of Johonnot's book was published by J. W. Schermerhorn & Company, one of the largest school supply dealers in the country. This version of the book recommended that in classrooms blackboards "should be about five feet wide, placed two feet above the floor, and should extend entirely around the walls."[49] Mindful of such advice, or eager to have the fullest possible space for lessons, many schools indeed invested in extensive blackboards. After the English school inspector and educational writer Joshua G. Fitch toured American schools and training colleges near the end of the century, he commented, "one very useful mechanical device . . . is to be found in nearly all the best American schools. It is the continuous blackboard, or blackened surface extending all round the room." Fitch thought that a large expanse of blackboard allowed teachers to refer to illustrations and ideas they had developed earlier in a lesson. It was also convenient for writing out assignments and for students' presentations of their work.[50]

Figure 2.3. (Opposite) Classroom blackboard of liquid slating on wood pulp (Hyloplate), 1923. Advertisement for a commercially made blackboard, sold in green and in black. By the time of this advertisement composition blackboards were expected to last for decades and compete with slate. Note that boards are shown on the side wall of a classroom, not just at the front. Photograph from American Seating Co., *School Furniture and School Supplies* (Chicago: American Seating, 1923), 54. Courtesy of Smithsonian Institution Libraries.

At many schools use of blackboards for recitations remained an entrenched ritual. Students at the United States Naval Academy, for example, received daily grades on presentations they made at the blackboard in each of their classes. This practice may well have originated in the 1840s, the earliest years of the academy, and was certainly in effect by the 1890s.[51] Indeed, by the beginning of the twentieth century mathematics educators took the blackboard for granted. In 1915, when H. C. Wright reported on mathematical equipment and its uses in the high schools of Chicago, he saw no reason to say anything about blackboards, although he did comment on the profusion of blackboard drawing instruments available at the University of Chicago High School. Wright argued that mathematics education should include "the education of the hand along with the training of the mind."[52] He commended the use of diverse equipment for teaching mathematics, including the slated globe (he called it a spherical blackboard), geometric models, graph paper, and surveying apparatus. The blackboard did not require mention.

In the twentieth century chalkboards made from chemical composites baked onto wood improved greatly. Several firms also developed new forms of board made from materials ranging from glass to porcelain enamel to plastic. Indeed, the felt-tipped marker replaced chalk in some schools. Nonetheless, the idea that large erasable surfaces should play a central role in the work of mathematics teachers and students endured.

CHAPTER THREE

Standardized Tests
The Many Guises of Efficiency

> The birth of physical science was a world-shaping event, inasmuch as it gave control over the material universe. The application of measurement and experimentation to mental products is as remarkable an occurrence, for it literally means control of the human machine. The last factor in efficiency to be controlled, the human factor[,] will in all likelihood yield the richest results of all.

The second half of the nineteenth century saw major changes in the production and distribution of school supplies. We have already noted improvements in the blackboard. Other classroom goods that came to be produced and distributed much more efficiently included such writing tools as pens, pencils, and paper. Quill pens gave way to pens fitted with mass-produced steel nibs and then to other forms of pen. The production of artificial dyes transformed the manufacture of ink. Wooden pencils with leads made from compounds of graphite came to sell cheaply enough for school use. Paper also became much cheaper, available not only blank but ruled with various lines. Thus, the slate and slate pencil gave way to paper and graphite pencil. Many of these products were first imported from Britain and Germany and then made in the United States.[1] All of these mass-produced goods entered the classroom. As one would expect, they fundamentally altered ways of teaching writing. The new tools also made possible much wider use of notebooks, written examinations, and written homework. Teachers of even elementary classes could teach large groups inexpensively, without relying primarily on slates and oral examinations.[2] Specially ruled paper also found a place in the mathematics classroom, as shown in chapter 12.

The success of new methods of mass production also led to questions about methods employed in schooling. Might one improve the efficiency of schools so that they kept pace with the newly efficient factories? Here we examine one method proposed for this purpose, the paper and pencil standardized test. Asked to think of objects used in mathematics teaching, few historians, teachers, or students would name standardized tests. The objects themselves—until recently, paper booklets printed on highly acidic paper, with associated manuals and score sheets—are flimsy, colorless, and ephemeral. Compared to textbooks and blackboards, the tests were not a part of daily school routine. Indeed, they were and are often administered separately, outside the regular classroom.

Nonetheless, standardized tests had an important influence on educational practice. The tests were part of a more general attempt to bring uniformity to diverse American certification procedures. Such efforts were especially important to education in the United States, as schools were administered by disparate local and private authorities, not by the central government or a single religious body. Tests also found important advocates among those who sought to bring the efficiency of business to education. This included advocates of the new discipline of educational psychology, particularly faculty and students at Teachers College of Columbia University.

Preparation of standardized tests, especially at the elementary level, proved more interesting to psychologists than to most mathematicians. By the 1920s, however, testing loomed large in mathematics education. Consider, for example, the Mathematical Association of America's 1923 report *The Reorganization of Mathematics in Secondary Education*. The committee of mathematicians and educators who prepared this volume devoted almost a quarter of the text to Clifford B. Upton's account of standardized tests.[3]

Those who introduced tests often distinguished between tests of arithmetic and tests of mathematics—the latter covered plane geometry and high school algebra. We refer to all of these tests as tests of mathematics for two reasons. First, the subject content of the tests overlapped. Many problems on arithmetic tests, for example, could be solved much more easily using algebra—and some students tried to do this. Several algebra tests contained substantial sections on arithmetic. Second, the types of tests developed were quite similar. Tests were prepared in both arithmetic and algebra for comparing the work of different school systems, for measuring the achievement of individual students, and for determining what precise topics required further review. Other examinations were used to gauge student aptitude for mathematical subjects. One should emphasize, however, that not even these "prognostic" tests were designed to single out mathematical genius. All of them sought to compare aspects of the learning and potential of large numbers of students.

Two further aspects of standardized tests merit mention. First, their claim to validity—and to any role in education—was the statistical analysis of quantitative data. In that sense their justification was fundamentally mathematical. Here they differed markedly from objects such as the blackboard or textbook. Second, those who made up tests tried to demonstrate scientifically that they were effective. Making up tests was not merely a source of income but a matter of professional prestige. Indeed, analysis of test results led to major revisions in test content and, for a few, to questions about the value of standardized testing generally.

The Regents Examination and the College Boards

American students took examinations long before the introduction of standardized tests. The faculty of individual schools quizzed both applicants and students. Textbooks, including those for subjects such as arithmetic, presented specific rules, gave numerous questions to be answered by students as applications of the rule, and included "promiscuous" questions that tested several topics. In addition to such work, some schools had outside examiners who came by and drilled students on an appointed day or days.

Toward the end of the nineteenth century there also were several attempts at a statewide and sometimes a national level to ensure greater uniformity in achievement among some school graduates. The most extensive state system of school examinations developed in New York. From the 1830s the state provided grants to private academies, although the schools themselves examined both applicants and those seeking promotions. In 1864 the regents of the University of the State of New York announced plans to offer examinations in such basic subjects as arithmetic, geography, grammar, and spelling. Students taking the exams could enter either a private academy or the academic department of one of the state's public district schools. The funding the state gave to these institutions depended on the number of enrolled students who had passed the exams. With the new standards the number of students enrolled in district schools and academies initially decreased.

By 1868 the regents also planned more advanced written examination as a way of standardizing the courses offered by academies, district schools, and high schools. These tests were given at the end of a student's work. Regulations of 1880–81 linked student performance on the exams to the state funding provided to the institutions, though not to graduation. The number of students attending these schools increased under this plan but not as rapidly as the number of children in New York state aged ten to nineteen. In other words, as historian of education Nancy Beadie has argued, the use of exams offered more uniform standards but may have discouraged some students from enrolling in school.[4]

For colleges who drew students from throughout the nation, scores on state examinations would not suffice. College faculty had long examined prospective students orally. In 1869, when Charles William Eliot was inaugurated as president of Harvard College, he urged that the school establish uniform written admissions examinations, an idea that was soon adopted.[5] Yet when each college offered its own tests, secondary schools had no single standard to which they could match their courses. As Nicholas Murray Butler of Columbia University put it, "No secondary school could adjust its work and program to the requirements of several colleges without a sort of competence as pedagogic acrobat that was rare to the point of nonexistence."[6] At the urging of Eliot and through the organizational efforts of Butler, twelve colleges joined together in 1900 to form the College Entrance Examination Board. The board initially gave entrance examinations in nine subjects, including mathematics. These written tests were devised by committees of college and secondary school teachers on the basis of previously announced requirements. They were first given in June 1901 to 978 students, over 700 of them applicants to Columbia. Tests were graded by readers from schools and colleges, and scores were passed on to colleges, which could use them as they wished in admissions. From this small beginning use of the "College Boards" spread throughout much of the United States.[7]

Admissions tests would have great influence on college entry and hence school curricula. They also were part of a more general trend to rationalize qualification procedures. Doctors and lawyers of the nineteenth century, for example, had learned their trade by private study with practitioners. Late in the century the better medical schools increasingly required that students have at least some undergraduate education. The Association of American Medical Colleges and the American Medical Association worked hard to establish more uniform medical school curricula. Moreover, licensing boards were established in most states to examine doctors and evaluate medical schools. These organizations would have their greatest effect after 1910, but their very establishment was a signal of a wider movement toward standardization in education. One sees a similar concern with establishing uniform credentials in the training of lawyers, civil servants, and actuaries.[8]

Standard examinations were by no means the only way that colleges sought to gain greater uniformity among their applicants. Some midwestern universities had programs that accredited schools whose graduates they would consider for admission. In the early twentieth century several institutions of the North Central Association joined together to establish common standards for accrediting high schools.[9]

Exposing Inefficiency in Elementary Education

While disparate reformers sought greater uniformity in admission to universities and to professions, a few authors worried about elementary education. One of the first attempts to measure and compare the achievements of groups of schoolchildren in arithmetic came from outside of academe. The New York City physician Joseph Mayer Rice had abandoned his practice in 1888 to spend two years studying psychology in Jena and Leipzig, Germany. On his return he planned and carried out extensive visits to public schools in cities across the United States. Between January and June 1892 Rice observed roughly twelve hundred students in thirty-six schools, reporting his observations in several articles published in the magazine the *Forum*. He concluded that "the characteristic feature of our school system may perhaps be best defined by the single word 'chaos.'"[10] Amid the welter of local school systems he found a distressing tendency for teachers to instruct by rote recitation of facts. He also pointed up problems of political interference in school administration, poor supervision of teachers, and overwhelming disregard for the happiness of students. Rice suspected that a great deal of the drill that took place in schools was a needless waste of time, and he set out to confirm this impression quantitatively.

To establish more precisely the role of teachers, school administrators, and more general social factors in learning and to determine whether time in school was used efficiently, Rice devised tests of spelling. He distributed the first of these tests, a list of fifty words to be dictated by teachers, to school superintendents. Twenty superintendents sent Rice the work of sixteen thousand children. Examining the results, he concluded that the differing enunciation of teachers undoubtedly had influenced the data. Rice then devised a second test, based on the spelling of words used in sentences, and supervised its administration to some thirteen thousand children. He found that the amount of class time devoted to spelling and the methods used to teach it had relatively little influence on test scores and concluded that "the old-fashioned spelling grind" that prevailed in most schools could be drastically curtailed.[11]

Rice had called on educators to establish standards for what children should know about not only spelling but also reading, penmanship, language, arithmetic, and other school subjects. Much of his later research focused on arithmetic. He began this work soon after his efforts on spelling, but editorial duties prevented him from analyzing and publishing the data.[12] At the 1897 meeting of the Department of Superintendence of the National Educational Association, he challenged school superintendents to answer two questions: How much time should be devoted to the study of specific subjects? What results should be accomplished? Rice hoped to find answers to these queries through scientific inquiry but found the superintendents unresponsive at best and sometimes openly hostile. Or, as he put it, "the ship of pedagogy, with respect to these two questions, has become waterlogged in a sea of opinions."[13]

A few years later, encouraged by the interest of Paul N. Hanus of Harvard University, Rice surveyed arithmetic teaching in eighteen schools in seven cities. He examined a total of six thousand children in grades 4 through 8. Each student took a test with eight word problems, with the problems differing for each grade. There was some repetition of questions from grade to grade "for the purpose of studying the growth in mental power from year to year."[14] Looking over the test results, Rice found that average scores for students in the same grade at different schools varied considerably, especially for seventh- and eighth-graders. This differed from scores on his spelling tests, in which results, especially for the more advanced grades, had been quite uniform across schools and school systems. Classes in the same city tended to group together as having scored well or poorly. This was true even if one compared students of differing home environments, ethnicity, and teachers. Test scores also did not increase with either the amount of time devoted to arithmetic teaching or with the assignment of homework in the subject. Rice concluded that supervisors played the major role in determining school achievement.[15]

Rice's tests required that students translate written prose into numerical terms and then carry out arithmetic operations. He realized that some of the operations he tested had not yet been taught to the students, although he argued that this factor accounted for only a small percentage of the variation among schools. Later testers not only would be more careful to test what students had been taught, but they would sometimes distinguish tests of arithmetic operations from tests that required knowledge of English prose. Even those who rejected Rice's tests remained most interested in the questions he raised. Were schools operating as efficiently as possible?[16] How could quantitative data be used to evaluate the quality of school systems?[17] What variables determined test outcomes?

Early Tests of Arithmetic

Rice urged educators to develop goals for the teaching of elementary subjects, devise tests to measure how well these goals were met, and use the tests to compare the efficiency of school systems. Several psychologists, particularly students of Edward Thorndike of Teachers College of Columbia University, took up the challenge. The tests they developed reflect several ideas of Thorndike that merit brief mention. First, Thorndike and his students, like Rice, had great faith in quantitative measurement. Thorndike's dissertation, his published papers, and his books were replete with graphs and numerical data. In the 1903 textbook *Educational Psychology* he wrote: "We conquer the facts of nature when we observe and experiment upon them. When we measure them we have made them our servants."[18] A pithier epigram that came to be associated with Thorndike and his colleagues was "All that exists, exists in some amount and can be measured."[19] Thorndike and his students not only made measurements but sought to present their data using the latest statistical techniques.[20]

Second, Thorndike believed that the human mind was best described as a combination of particular and independent capacities, associations, and responses. In a series of experiments carried out with his colleague Robert S. Woodworth, he found little transfer of training from one area of learning to another. In 1901 Thorndike and Woodworth argued that extensive study of one subject—such as Latin or mathematics—would not provide general mental discipline that improved learning in other areas. Educational authorities, particularly those who wished to expand science education, had argued for a half-century that mathematics did not deserve special status in the schools as a source of mental discipline that was central to a general education.[21] Thorndike and Woodward's results buttressed this view. They also led Thorndike to encourage those who made up tests to be careful in identifying the specific skills they were testing.[22]

Third, Thorndike would combine his faith in quantitative measurement and his belief in independent mental capacities to suggest that human traits could be measured upon a series of scales. In his book *Educational Psychology* (1910) he claimed that there were "scales for every thing in human nature . . . each person being recorded as *zero* in the case of things not appearing in his nature."[23] Both Thorndike and some of his students would try to develop such scales. Initially, they sought to describe groups, not to rank individuals.

In 1903 Thorndike and his student W. A. Fox, the superintendent of schools in Albion, Indiana, published a preliminary study of abilities involved in the study of arithmetic. They tested seventy-seven high school students on addition, multiplication, computations with fractions, and word problems. Scores on the four different tests showed relatively modest correlations. Girls scored slightly better than boys, with the variation in scores greater for boys.[24] A few years later Columbia University graduate student Cliff W. Stone extended this work, developing the first standardized test of arithmetic given to groups of children. Stone, a Wisconsin native, had attended the state normal school in Oshkosh, graduating in 1899. He then went on to Columbia University, earning a B.S. degree in education in 1904 and a Ph.D. degree at Teachers College in 1908. For his doctoral work Stone tested the arithmetical skills attained by sixth-graders. He prepared a set of fourteen problems in straightforward arithmetic operations as well as twelve written problems intended as a test of reasoning. Students were given twelve minutes for the first test and fifteen for the second. Stone administered the test, in conditions as nearly identical as possible, to three thousand sixth-graders in twenty-six school systems in six different states. Rice had not tried to say what learning arithmetic entailed. Stone, however, followed Thorndike and argued that arithmetic teaching was the training of several special abilities. Scientific teaching of the subject would develop these individual abilities as efficiently as possible.

Comparing test scores, Stone found considerable variability among school systems and even greater variation among children in the same school system. Like Rice, he found that schools differed markedly in the amount of time devoted to arithmetic—from 7 percent to 23 percent of class hours were devoted to the subject. Yet the time children spent learning arithmetic was not reflected in their test scores. If anything, students at schools devoting smaller portions of the day to arithmetic had higher scores on Stone's tests.[25]

In June 1909 Stone's dissertation received a laudatory review in the *Elementary School Teacher* from Chicago teacher and textbook author James F. Millis. Adopting the language of contemporary manufacturing, Millis suggested that "this scientific study provides the educational world with a means of beginning to standardize its product."[26] Stuart Appleton Courtis saw different possibilities in Stone's work. Courtis was a teacher at the Liggett School for Girls, a private institution in Detroit that had students in kindergarten through "grade thirteen." He had contemplated a career in electrical engineering but in 1898 abandoned his studies at the Massachusetts Institute of Technology to return to Michigan and be near the woman he loved. Courtis suggested that Stone's test could be given not just to sixth-graders but to students from third grade through high school. His school had recently introduced laboratory methods of instruction in arithmetic teaching. By comparing scores on tests taken by students over several years, Courtis thought he might arrive at "a rational basis for the estimation of the influence exerted by any method, material, or teacher."[27]

Courtis gave the Stone tests to students during the academic year 1908–9 and made a detailed study of the results. He quickly concluded, as Stone had suggested, that success in arithmetic depended on skills in several different areas. Finding that Stone's two tests did not distinguish these skills satisfactorily, Courtis abandoned the idea of using these examinations repeatedly and set out to develop his own tests. He enlisted the aid of all the girls in the school. An article he wrote for the school yearbook described the "strange exercise" in the assembly room: "A hundred heads bent over a hundred scurrying pencils, a silence vibrant with the energy of a hundred minds at work upon a single task; a signal, a pause, and a sigh of relaxation from a hundred lips as from the lips of one."[28]

Initially, Courtis introduced a set of four tests of speed in the four arithmetic operations that required use of whole numbers less than ten. Another speed test, added in the course of the year, measured accuracy in copying numbers. Courtis also included a reasoning test that required that students identify the arithmetic operation needed to solve a written problem. He rounded out his exam with tests of arithmetic operations and reasoning like those of Stone.[29] By April 1911 Courtis was sufficiently satisfied with his work to begin to distribute it nationally (Figure 3.1).[30]

Upon further consideration Courtis developed fundamental doubts about

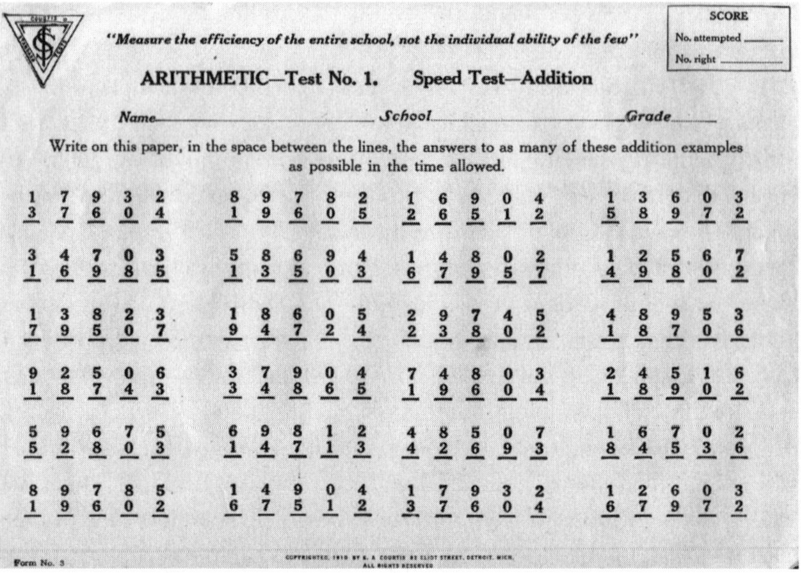

SCORE
No. attempted ______
No. right ______

"Measure the efficiency of the entire school, not the individual ability of the few"

ARITHMETIC—Test No. 1. Speed Test—Addition

Name ______ *School* ______ *Grade* ______

Write on this paper, in the space between the lines, the answers to as many of these addition examples as possible in the time allowed.

1 7 9 3 2   8 9 7 8 2   1 6 9 0 4   1 3 6 0 3
3 7 6 0 4   1 9 6 0 5   2 6 5 1 2   5 8 9 7 2

3 4 7 0 3   5 8 6 9 4   1 4 8 0 2   1 2 5 6 7
1 6 9 8 5   1 3 5 0 3   6 7 9 5 7   4 9 8 0 2

1 3 8 2 3   1 8 6 0 5   2 9 7 4 5   4 8 9 5 3
7 9 5 0 7   9 4 7 2 4   2 3 8 0 2   1 8 7 0 6

9 2 5 0 6   3 7 9 0 4   7 4 8 0 3   2 4 5 1 6
1 8 7 4 3   3 4 8 6 5   1 9 6 0 4   1 8 9 0 2

5 9 6 7 5   6 9 8 1 2   4 8 5 0 7   1 6 7 0 2
5 2 8 0 3   1 4 7 1 3   4 2 6 9 3   8 4 5 3 6

8 9 7 8 5   1 4 9 0 4   1 7 9 3 2   1 2 6 0 3
1 9 6 0 2   6 7 5 1 2   3 7 6 0 4   6 7 9 7 2

Form No. 3

Figure 3.1. Courtis Arithmetic Test Number 1, copyright 1910. This was the first of a set of eight tests on various aspects of the subject. Courtesy of Archives of the History of American Psychology, The University of Akron.

the validity of his examination. On the one hand, he found that the ability to add, subtract, multiply, and divide small numbers was not closely tied to success with more complex problems. Hence, he replaced his four speed tests with others that had more complex problems. On the other hand, he discovered that success on the reasoning test depended more on reading ability than on arithmetical skill and dropped this section. The second form of Courtis's product, called the Courtis Standard Research Tests, Series B, was the first widely available standardized test of arithmetic.[31] Courtis himself was hired to head the newly established Bureau of School Efficiency in the office of the superintendent of the Detroit Public Schools. He also participated in surveys of the schools of Boston, Cleveland, and New York City, attracting considerable publicity.[32]

Courtis distributed his tests under the motto "Measure the efficiency of the entire school, not the individual ability of the few."[33] He hoped that his tests could be used to show the strengths and weaknesses of individual students, yet he found that the scores of girls who took his early tests repeatedly fluctuated too much for him to rely on them as indicators of student ability or achievement. He eventually concluded that results were best used only by giving exams several times during any one period in a child's life and comparing them to scores obtained by the same child at a later time. In other words, tests were as much a measure of maturity, attention, and good health as of knowledge.[34] This was not the efficient tool that many had envisioned.

Despite Courtis's doubts, his tests spread widely. Between August 1914 and August 1915 over four hundred thousand of his examinations were sent to educators in forty-two states. Courtis sold tests from 1909 until 1938, when he withdrew them from the market because he no longer believed that they were satisfactory measures of student knowledge. During this period over twenty million of his tests sold around the world.[35]

Testing Individuals in Arithmetic

The early Courtis tests were not aimed at individuals. Test makers soon tried, however, to prepare standard examinations that would gauge both how much students had learned and what specific difficulties they had. In 1909 and 1910 Thorndike introduced scales for evaluating the handwriting of elementary school students that offered a model for early achievement tests. He assumed that student handwriting could be ranked along a single scale of steadily improving quality. The success of this handwriting scale led Daniel Starch of the University of Wisconsin and Clifford Woody, a graduate student at Teachers College, to propose scales for arithmetic. Using the results of tests given to 2,515 pupils in grades 3 through 8 who attended eighteen different schools, Starch prepared a scale with twelve word problems of increasing difficulty. He described his scale in publications and distributed it himself.[36] Woody made up a set of problems for each of the four arithmetic operations that increased steadily from the simplest examples to those that were so challenging that no more than half of eighth-grade children could solve them, even if they were given unlimited time. He presented his results in a 1916 doctoral dissertation, arguing that the range of problems he presented made it possible actually to test the work of children in the lower grades.[37] Woody's test was distributed by Columbia, first as the Woody Scales and then, in a revised form coauthored by William A. McCall, as the Woody-McCall Mixed Fundamentals.[38]

Starch and Woody hoped to measure what children had learned compared to other children. At about the same time, both Courtis and University of Chicago doctorate Walter Scott Monroe proposed tests that would show more precisely where individual students needed to do further classroom drill. Following medical language, such tests were called "diagnostic."[39] Both Courtis and Monroe assumed that skill in arithmetic was the result of training in separate specific habits, each of which a student needed to establish. Courtis distinguished seven skills associated with addition, ranging from adding two one-digit numbers to adding with carrying over to adding numbers of different lengths. Considering the other operations of arithmetic, he came up with thirty specific habits that needed to be tested.[40] For surveys of the schools of Cleveland, Ohio, and Grand Rapids, Michigan, Courtis devised a set of fifteen tests of various aspects of arithmetic, each of which took twenty-two minutes.[41] To help improve student skills in the different areas being tested, Courtis also designed a set of cards with practice exercises.[42]

As Monroe commented, "Any series of classroom tests must not require a large amount of time if they are to be used by any besides the most enthusiastic workers."[43] The five and a half hours required to give the survey tests that Courtis had developed for Cleveland was clearly excessive. Monroe proposed a set of twenty-one tests, each including only one or two examples, which could be given in a total of thirty-one minutes. Over the next decade several other psychologists developed examinations designed to identify areas of arithmetic in which children needed to do special practice. Teachers College graduate student John W. Studebaker also proposed additional materials for drill.[44] In general, however, teachers continued to rely on their own observations, quizzes, and recitations to pinpoint children's difficulties.

Testing in Algebra and Geometry

The same kinds of tests developed for arithmetic also were proposed for high school algebra. Tests of the ability of groups of students, scales of individual achievement, diagnostic tests, and remedial exercises all had their advocates. In 1915 Monroe, who was then at the Kansas State Normal School in Emporia, proposed a test of attainments of first-year high school algebra students. He hoped to set standards for ability in this subject, as Stone and Courtis had sought standards for arithmetic.[45] The previous year Thorndike had tried to show how teacher rankings of the difficulty of algebra problems could serve as the basis of a scale of problems of increasing difficulty. His paper was more an explanation of what a scale was than a proposed test.[46] In Illinois Harold O. Rugg of the University of Chicago and his collaborators built on Monroe's ideas and made a mind-numbing analysis of skills to be mastered in the first year of algebra. They also examined Thorndike's scale and found it entirely inadequate as a measure of student accomplishment. Teacher rankings of the relative difficulty of problems did not compare well with what students actually did. It also proved difficult to construct a scale that was evenly graduated as problems became harder. Moreover, solving the problems Thorndike had set up required the use of several different skills simultaneously. Hence the test would not reveal specific student difficulties.[47] Dropping the idea of a scale, Rugg and his associates prepared a series of tests of specific skills needed in algebra; it would be distributed through the University of Chicago bookstore.[48] They also developed a set of remedial exercises for drill in specific algebraic manipulations, which teachers could use according to the needs of students. By efficiently mastering routine processes such as factoring equations, they argued, students would have more time and energy to spend on the more important topic of understanding how these processes could be applied.[49] Thorndike and his associates at Columbia were undaunted by this attack on scales of achievement. In 1918 Teachers College graduate student Henry G. Hotz prepared a new set of scales for achievement in first-year algebra. Hotz's algebra scales would be distributed by the Bureau of Publications of Teachers College for at least a decade.[50]

Standard tests of plane geometry developed more slowly than those of arithmetic and algebra. The subject was not as widely taught; hence, there was less pressure to teach it as efficiently as possible. Furthermore, the skills required for success in geometry—spatial perception, precise statements of theorems, logical reasoning from hypotheses to conclusions, and coherent presentation of deductions—were not easily represented by short-answer tests. John H. Minnick, a former mathematics teacher at the Horace Mann School of Teachers College and a graduate student at the University of Pennsylvania, developed the first widely used standardized geometry examination. Minnick focused on four aspects of the subject—drawing a figure, stating hypotheses and conclusions in terms of the figure, recalling results that followed from given facts, and selecting and arranging information needed to arrive at a conclusion. He prepared tests for each of these skills, tried the examinations in the classroom, and presented his results in his 1918 dissertation.[51] A form of the test was distributed by Public School Publishing Company of Bloomington, Illinois, at least as late as 1929.[52]

Detailed studies of the marking of conventional geometry tests led to new forms of examination. In 1913 Daniel Starch and Edward C. Elliott had compared the scores assigned by different high school teachers to the same test paper. They found that when 116 geometry teachers were given the same paper to grade, they assigned it marks that ranged from under 30 to over 90 out of a possible 100.[53] Starch also reported that teachers were not consistent in grading. If the same teacher evaluated a paper a second time, the two sets of marks often varied considerably.[54]

Starch and Elliott's findings, as well as other studies of school marks, were used to justify the adoption of standardized tests.[55] Even when tests were uniform, however, evaluating achievement in geometry proved difficult. By the early 1920s sufficient numbers of students were taking the College Entrance Examination Board achievement test in plane geometry to justify statistical analysis of the results. Teachers College graduate student Ben D. Wood analyzed results from the 1921 exam, which consisted of six questions, four proofs, and two computations requiring knowledge of geometry. Wood found that when two different readers scored the tests, results were quite uniform—these teachers did not suffer from the problems of inconsistency noted by Starch and Elliott. Yet students had much more difficulty with some parts of the examination than with others. Comparing scores on the three even-numbered problems with results from the three odd-numbered problems, Wood found a low correlation between scores on the two sections. A considerable number of students judged on only half of the exam questions would pass or fail depending on the part of the exam that was selected.

This conclusion does not seem surprising, particularly as all three of the odd problems were proofs, while the even problems consisted of one proof and two computations. Wood's deductions from his observation are perhaps more curious. He claimed that achievement tests based only on a small number of questions were inherently unreliable. He noted that the portion of the students who had passed the College Boards plane geometry examination varied considerably from year to year and suggested that problems requiring the full presentation of proofs should be replaced with a much larger number of true-false, fill-in-the-blank, and short-answer questions.[56] In a paper read to the American Mathematical Society in October 1922,

William L. Crum of the Yale University mathematics department argued that Wood's use of statistics was highly questionable, but Crum's paper was not published until the fall of 1923.[57] By then Wood had republished his conclusions about the reliability of the College Board examinations in geometry in his influential doctoral dissertation *Measurement in Higher Education*.[58] They would be reprinted without question by later authors.[59]

In 1921 Raleigh Schorling, who taught mathematics at the Lincoln School of Teachers College, had prepared a geometry examination in the general form favored by Wood, emphasizing numerous short questions rather than lengthy proofs. Schorling soon left New York, but Vera Sanford, also of the Lincoln School, prepared revisions of the exam and tested their reliability. Sanford suggested that the part of the test that required filling in blanks might be better given as a multiple-choice test, as it would then be less of a test of linguistic ability.[60] This is the first mention of a multiple-choice question in a standardized test of mathematics that we have found, although multiple-choice (also called "multiple-answer") questions had been used in tests of reading since at least 1915.[61]

Sanford hoped that the new geometry test would not only be more fair than its predecessors but would assist in uncovering student difficulties. Also, as Teachers College professor Clifford Upton pointed out, it could save the time of skilled teachers. Upton wrote in 1923: "The College Entrance Board in its June examinations uses the energy of approximately twenty-five to thirty well-trained mathematicians, many of them college professors, for a period of about two weeks, in grading the papers of the plane geometry examination. It seems that this energy could serve society better than to be used in grading papers of doubtful value."[62] Others at Columbia also continued to work on geometry tests. Ben Wood and Columbia College dean Herbert E. Hawkes published an achievement test in geometry. Following the style that Wood preferred, they omitted lengthy proofs, testing the essence of plane geometry in sixty-five true-false and thirty-five fill-in-the-blank questions.[63] More generally, "new style objective tests," designed especially for easy marking and made up of numerous short-answer questions, were recommended for classroom use.[64] The College Entrance Examination Board started using the format for its aptitude tests in 1926 and for its achievement tests in 1937.[65] The New York Regents Examinations took this form from the mid-1920s. Objective tests were soon introduced for admission to law and medical schools and for the New York bar examination. By the 1930s Wood was working with Reynold B. Johnson and the International Business Machines Corporation on machines that could score such tests automatically.[66] By the late 1940s even the Society of Actuaries had adopted this style of examination for some of its preliminary tests.[67]

Prognostic Tests

From the nineteenth century physicians and psychologists had sought ways to predict how well students would do in school. In 1905 the Frenchmen Alfred Binet and Theodore Simon introduced a set of simple tests designed to determine whether young children who were mentally disabled would benefit from being placed in regular schools. These tests, unlike the mathematics examinations we have described, were given to students individually. Stanford psychologist Lewis M. Terman prepared his own version of the Binet-Simon test and represented the results as one hundred times the ratio of a child's mental and chronological ages—that is to say, as an intelligence quotient. During World War I Edward Thorndike, Ben Wood, and other psychologists volunteered their services to the United States Army. As officers in the Sanitary Corps, they developed group tests of intelligence that could be given much more rapidly than the tests of individuals that had been carried out previously. These tests were meant to weed out those recruits who were mentally unfit for overseas service.[68]

In this same period Agnes Low Rogers, a Scottish-born graduate student at Teachers College, worried about devising tests that would weed out high school students who were unfit to study advanced mathematics. Twenty years earlier it had been assumed that all high school students required the mental discipline associated with the study of subjects such as mathematics and Latin. Thorndike and Woodworth's research had suggested that mental training in one area was not necessarily transferred to work in other disciplines. Learning mathematics was not necessary to build the mind. Moreover, high schools were accepting a much wider range of students, some of whom were more interested in clerical or other vocational training than academics. Might it be possible to test a student's aptitude for mathematics so as to give a "prognosis" for success in the subject? Those who lacked mathematical ability could then be directed to more fruitful areas of study.

Further incentive for such testing came from the educator Leonard Ayres. In 1909 Ayres published a book entitled *Laggards in Our Schools,* in which he argued that schools were not promoting students efficiently. Many students were older than their class grade suggested they should be. Ayres had not collected data on how long children had actually been in school, but he argued that training slow pupils was an enormous waste of taxpayer dollars.[69]

The demand for greater efficiency in promoting students, combined with the growing diversity of student interests, led Rogers to propose a new test of mathematical ability. This prognostic test was to be given to students after they had studied both algebra and geometry for a few months. As Rogers explained: "It would seem right in a democracy that no child of normal intelligence should be deprived of the opportunity of becoming acquainted with a realm of thought, which has meant so much for the advancement of science and of civilization as mathematics. Nor should he be deprived of a training in the use of the tools of quantitative thinking, except where he lacks the ability to profit by it."[70]

In her research Rogers tried out seventeen examinations that tested algebra, geometry, reasoning, arithmetic, and linguistic ability. Comparing test scores, she judged the interdependence of the traits they measured. Success on algebra and geometry tests proved to be no more closely correlated than success in either of these subjects and language. This revelation not only suggested the importance of a command of the vernacular in mathematics but also indicated that the skills required to learn the two mathematical subjects were not closely related. The result, Rogers commented, might be problematic for those mathematicians who hoped to unite the teaching of the two subjects. Rogers concluded that by combining student scores on tests of algebraic computation, interpolation, adding missing steps in a series, geometry, superposition of geometric figures, and matching problems and equations, she could make satisfactory predictions about how well students at Teachers College's Horace Mann School would do in mathematics. Rogers stressed that her tests said nothing about the sources of mathematical success or failure but merely reflected correlations. Her "sextet" of tests was distributed for classroom use by the Teachers College Bureau of Publications.[71] She continued to improve and revise the exam, introducing a second form in 1921.[72]

William David Reeve, a professor at Teachers College, commended Rogers's test in a 1924 article in the *Mathematics Teacher.* He even went a step farther, suggesting that students need not be exposed to even a smattering of algebra and geometry. Reeve suggested that other measures of "abstract intellect" might work even better than the tests that Rogers had chosen. Those who failed such tests would not be required to take any courses beyond arithmetic, "doing justice to a large number of pupils who are forced to study mathematics against their will and in many cases without profit."[73] Reeve's colleague, the historian of mathematics David Eugene Smith, was more cautious. In March 1923 he wrote in the *Teacher's College Record,* "The prognostic test at its best achieves quickly and with improved results that which the schools have heretofore discovered after a loss of valuable time; at its worst it leads into a determinism that is more dangerous than the extreme form of Calvinism which left each individual without hope."[74] Tests of mathematical aptitude would soon be incorporated into test batteries given to students entering high school and used both in guidance and in grouping children into different sections. Thus, by the 1920s advocates of testing believed that well-designed short-answer tests could provide objective, efficient standards for evaluating students from elementary school through college. These new educational experts, buttressed by scientific demonstrations and statistical authority, shaped both the practice of teaching and the direction of student lives.

CHAPTER FOUR

The Overhead Projector
Snapping the Class to Attention

The usual way is to work from individual textbooks, use the blackboard extensively, and pass around one-of-a-kind examples. The result: time is wasted and students are distracted. A better way is to snap the class to attention with a big bright image that gives every student a single focal point.

As a teacher first pointed out to us and subsequent study amply confirmed, the overhead projector has been ubiquitous in mathematics classrooms. The device originated in the nineteenth century as a tool of science lecturers. Its general use in the classroom emerged from a new group of experts, authorities on what was called "audiovisual instruction." During the 1920s and 1930s various staff members associated with film companies, urban school districts, museums, and state departments of education explored the possibilities of using "visual instruction." They were particularly interested in expanding the role of photography in exhibits, slides, filmstrips, and movies. Visual aspects of geography, history, literature, science, and industrial arts received special attention. Research in the area was also carried out at schools of education associated with universities such as Columbia, Wisconsin, Yale, and Chicago.[1]

With the outbreak of World War II, the U.S. armed forces needed to train troops quickly and efficiently, often with inexperienced instructors. In addition, as one later account put it, "The majority of men training for military duty were not accustomed to serious study and prolonged mental concentration."[2] To assist them, the armed forces greatly extended their use of "training aids" such as movies, filmstrips, charts, and, to a lesser extent, overhead projectors. Such programs required not only people with subject area knowledge but those specially trained in using media for educational purposes. Thus, to make up for a lack of expert teachers, the armed forces called for experts in instructional technology.

After the war audiovisual instruction not only continued to be of military interest but also began to be widely used in schools. Military contracts offered an incentive for private firms to develop new apparatus and often helped determine the capabilities of machines produced. After passage of the National Defense Education Act in 1958 and the Elementary and Secondary Education Act of 1965, federal matching funds became available for schools to purchase such equipment. By the 1960s schools of education had entire departments devoted to what had come to be called "instructional technology," or "educational technology."

Early advocates of educational technology emphasized the benefits of replacing verbal instruction by teachers with materials that appealed to a wider range of senses.[3] The overhead projector at best enhanced rather than replaced the lecture. Nonetheless, the story of the instrument well illustrates the impact of audiovisual educators. U.S. Army and Navy trainers adopted the device on a limited basis during World War II and encouraged its improvement in the 1940s. Their specifications recast the product into one that proved to be better suited to general classroom use. By the 1960s lower prices, new methods of producing transparencies and projectors, federal funds, and promotional efforts by manufacturers and educators combined to make the overhead projector a common tool in many mathematics classrooms. Manufacture and use of these instruments required sophisticated technologies—a power grid supplying electricity uniformly; methods for producing cheap, precise, lightweight lenses; and new techniques for making transparencies. These technologies succeeded so well that overhead projectors became an unremarkable classroom commonplace. During the last forty years of the twentieth century transparencies and objects were introduced specifically to teach mathematics with overhead projectors, including slide rules, coordinate grids for graphing, and electronic calculators as well as devices for elementary school teaching. Hence, the use of overhead projectors recast familiar mathematical teaching tools.

Despite the widespread diffusion of overhead projectors, there is very little basic historical literature about them. Existing accounts tend to reflect the view of specific manufacturers.[4] Here we explore the history of this tool as an illustration of the rise of a new group of educational authorities, placing particular emphasis on mathematics teaching.

Small Stage Overhead Projectors

Images projected onto or through a screen were used in popular scientific lectures from the nineteenth century. The oil or gas flame of a "magic lantern" produced dramatic views from photographs or paintings on glass. Magic lanterns also could provide enlarged images of biological specimens, most notably microscopic organisms.[5] A few scientists and opticians wished to project horizontal arrangements of objects, such as waves propagating in a tank of water or iron filings around a magnet. By the 1870s the French optician Jules Duboscq, the English physicist John Tyndall, and the American chemist, lecturer, and later university president Henry Morton had devised ways to project images from horizontal surfaces onto a vertical screen. These images could then be studied by an audience of students or the general public. Morton's "vertical lantern" was manufactured by a small firm in Hoboken, New Jersey, and distributed by it and by other lantern slide supply companies (Figure 4.1).[6]

In the early twentieth century the invention of new illumination systems,

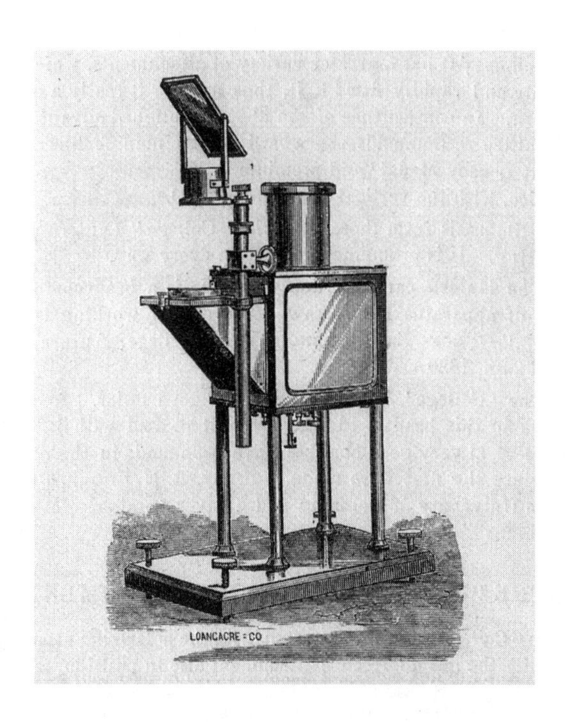

Figure 4.1. The College Lantern, an early overhead projector, 1881. From Edward L. Wilson, *Illustrated and Descriptive Catalogue of Magic Lantern Slides, Magic Lanterns and Appliances* (Philadelphia: E. L. Wilson, 1881), 145. Courtesy of Smithsonian Institution Libraries.

and the growing availability of electrical power encouraged improvement in projectors, including overhead projectors. The Spencer Lens Company of Buffalo, New York, offered a "College Bench Lantern," which it sold directly and through dealers such as Central Scientific Company of Chicago. From as early as 1909, customers could purchase a "vertical attachment" for this projector that was "designed expressly for the illustration and exhibition of the various phenomena attending the scientific experiments relating to heat, light, sound, electricity, magnetism, cohesion, figures and crystallization." The same attachment could be used to project images of the sun captured with the Spencer Company's Heliopticon.[7] By 1912 the firm was manufacturing an enclosed projector, illuminated with a carbon arc light instead of limelight. This machine was called a Delineascope and came with a vertical attachment.[8]

In the 1920s Charles A. Kofoid, a zoologist at the University of California at Berkeley, suggested to the Spencer Lens Company that professors would value a projector that they could operate themselves while still facing the class.[9] By 1929 the company was advertising the "Spencer Lecture Table Delineascope" as "the only lantern in the world permitting the lecturer to face his class and the lantern." The instrument was designed for projecting either 3¼ × 4 inch glass slides or transparent specimens placed on glass. The image appeared on a screen over the head of the seated speaker, and a sharp pencil could be used to point out specific features on the slide. This Delineascope could also be equipped with an attachment for projecting "Filmslides" (filmstrips). The apparatus was relatively compact (5 in. [w] × 13 in. [d] x 23 in. [h]) and weighed 14 pounds. According to advertisements, it deserved a "permanent place on every lecture table, ready for instant use." It sold for $75, or $113.50 with the Filmslide attachment.[10]

The Spencer Lens Company soon had rivals in the area of overhead projection. Bausch & Lomb Optical Company of Rochester, New York, had offered a "Balopticon," or "projection lantern," for lantern slides since 1908.[11] In about 1930 it began selling an attachment to the Balopticon that allowed a speaker to sit at a desk facing the audience, place glass slides on the stage of the projector, and have images appear above his or her head (Figure 4.2). Bausch & Lomb actually called the instrument an "overhead projector," the first use of the term we have encountered.[12]

Advertising for this instrument, like that for the Spencer Lecture Table Delineascope, stressed that it permitted a speaker to maintain eye contact with listeners, to change slides without an assistant, and to point out details

Figure 4.2. Bausch & Lomb Overhead Projector, 1934. From Bausch & Lomb, *Balopticons and Accessories* (Rochester: Bausch & Lomb Optical Co., 1934), 34. Courtesy of Smithsonian Institution Libraries.

on the image using a pencil or similar object, without needing a special pointer. There is no specific mention of projecting anything but glass slides and no explicit reference to educational use. The Balopticon overhead projector cost $75 to $80 in 1930; the projector attachment alone was $40. These items were listed in Bausch & Lomb catalogs through the 1930s, with somewhat different models introduced toward the end of the decade and further modification after World War II.[13]

These early overhead projectors had even less influence in the mathematics classroom than other projectors. A 1941 bibliography prepared by the Committee on Multi-Sensory Aids of the National Council of Teachers of Mathematics (NCTM) made no reference whatsoever to overhead projectors. This report did note the availability of several sets of slides, a few "film slides," a small number of films, and projectors for them. It also listed articles describing how teachers might make their own glass slides.[14] The committee went on to compile a book entitled *Multi-Sensory Aids in the Teaching of Mathematics,* which appeared in 1945 as the eighteenth yearbook of the NCTM. The section of the book devoted to "motion pictures" included discussion of opaque projectors, glass and 35 mm slides, filmstrips, stereographs, and motion pictures. There was no mention, however, of overhead projectors.[15]

From the Bowling Alley to the U.S. Armed Forces

The overhead projector also found use outside of academe. In the late 1930s the Brunswick-Balke-Collender Company of Chicago, a manufacturer of billiards tables and bowling equipment, began selling "Tel-E-Score" units for use in bowling alleys.[16] These devices projected written scores of bowling games onto a screen at the head of the alley. They were both more compact and easier to update than the blackboards previously used for the purpose and apparently proved sufficiently successful to be purchased quite widely. The units occupied the floor space of a small desk and projected an image in front, rather than over the head, of the user. Plans for this projector initially came from Harold G. FitzGerald, a Los Angeles inventor. A modified form of FitzGerald's machine was reportedly used by a mathematics teacher, but the practice did not become widespread.[17] In July 1940 FitzGerald and John B. Coker of San Diego applied for a patent for "score projecting apparatus" that would be granted three years later.[18] It was sufficiently popular that in April 1942 Coker applied for a patent for a projector that allowed two groups of scores written on a single sheet to appear on adjacent screens, rather than one above the other, which made it possible to use the score projector in bowling alleys with low ceilings. Coker received his patent in 1945 and promptly assigned it to Brunswick-Balke-Collender.[19]

During World War II the United States armed forces made extensive use of audiovisual materials in training troops. The effort centered on film, a wide variety of graphics, and manuals.[20] To teach soldiers about the inner workings of mechanisms, however, the Army Ordnance School at Santa Anita, California, asked FitzGerald to design a new form of overhead projector. A wartime project, it was called the Victorlite. By late 1944 Victorlites were used by the army and navy, had been tried in the Los Angeles city government, and were owned by a few local schools.[21]

Postwar discussion of military use of overhead projectors described them

Figure 4.3. Overhead projector and classroom, United States Military Academy, 1961. Courtesy of Frederick Rickey and the United States Military Academy Archives.

as "bowling alley projectors" and "visual aid projectors." They were mentioned in a 1947 publication as devices developed for military training that might be introduced into American education more generally.[22] Appropriately enough, such early use of overhead projectors occurred at the U.S. Military Academy (USMA) at West Point (Figure 4.3). After the war modified Victorlites sold as Visual Cast projectors (transparencies were dubbed visualcasts, or VCs). The 1949 report of the USMA superintendent mentioned that the mathematics department had tried out two of these projectors. The trial proved satisfactory enough that soon the machines were being used by other departments.[23] Other schools would adopt the new technology more slowly.

Large Stage Overhead Projectors

In the postwar years technical improvements transformed the overhead projector and the production of transparencies. These changes reflected military training needs, as spelled out in government contracts. One particularly enthusiastic and influential advocate of the overhead projector was Allan Finstad. Finstad had received an M.A. degree in education in 1943 from the University of Washington. His thesis was representative of prewar advocates of visual instruction; he wrote about the use of motion pictures, particularly in science teaching. From 1943 to 1946 Finstad was a training aids officer in the U.S. Navy. After the war he stayed on as the civilian head of the Navy's Training Aids Division. In 1948, amid renewed concern about national security, he persuaded the engineer and former fellow training aids officer George Beckwith to join him in his work. Finstad and Beckwith wished to produce transparencies from illustrations in 8 × 10 inch printed manuals. The Ozalid Division of General Analine and Film Corporation in Johnson City, New York, had imported and developed chemical processes for making copies using diazonium salts. The dry print machines could produce the color transparencies needed by the navy at a cost of about fifty cents, compared to the ten to fifteen dollar expense of color photographs of the same size. Mindful of the size of the prints they wished to copy, Finstad and Beckwith wanted an overhead projector with a larger stage than had previously been built. Specification for a U.S. Navy contract prepared in 1948 described the overhead projector they wanted. Features were to include a 10 x 10 inch stage for the transparencies and a sufficiently bright image so that the projector could be used with ordinary lighting in an undarkened room.[24]

Sixteen firms bid on the contract for five hundred of the new machines, most of them planning to use a system of glass optical components. Charles Beseler Company, a maker of opaque projectors since 1898 and of overhead projectors since 1943, proposed, however, to incorporate a plastic lens into the new overhead. This approach not only reduced the weight of the machine but also made it much cheaper to build. Beseler offered to sell its projectors for $145 apiece, while other firms listed prices of up to $1,000 per unit. The company not only won the initial bid but soon was supplying further machines to the navy as well as overhead projectors for the army and air force.[25]

Beseler was also happy to sell projectors to civilians. An advertisement for its Vu-Graph from about 1952 indicated that it could be adjusted to project 3¼ × 4 inch slides, 2 × 2 inch slides, or 35 mm filmstrips. The device also had an attachment that carried a roll of cellophane across the 7 × 7 inch lighted stage of the projector on which a speaker could write directly with a wax pencil. The Vu-Graph allowed lecturers to face their audiences to project images above and behind them and to speak in a fully lighted room.

Preparing images did not require purchasing or making slides or transparencies.[26]

Allan Finstad did much to publicize the new instrument, beginning with his work for the navy and then, from 1951 to 1956, as an "educational representative" for Charles Beseler Company. He initially prepared demonstrations and manuals primarily for army and air force centers but soon expanded his reach to include universities, colleges, and large city school systems. By 1957 Beseler was advertising a Vu-Graph Teaching Center, a full-sized desk with an inset overhead projector that displayed images as large as 10 × 10 inches (Figure 4.4). The desk and projector combined cost

Figure 4.4. Brochure showing Beseler Vu-Graph Teaching Center, ca. 1957. Courtesy of Smithsonian Institution Libraries.

$550, a substantial sum for most school systems. Nonetheless, Beseler clearly had a broad market in mind. Sitting at its teaching center was a woman with her sleeves rolled up, not a man in suit and tie or navy uniform.[27]

By 1960 overhead projectors took several forms. Spencer (by then the Instrument Division of American Optical Company) had modified its Delineascope to project images from 10 × 10 inch transparencies as well as glass slides. One could also scroll a roll of acetate across the projection stage of the machine. Central Scientific offered this Delineascope for $315.[28] A challenge to Ozalid transparencies came from Minnesota Mining and Manufacturing Company (3M) in St. Paul, Minnesota. Scientists there invented thermal processes to produce copies from text. Staff of the Thermo-Fax Division envisioned a large educational market for overhead projectors and hence for the firm's transparencies and Thermo-Fax copiers. In December 1960 the company announced its first overhead projector, a 40-pound machine that cost $395.[29] By September 1962, 3M also was selling a special material for making transparencies.[30]

During this period a team of 3M scientists, led by Roger H. Appeldorn, developed a method for producing plastic Fresnel lenses in quantity or, as he later put it, "by the yard."[31] In February 1963, the company announced its Model 66 overhead projector, which was relatively small, simple to operate, reliable, and cost only $159.[32] Some other overhead projector makers lowered their prices to compete, making the product within the means of more school systems.[33] For those without $159, Edmund Scientific Company of Barrington, New Jersey, offered a kit for assembling projectors. An advertisement in the November 10, 1961, issue of *Science* urged readers to "make your own $300 overhead projector for less than $50." Edmund Scientific supplied the projection lens, condenser lens, lamp, socket, and directions for $45. The case and wiring would be purchased locally; apparently, there was no fan.[34] The product was sufficiently successful to be offered at least through 1968—by then the total cost was estimated as less than $40. Edmund sold the lenses and instructions for $17, offered several other components separately (including a fan and motor), and expected purchasers to come up with their own lumber and wire.[35]

The Overhead Projector in the Cold War Classroom

Around 1960 novelty alone may have led some teachers to try overhead projectors. Appealing new devices have usually found at least a few customers.

General acceptance of the new tool depended on several other developments. The first was the general expansion of the school-age population. Numerous communities were building and furnishing new schools and sought to equip them with the latest improvements. The overhead projector was sold both as a replacement for the blackboard in temporary classrooms and as a modern instructional aid. Some teachers tried the machine on an interim basis while teaching in temporary classrooms and continued to use it once chalkboards were available again.[36] Others adopted the device more directly. At a few schools, such as those in Newton, Massachusetts, overhead projectors were hailed as an answer to the shortage of qualified teachers. The instrument allowed teachers to write text that could easily be seen by 65 to 400 students at a time.[37]

Similarly, instructors at Millersville State College in Pennsylvania tried large group instruction in elementary college mathematics. They compared test scores from classes of 40 students with those from classes of over 140 students. To handle the larger classes, they used a microphone, an overhead projector, and student assistants, who taught help sessions. The authors reported in 1965 that students in the larger classes appeared to do somewhat better than those in the small classes, although they declined to draw broad conclusions.[38] This research might have been undertaken in response to a 1963 report that had been prepared under the direction of J. Sutherland Frame of Michigan State University for the Conference Board of the Mathematical Sciences. This monograph, published under the title *Building and Facilities for the Mathematical Sciences,* anticipated rapid increases in mathematics enrollments. It recommended that introductory mathematics classes at universities increase in size from about twenty-five students to a few hundred, if need be, with projection apparatus available as required.[39]

A second aid to the spread of overheads was the U.S. federal government. The National Defense Education Act of 1958 and the Elementary and Secondary Education Act of 1965 offered money for both research on the use of audiovisual equipment and the purchase of such equipment by schools. Journals for mathematics teachers and audiovisual specialists as well as manufacturers publicized the program.[40] Publications prepared for the U.S. government reported on the present and possible future use of a wide range of audiovisual materials, noting the spread of overhead projectors in particular.[41]

Of course, those making overhead projectors also publicized their wares. In addition to advertisements and special classes, there were publications on the overhead projector such as those by Horace C. Hartsell of the education department at Michigan State University. Hartsell was one of the first to advocate widespread use of the overhead projector in mathematics teaching. He wrote in a 1958 article in *Audiovisual Instruction* that "probably the greatest innovation for the teaching of arithmetic in the last 10 years has been the introduction of the overhead projector."[42] Here he particularly commended the open face of the machine, which allowed teachers to superimpose projected materials, and the large work surface for writing. Calhoun C. Collier, also of Michigan State University, quoted Hartsell's comments approvingly in the November 1959 issue of *Arithmetic Teacher.*[43] The next year Hartsell and Wilfred L. Veenendaal wrote a booklet entitled *Overhead Projection,* which was published for the Instrument Division of American Optical Company, maker of the Delineascope.[44]

A more extensive campaign to publicize the overhead projector in the schools came from 3M. To be sure, the marketing department reportedly first left it to Roger Appeldorn and others designing overhead projectors to make sales calls on schoolteachers.[45] These efforts proved sufficiently successful that in 1962 the company produced a twenty-five-minute film entitled *A New Dimension in Teaching,* which described how the overhead projector could be used in the classroom. As one might expect, the film discussed not only techniques of overhead projection but the making of transparencies.[46] The company also took steps to train teachers in the use of overheads. The Detroit school system, for example, received equipment for a "mobile training center" housed in the trailer of a truck. The air-conditioned unit seated twenty-one and came equipped with overhead projector, Thermo-Fax copier, and screen.[47] More generally, in August 1963, 3M president Bert S. Cross announced plans to give $1.5 million in overhead projectors and related materials to schools across the country.[48] The following May the company presented five hundred schools with identical gifts of equipment: eleven Thermo-Fax overhead projectors, two copiers, and supplies for making transparencies.[49]

These efforts to promote the overhead projector had only a modest impact on the literature of mathematics teachers. In the 1959 article from *Arithmetic Teacher* mentioned previously, Calhoun Collier described the instrument as only one of several aids to arithmetical understanding.[50] A 1960 article in *Mathematics Teacher* by Viggo P. Hansen of the University of Minnesota High School in Minneapolis was more focused, describing in detail transparencies that might be useful in teaching set theory, geometry, and arithmetic.[51] Similarly, the February 1962 issue of the magazine included an article by Alan R. Osborne of the University School at the University of Michigan that discussed the use of the overhead projector in algebra teaching, with particular emphasis on graphing.[52]

Mathematicians and mathematics teachers may well have concluded that use of the overhead projector required little explanation. They took steps to ensure that the instruments would be available to those who wished to use them. The 1963 report *Building and Facilities for the Mathematical Sciences* discussed both chalkboards and overhead projectors at some length, suggesting that mathematics buildings should be equipped with both and with the necessary screens, electrical wiring, and facilities for making transparencies.[53] A summary of relevant sections of this report appeared in *Mathematics Teacher* in 1964, and overhead projector manufacturers began advertising in the journal the following year.[54] Thus, it is hardly surprising that in 1966, when the School Mathematics Study Group prepared a revised edition of its experimental textbook *A Brief Course in Mathematics for Junior High School Teachers,* it included a supplement with 214 masters to be used in the preparation of transparencies.[55]

A more general perspective on the diffusion of the overhead projector in mathematics teaching during the 1960s and early 1970s comes from examining *Instructional Aids in Mathematics,* the thirty-fourth yearbook of the NCTM, published in 1973. This volume included discussion of textbooks and other printed materials, teaching machines and programmed instruction, electronic calculators and computers, models, and elementary school manipulative devices. In a lengthy article on "projection devices" Donovan R. Lichtenberg of the Department of Secondary Education at the University of South Florida commented that motion pictures and television "have not been enthusiastically received by mathematics teachers." In contrast, the overhead projector "has seen overwhelming acceptance in the past decade." Indeed, in some classrooms students were "so accustomed to the overhead projector that it no longer commands their attention."[56]

Lichtenberg attributed the success of the overhead projector to factors long mentioned by projector manufacturers—the ease of writing on a horizontal surface rather than a vertical blackboard, the control a teacher gained by facing a class, and the facility with which students could take notes in a room that was not darkened. It seems equally likely, however, that the overhead projector, like the chalkboard, became popular with teachers precisely because it did not "command"—in the sense of require—attention. The machine was simple, flexible, durable, sturdy, and relatively unchanging. It required little maintenance and had few parts that could malfunction. Indeed, the Model 66 overhead projector introduced by 3M in 1963 remained in production through 1975. Changes made to types of projectors and methods of producing transparencies did not require extensive retraining. In other words, the overhead projector allowed the teacher to concentrate on teaching and to accumulate skill—and transparencies—from year to year.

Mathematical Instruments for the Overhead Projector

Decades passed between the introduction of the blackboard into the American mathematics classroom in the early nineteenth century and the preparation of special blackboard drawing instruments after the Civil War. Once the overhead projector entered the classroom around 1960, materials to be used with it came more quickly. The first of them tended to be especially suited for use in high schools and elementary college courses.

"Projectuals" for the mathematics classroom included special slide rules. In 1955 the U.S. patent office received applications for two related patents for slide rule demonstration devices to be used with an overhead projector. The first proposal was from Paul D. Grimmer of Jackson Heights, New York. The second came from two other New Yorkers, Thomas C. Coale of Levittown and Wesley F. Heyman of Glen Cove. Both patents were granted on July 8, 1958. Large demonstration slide rules had long been sold for classrooms but were awkward to use and showed only one side of the instrument. The transparent slide rules used with overhead projectors were cheaper, easier to manipulate, and more convenient. The two sides of a duplex slide rule could be reproduced in transparent plastic, with scales from the front and back of the rule placed parallel to one another. The scales of the front and back of the slide were drawn on another plastic sheet that slid over the front, lining up with appropriate scales below. There was also a transparent sliding indicator that fit over the entire instrument.[57] In the course of the 1960s slide rules for use with an overhead projector were sold by overhead projector manufacturer Charles Beseler Company of East Orange, New Jersey, as well as by such American slide rule dealers as Keuffel & Esser Company of New Jersey and New York, Frederick W. Post Company of Chicago, and Pickett of California. German slide rule makers such as Nestler, Aristo, and Faber-Castell also sold the product. Other vendors included educational supply houses such as Welch Scientific Company.[58]

The overhead projector also offered a new way to present graphs and other visual information. Cloth painted with liquid slating and then marked with a rectangular grid of lines had been sold for kindergarten use at least as early as 1888. By the 1920s such "graph charts" were available in several sizes—and continued to sell for decades.[59] Early articles on the overhead projector discussed how grids of lines might be transferred to a transparency for graphs done in either rectangular or polar coordinates. The ambitious teacher might even use a series of overlays to give a dramatic view of geometric theorems or demonstrate the intersection of sets through Venn diagrams.[60] Prepared transparencies showing a grid of lines were not widely mentioned, although they were on sale at least by 1963.[61]

In the 1970s, when the electronic calculator displaced the slide rule as a routine tool of computation, overhead versions of this device were made.[62] Yet it was only when calculators became more complex, capable of both symbolic manipulation and graphing, that projecting calculator screens attracted considerable pedagogical interest. By May 1989 teachers using the HP-28S calculator could purchase a separate overhead display that allowed them to show an entire class equations and results. This device received the output of the calculator into a separate viewing screen, which was then projected.[63] Similarly, when Texas Instruments introduced its TI-81 "graphics calculator," it also sold a special overhead projection unit to display the calculator output.[64] Another calculator manufacturer, Casio, made rather different provisions for teachers. In addition to its fx-7000G graphing calculator, it sold a form of the instrument for teachers known as the OH-7000G. This came with a separate special lens that projected and enlarged the display onto a wall or screen.[65]

By the mid-1990s overhead projectors had also become commonplace in many elementary and middle school classrooms. Catalogs of mathematics education material contained separate sections devoted to "overhead materials." A 1997 catalog of Delta Education, for example, included overhead geoboards, clocks, thermometers, pattern blocks, attribute blocks, tangrams, Cuisenaire rods, number lines, dice, bills and coins, and rulers. For more advanced students there were algebra tiles and transparent protractors as well.[66]

At the end of the twentieth century those who discussed the use of "technology" in the classroom rarely had overhead projectors in mind. Those authors who mentioned it tended to joke that the overhead projector had reached the bowling alley decades before it was in the classroom. Seeing computers in bowling alleys, they predicted that they might soon be in academia as well.[67] Closer study reveals that overhead projection emerged in the nineteenth-century lecture hall, long before its application in popular sports. By the 1960s military patronage, sophisticated design and manufacturing techniques, and new federal funding for classroom equipment transformed the technique into a practical teaching tool. It received little of the attention of contemporary novelties such as teaching machines, language laboratories, and educational television but was adopted to teach a wide range of subjects, including mathematics. Indeed, the development of the overhead projector into a robust, user-friendly, relatively inexpensive, stable instrument, based on sophisticated materials science, offers a model that advocates of other educational technologies might do well to study.

CHAPTER FIVE

Teaching Machines and Programmed Instruction
A Lifeline in a Sea of Students

The programming movement has a very broad appeal. It offers to psychologists a new means for controlled experimentation, to educational administrators a lifeline in a sea of students, to audiovisual specialists an additional medium, and to academic or commercial promoters profitable possibilities.

Early-twentieth-century concerns about efficiency in education led to the introduction of standardized tests. During the 1940s and 1950s concerns about efficient military training heightened interest in audiovisual instruction, contributing to the widespread use of devices such as the overhead projector. Other mid-century suggestions about efficiency went a step farther. Adopting ideas of psychologists Sidney L. Pressey, B. F. Skinner, and Norman Crowder, educators, mathematicians, and businessmen proposed a radical transformation of the American classroom. They suggested that long-established patterns of lectures, recitations, and textbook exercises should be supplemented or even replaced by self-paced programmed instruction using teaching machines or specially organized printed texts. Classrooms might become rooms of machines. Much of the routine work of teachers could be automated, and apparatus for group instruction such as the blackboard would be largely superseded by devices used by individuals.

Programmed instruction emerged from laboratory experiments on human and animal learning. Advocates claimed that students could learn efficiently by studying material that had been broken into carefully ordered short items, or "frames." Each frame provided new information and also required a written or push-button answer. The student learned immediately whether his or her response was correct and, thus reinforced, moved to the next frame. The mathematics educator James T. Fey later described the technique more pithily in an oral history interview. Given a subject, Fey said, "all you have to do is break it into a million behavioral objectives and you just give the kid the packet and they learn on their own."[1]

In the years following the October 1957 launch of the Soviet *Sputnik* satellite, improved teaching of mathematics and science was seen as a national priority. The number of adequately trained mathematics teachers seemed woefully small. Techniques that promised to allow students to learn on their own were most welcome. Trials of programmed materials received generous funds from foundations, private companies, and the federal government. Although programs were envisioned to teach everything from psychology to poetry to Sunday school, the most extensive use of the technique was in mathematics teaching.[2] Programs were offered that taught topics from arithmetic to abstract algebra. At least two major projects for mathematics curriculum reform, the School Mathematics Study Group (SMSG) at Yale and then Stanford and the University of Illinois Committee on School Mathematics (UICSM), produced programmed textbooks for high school use. A committee of the Mathematical Association of America (MAA) developed a programmed course in calculus. At about the same time, Stanford and Illinois also hosted early projects for using electronic computers in classroom teaching. Participants in these attempts at computer-based instruction were aware of work in programmed instruction, although they took rather different approaches. In the mid-1960s, as federal funding increasingly focused on equity in education, programmed materials for remedial training received considerable attention. The radical improvements promised by the new technique proved elusive, and it largely disappeared from the schools. Nonetheless, the view that students might learn from individual study of carefully organized material presented to them at small machines proved to be a lasting legacy.[3]

Origins

The first teaching machines were an indirect consequence of the boom in mental testing described in chapter 3. Several inventive Americans envisioned more efficient ways of administering and scoring objective tests. One of them, Sidney L. Pressey, began thinking about such a machine in 1915 as a graduate student at Harvard. He was diverted to other endeavors by World War I. After the war Pressey taught at Indiana University and then Ohio State University. In Ohio he developed a device that could be used to administer and score tests and also to reward correct responses. Pressey exhibited his teaching machine at two annual meetings of the American Psychological Association in the mid-1920s. In 1926 he applied for a patent for his improvement in "machines for intelligence tests" and was granted it two years later. Pressey also published several articles about his machine in *School and Society*.[4]

An early example of Pressey's machine appears in Figure 5.1. Multiple-choice questions with four possible answers were written on a sheet of paper attached to a rotating drum. A narrow slot in the cylinder that surrounded the drum revealed one multiple choice question at a time. The four keys on the right were numbered to correspond to possible answers; a student pushed down a lever to indicate an answer. If the machine was set on "test," it then advanced automatically to the next question, tallying the number of correct answers as it went. The machine also could be used in "teach mode." In this case the drum would not advance until a question was an-

Figure 5.1. Sidney Pressey with his early teaching machine, 1960. National Museum of American History collections, gift of Science Service. Smithsonian Negative no. 2005-28923.

swered correctly. By 1927 Pressey had modified his machine so that those questions that had been answered correctly twice no longer appeared before the student. This feature made it possible for a student to concentrate on material that required further drill.

At the close of a 1933 book entitled *Psychology and the New Education,* Pressey envisioned an "industrial revolution" in education in which "educational science and the ingenuity of educational technology combine to modernize the grossly inefficient and clumsy procedures of conventional education."[5] Coming in the midst of the Great Depression, this call for increased expenditures for classroom automation fell on deaf ears. In 1929 the William M. Welch Scientific Company of Chicago offered for sale a commercial version of Pressey's machine, but it did not prove successful.[6]

By the 1950s a burgeoning school population and new prosperity encouraged educators and psychologists—as well as mathematicians—to take a new look at possible improvements in education. Harvard psychologist Burrhus Frederic Skinner had long been interested in devices for training rats and pigeons through a careful system of rewards and punishments. According to Skinner's own account, a 1953 visit to his daughter's arithmetic class persuaded him that children also might benefit from automated instruction. At a conference on current trends in psychology held at the University of Pittsburgh the following year, Skinner exhibited a prototype machine for arithmetic training. This was a wooden box that contained a roll of paper tape with arithmetic problems written on it. The tape rolled crosswise so that one problem at a time appeared in front of the child. The levers beneath the problem were moved to set the answer. If the answer was correct, a knob at the front of the machine could be turned to reveal a new question. An improved version of this machine, specifically intended to teach arithmetic and spelling to grade-school students, was made by the Typewriter Division of International Business Machines (IBM).

Over the next few years Skinner developed his instruments for providing drill into machines that presented new material. By the spring of 1958 he and his colleagues had designed and programmed a machine used in teaching a course on introductory natural science at Harvard and Radcliffe. This work was sponsored by the Fund for the Advancement of Education of the Ford Foundation. By this time Skinner had concluded that most information conveyed in the classroom could be divided into short segments presented in a carefully thought-out linear sequence. Students would learn new

Figure 5.2. B. F. Skinner teaching machine, 1956. National Museum of American History collections, gift of B. F. Skinner. Smithsonian Negative no. 85-3334.

material by responding to information presented to them—filling in blanks or writing short answers.[7]

Skinner's 1958 machine looked somewhat like a record player, with questions written out on sectors, or "frames," in a paper disc (Figure 5.2). The discs fit onto a rotating turntable. To the right of the turntable was a paper tape for answers. The machine's lid covered both turntable and paper. One slit in the lid revealed a frame at a time, another the section of paper on which the student wrote out his or her answer. Advancing the disc revealed the answer and also advanced the paper tape so that it could not be changed (to prevent cheating). If a student found the answer was correct, he or she moved a lever, ensuring that the question would be skipped if and when the student went through the disc a second time. Skinner applied for a patent for this machine, and one was granted in August 1958, with rights assigned to IBM.[8] That October he published an article on teaching machines in *Science* magazine. Here he declared that growing demand for education called for "the invention of labor-saving capital equipment" and suggested that his device might supply such equipment.[9]

Skinner argued that students would learn efficiently by reading a series of short frames that required their immediate response, and they should know at once whether their responses were correct. Both of these ideas were widely accepted by advocates of programmed instruction. Skinner's claims that frames should be ordered in a linear sequence and that machines were needed to control the presentation, entry, and review of material were not as widely accepted. The author and inventor Norman A. Crowder believed, for example, that it was more useful to prepare somewhat longer frames giving information and then ask a question with multiple-choice answers. A student was directed to a new frame according to the answer given. If the answer was correct, the new frame gave new information. If not, the new frame presented the idea another way and gave new questions. Eventually, the student was sent once again to the frame that he or she had missed. Crowder called this arrangement of text and questions an "intrinsic program"—others would refer to it as a "branched program." He used intrinsic programming to prepare a course for electronics technicians in the U.S. Air Force. A 1961 report indicates that the air force had purchased eighteen of the Autotutor Mark I teaching machines designed by Crowder. The devices, which sold for $5,000 to $7,000, were used for electronics training at Keesler Air Force Base.[10]

Crowder also pioneered the use of programmed instruction in far less expensive printed books. His volumes included texts on number systems (1958), binary and octal arithmetic (1960), algebra (1960), and trigonometry (1962) as well as games such as bridge. From 1960 through 1967 they were sold by Doubleday in the United States, with various foreign editions published in England, France, and South America.[11] Programmed textbooks had none of the guarantees against cheating or the elaborate record keeping of teaching machines, and when they were arranged in Crowder's "scrambled text" form, it was especially difficult to review material. Yet they sold for only a few dollars, could be used anywhere one could read a book, and required no maintenance. They also could be produced and distributed by existing publishers and became the preferred form of much programmed instruction. Indeed, when B. F. Skinner had difficulty marketing commercial forms of his teaching machines, he resorted to publishing a programmed textbook.[12]

Early Programmed Instruction in Mathematics

Attempts to program mathematics instruction followed promptly in the wake of Skinner's 1958 article. That fall John W. Blyth, a professor of philosophy at Skinner's alma mater, Hamilton College in Clinton, New York, received a grant from the Ford Foundation's Fund for the Advancement of Education to develop a programmed course in logic. Results were sufficiently encouraging for Blyth to request and receive further funds in the spring of 1959. He used the money to develop an undergraduate course in analytic geometry and calculus as well as elementary instruction in French, German, and psychology. Blyth and his associates, including mathematics professor B. F. Gere, arranged to have teaching machines manufactured locally by a firm called, appropriately enough, Hamilton Research Associates. Frames of the program were presented either on microfilm (in a machine called the Visitutor), on flash cards stored in a projector (in the Visitutor Card Model 200), or on a push button–operated tape recorder (in the Auditutor, used especially for language instruction).[13]

In a March 1960 article in the *American Mathematical Monthly* Blyth claimed that the use of teaching machines in logic teaching allowed him and a colleague in the philosophy department to reduce class contact hours by a third, increase the level of student mastery of the subject, and cover more material. Moreover, the professors were freed from routine checking and drill and could devote class time to further development and application of skills and ideas introduced in the lessons. Blyth believed that teaching machines promised "superior private tutoring on a massive scale."[14]

Other optimistic reports came from Roanoke, Virginia. Allen Calvin, a psychologist at Hollins College, developed a program for teaching ninth-grade algebra. In the spring of 1960 he used it with a group of thirty-four eighth-grade students. Without teacher, textbook, or homework students were able to cover the year's work in a single semester.[15] In these early experiments the students used a variation on Skinner's teaching machine developed by Foringer & Company of Rockville, Maryland, and later sold by Programmed Teaching Aids, Inc., of Arlington, Virginia.[16] Staff at Hollins College went on to work with Encyclopaedia Britannica Films, Inc., of Wilmette, Illinois, developing programmed courses in first- and second-year high school algebra, plane geometry, trigonometry, and calculus. By 1961 courses were also planned in solid geometry, intermediate calculus, and differential equations.[17] The form of these materials quickly moved away from machines toward less expensive sets of loose-leaf notebooks sold under the name TEMAC. Frames were printed on pages stored in the notebooks and viewed in a clear plastic sheet holder that had slots cut into it for recording answers. An opaque sliding shield covered the correct answers. Students wrote down their answer and then pulled down the slide to reveal the correct response. There was no mechanism to tally results or allow for easy repetition of any questions that had been missed.[18] Use of programmed instruction in the Roanoke schools would continue at least through 1965.[19] Field testing of more advanced programs took place through the Britannica Center for Studies in Learning and Motivation at Stanford University.[20]

Evan Keislar of the University of California at Los Angeles and Patrick Suppes of Stanford suggested that programmed learning also had a place in elementary education. In the spring of 1959 Keislar wrote a short multiple-choice program designed to teach elementary schoolchildren to calculate areas of squares and rectangles and to apply such calculations to practical arithmetic problems. For the task he adopted a Film Rater used by the U.S. Navy to teach aircraft identification. Multiple-choice questions were projected from film onto a viewing plate, and a student entered the answer by pushing a button. If the answer was correct, a green light shone, and the student moved on to another problem. If not, a red light turned on, and the student tried again until the correct button was pushed. Keislar's teaching machine was an automated version of Pressey's earlier device. He tried out his program with fourteen students at UCLA's University Elementary School. Comparing their performance with that of fourteen students who had not spent time at the machine, he concluded that the program did in fact teach something but could be improved.[21] Mindful of contemporary interest in introducing children to more general algebraic structures, Keislar then obtained a grant from the Ford Foundation to use programmed learning to teach first-graders about algebraic structures. He employed blocks and games in this teaching as well as a machine for projecting several images simultaneously. His results were published by the U.S. Office of Education, although the apparatus was not widely adopted.[22] During this same period Patrick Suppes and Newton Hawley published a programmed textbook for teaching elementary school geometry.[23]

Thus, by the close of 1961 programmed instruction had been used experimentally to teach mathematics to American students from first grade through college.[24] The devices ranged from machines selling for substantial sums to textbooks that cost a few dollars. *Mathematics Teacher,* the journal of the National Council of Teachers of Mathematics (NCTM), had published its first article on the new teaching tools that May. Here Ralph T. Heimer, a graduate student in education at Pennsylvania State University, sought to inform high school mathematics teachers about the recent "eruption of interest in the technique of automated instruction." He briefly described linear and branched programs, noted that they were presented in both machines and printed textbooks, and described a few of the possible advantages of the approach. Programmed materials continuously involved the student, provided immediate reinforcement of correct answers, and allowed each student to proceed at his or her own rate. Heimer made no direct reference to any possibility that programmed material might replace teachers but described plans under way at Penn State for using closed-circuit television as a teaching machine.[25]

Later discussion of programmed instruction in *Mathematics Teacher* took a more cautious approach. J. F. Clark, a teacher in Langley, British Columbia, published an article in November 1962 entitled "Programmed Learning: My First Six Months." Clark had used the TEMAC first course in ninth-grade algebra published by Britannica Films. She reported that both her students and her own daughter (an eighth-grader) could learn from the material. Indeed, she found that "a programmed learning text is possibly a refinement of some of the best work sheets ever constructed."[26] At the same time, she warned that teaching machines were part of "one of the greatest commercial invasions of education since textbooks and audio-visual aids became standard equipment."[27] Manufacturer's rhetoric, she wrote, should not prevent a critical evaluation of materials. Clark also feared that a programmed course could reduce a teacher to a clerk. She acknowledged that it might give a teacher time to work with students individually and to provide special projects for those who completed the program swiftly. Programmed materials also might suit the needs of those studying independently through night schools, correspondence courses, or entirely on their own.

In the same issue of *Mathematics Teacher* Paul McGarvey of the Lakewood Public Schools in Lakewood, Ohio, reported on his experiences using the TEMAC materials in a summer course in ninth-grade algebra. McGarvey found that most students found the materials interesting and useful and that the eleven students who were repeating algebra significantly improved their scores. Programmed exercises, however, left little opportunity for discovery learning.[28]

Programmed Instruction and the "New Math"

From the early 1950s a number of American mathematicians and educators had sought to improve school mathematics by revising the curriculum, preparing new textbooks, and training teachers in their use. The "New Mathematics" emphasized the theoretical foundations underlying school subjects, introducing ideas and notation of set theory and abstract algebra to much younger students than had previously been done. After the launch of *Sputnik* in October 1957, the U.S. Congress appropriated funds specifically intended to improve teaching in science and mathematics. Some of this money was spent on programmed textbooks for mathematics.[29]

Work on the Programmed Instruction Project at the University of Illinois began in April 1961, when the U.S. Office of Education awarded a grant to psychologist Lawrence M. Stolurow of Illinois and mathematics educator Max Beberman of the UICSM. Funds were used to develop and test two new forms of the UICSM ninth-grade algebra textbook. One was in the form of a linear program, the other branched. As the positions of the principal investigators suggest, these programs sought to incorporate the knowledge of both mathematicians and psychologists. The textbooks also attempted to convey some of the dynamism produced by the motion and intonation of a good teacher through the layout and typefaces used in the text. Classroom use of programmed textbooks alternated with presentations by teachers to prevent monotony.[30] Early UICSM materials were designed especially to encourage students to pursue more advanced mathematics. In the mid-1960s, as the focus of federal funding shifted to remedial work, so did the aims of the project.

A second effort to improve mathematics teaching, the School Mathematics Study Group, was directed by Edward G. Begle at Yale University from its beginning in 1958 and moved to Stanford when Begle joined the faculty there in 1961. One major task of the SMSG was preparing classroom textbooks that had been developed at summer conferences of mathematicians and mathematics teachers, tested at schools around the country, and revised in further summer sessions. Several of these texts were published by Yale University Press. In 1961 the Panel on Programed Learning of the SMSG received funds from the National Science Foundation to produce its ninth-grade algebra textbook in programmed form. Using a manual prepared by former mathematics teacher Leander W. Smith of the SMSG staff, a group of twelve people with mathematical training met at Yale in August 1961 for training in programming. By the fall of 1962 preliminary textbooks were available that had been written in both branched (form MC, or multiple-choice) and linear (form CR, or constructed-response) form. Like their counterparts at Illinois, the Yale group suggested ways in which usual programmed texts might be improved. In particular, they recommended a "hybrid" form of programming. Their hybrid programmed text contained more straightforward explanation than a usual programmed book as well as an index. By 1964 all three of these forms of programmed textbook had been tested and were published in paperback form by Stanford University.[31]

A third project, aimed at college students, grew out of efforts of the Mathematical Association of America. At its January 1962 meeting the association created a standing Committee on Educational Media, with subcommittees on television, film, and programmed learning.[32] The subcommittee on programmed learning, headed by Brewster H. Gere of Hamilton College, applied for and received funds from the National Science Foundation to prepare a programmed textbook on first-year calculus. Those who were chosen to write the book first assembled at Stanford in the summer of 1964.[33]

To do their work, the authors first needed to decide what kind of program would be most suitable and how it should be used in the classroom. One subcommittee member, Kenneth O. May of Carleton College and the University of California at Berkeley, prepared a provocative outline of existing programmed learning. It is unclear how much this paper was used by the textbook authors, but it was widely circulated by the MAA in 1965 and revised for publication in the NCTM's *Mathematics Teacher* the following year.

May's paper first outlined the approaches to programmed learning of Skinner, Crowder, Pressey, and the SMSG. He noted that an increasing shortage of teachers forced some students to learn on their own but that the traditional printed material available was inadequate. May thought that programmed learning appealed broadly to psychologists, administrators, audiovisual experts, and entrepreneurs. Sorting through the studies available to him, he concluded that students indeed could learn from programmed materials. There was no significant difference between the amount of learning that occurred with them than with traditional teacher and textbook, and learning did not take place more efficiently. Indeed, typical programs did not teach such important skills as careful, sustained reading and independent problem solving, which were most desirable in mathematics education. May then examined and rejected many of the more specific claims of programmed instruction. It was not equivalent to a private tutor, was not the only path to self-paced instruction, and might motivate pigeons but in the long run bored people. It also did not save teachers' time or school systems' money. Some texts, such as those of the SMSG, might provide useful supplements to teacher presentations but were not satisfactory replacements.[34]

May's blunt evaluation of programmed learning was sent to every member of the MAA with a disclaimer explaining that the views expressed were those of the author and not necessarily those of either the Committee on Educational Media or the association as a whole. A notice of the work published in the *American Mathematical Monthly* stressed this point and drew the reader's attention to May's statement that programmed learning might prove to be a useful supplement to usual mathematics instruction.[35]

Meanwhile, writing of the programmed calculus book continued. Following the spirit of the New Math, it emphasized rigor and theory, attempting to offer a proof for every theorem included. *A Programed Course in Calculus,* published in 1968, was a text in five paperback parts, with a total length of over a thousand pages. It soon was followed by a supplement that included a summary of each section and further exercises. The text was adopted for supplementary work at such schools as University of Ottawa and Amherst College, with both faculty members and students offering rather mixed reviews.[36] Yet even this minor popularity did not endure.

Reactions

Kenneth May was by no means the only mathematician or educator to ask how the techniques of programmed instruction might influence mathematics teaching. The claims of programmed learning left a vivid impression on several graduate students in mathematics education, who wondered where they would fit within a world of teaching machines. Recounting that time, F. Joe Crosswhite, who had done graduate work in mathematics education at Ohio State University from 1961 to 1964 and then taught there for many years, described the era of programmed learning as "when we were hoping to get by without teachers." He found that careful task analysis for training specific skills, such as that recommended by Robert Gagné of Princeton University, was very useful. Yet the technique seemed ill suited to providing a broad mathematical education.[37] Shirley A. Hill obtained her Ph.D. degree from Stanford in 1961, studying under Patrick Suppes, and went on to teach mathematics education at the University of Missouri at Kansas City. Hill later described programmed learning "as much too rote, too cut and dried, lacking deep meaning."[38] James T. Fey, who had earned his bachelor's and master's degrees at the University of Wisconsin and then taught high school math for a year, went to Teachers College at Columbia University in 1964 to work on his doctorate. There he discovered that some scholars thought that student interaction with a dynamic teacher might be replaced by programmed instruction. Fey asked one of his professors, Howard Fehr: "Is there a future in this business? I'm in here, and maybe we're all going to be replaced by machines." According to Fey, Fehr replied: "Don't worry. When I started out in education, it was a radio that was going to do it."[39] Fey was apparently reassured and went on to have a successful career at the University of Maryland.

Student reaction to programmed instruction is more difficult to trace. Early reports suggested that students enjoyed the novelty—as they might enjoy any break from routine.[40] Even then, however, the prospect of long-term use of programmed instruction was unwelcome. In the spring of 1961 George L. Henderson of Moline Senior High School in Moline, Illinois, used programmed materials for six weeks in his Algebra II course. He described his procedures and results in an article in *Mathematics Teacher.* As an appendix to the article, he presented three essays in which students described their reactions to the new approach. All three students commented that reading chunks of information and responding to them had its tedious moments. As one wrote, "Doing the same thing more or less day after day does get monotonous." The student whom Henderson judged to be the best of the three in algebra was most critical of programmed instruction. Because of the "slow step by step procedure" used in programmed material, the boy wrote, "the student does not have to think or reason things out which is usually considered one of the prime objectives in algebra or any math course."[41]

Later programmed instruction answered some of these criticisms by using a greater variety of programming styles. The SMSG's "hybrid" program illustrates this development. More "flexible" programming also appeared in a UICSM unit on solid geometry. There was also a more general call for programmed materials that covered specific topics that could be used as supplements to regular class work.[42] The teaching machine was relegated to the sidelines, not envisioned as being at the center of the classroom.

The interaction between the rhetoric of programmed instruction and the reality of children learning is perhaps most vividly illustrated by a simple set of cards. Printed cards with arithmetic problems written on them had been used in classroom drill from at least the late nineteenth century. By the 1920s they were sold as "flash cards." They were most commonly used to teach elementary arithmetic but were also proposed as aids for teaching geometry, trigonometry, and factoring.[43] In the early 1960s New Jersey's Edmund Scientific Company revived the traditional cards for teaching arithmetic operations and also published a card game designed to teach basic identities in trigonometry and formulas in calculus through a game similar to solitaire.[44] Later in the decade the New York firm of Ed-U-Cards Manufacturing Corporation began selling "New Math Flash Cards," which reflected more contemporary views about both mathematics and the psychology of learning. Both sides of each card had a simple sum, written horizontally. A cardboard "sliding number cover" fit over part of the card and could be used to cover any number in the equation with a blank square. A window in the other side of the cover revealed the full equation (Figure 5.3). Using the cards, a parent and child, or a child alone, could set up equations for adding numbers from 0 to 9. The solution provided instant reinforcement. Moreover, like missed frames in the program of a teaching machine, cards that had been answered incorrectly could be singled out for further review. As the instructions for the flash cards put it: "Drill now becomes an exciting game that reinforces learning. Both the child and the person testing him know at once when basic number facts have been mastered."[45] Ed-U-Cards made similar cards, with identical instructions, for teaching rote multiplication. The paper sliding number cover was not terribly sturdy, and it seems unlikely that the cards were used as described. Nonetheless, it was in this simple and unstructured form that many children and parents first encountered programmed instruction.

The trademarked logo of Ed-U-Cards Manufacturing Corporation is a stylized child wearing a mortar board. By the 1970s the number of college

Figure 5.3. New Math Flash Cards, ca. 1965. National Museum of American History collections, gift of Sherman L. and Marjorie A. Naidorf. Smithsonian Negative no. 2006-0197.

graduates with degrees in mathematics increased, and the boom in school-age population began to dwindle. Mathematics education also seemed to be a less urgent national priority. All of these factors, combined with new psychological theories, worked against the use of programmed learning. Both the promise that machines might be used for teaching and the failure to realize this promise would be recurring themes, however, in late-twentieth-century American mathematics education.

PART TWO

Tools of Calculation

CHAPTER SIX

The Abacus
Palpable Arithmetic

> To do anything like justice to this instrument, would form a volume in itself; suffice it to say, that it was one of the best instruments that was ever introduced into an infant school, and I do sincerely hope that no nursery will be without it.

The abacus, or numeral frame, was introduced into American schools in the 1820s and 1830s as a tool for teaching arithmetic to young children. Like the blackboard, it was adopted from diverse European sources, particularly in Great Britain and France. Although the abacus had a long history as an aid to calculation, its use in Western education was quite recent. It had been brought to France from Russia at the time of the Napoleonic Wars and introduced in primary school teaching. It was also developed in the early 1820s, apparently independently, by British teachers at schools established for very young children. These "infant schools," generally intended for the poor, taught children from one to seven years of age.

The idea that most young children could and should learn arithmetic—or indeed attend school—was itself relatively new. Arithmetic had traditionally been taught to only a small number of children—most of them boys—beginning at age twelve or thirteen. Textbooks started by teaching written numerals and then required memorization of numerous rules used to solve separate classes of problems. The first British advocates of the abacus retained this emphasis on written arithmetic. The influential Swiss educator J. H. Pestalozzi had argued, however, that young children would do much better to learn to do arithmetic in their heads before they worried about learning written numbers. He and his followers not only encouraged exercises in mental arithmetic in their own classes but urged others to do likewise. New textbooks for young children also adopted the approach.

Pestalozzi advocated the use of objects in arithmetic teaching. His admirers in both Britain and the United States appreciated his idea that children should spend time counting and grouping simple objects as part of their study of arithmetic, and they found that the abacus was most useful for this task. Thus, the instrument came to be associated with a major reform in arithmetic teaching. Even when Pestalozzi's influence faded, the abacus remained as a tool for introducing children to counting and simple arithmetic.

The Early History of the Teaching Abacus

Although the abacus has been used in the Western classroom for only about two hundred years, it has a much longer history as a practical aid to computation. The ancient Greeks and Romans used pebbles and other counters moved along lines to carry out routine arithmetic. Medieval and early modern European clerks and merchants continued the practice, using counters that were moved on lines marked on special tables.[1] The Chinese and the Japanese developed distinctive portable versions of the abacus. By the end of the seventeenth century a form of the instrument on which beads were moved crosswise on horizontal wires had become popular in Russia (Figure 6.1). It was this style of the abacus that would be adopted in the American classroom.[2]

According to an 1843 account by the French mathematician and historian of mathematics Michel Chasles, the Russian abacus came to the French schoolroom through the efforts of the French soldier, engineer, and geometer Jean Victor Poncelet. Poncelet was a pupil of Gaspard Monge at the École Polytechnique in Paris, graduating in 1809. He continued his studies in his hometown of Metz in Alsace-Lorraine. In March 1812 Poncelet was called from the classroom to serve as a lieutenant in Napoléon's ill-fated attack on Russia. That November he was taken prisoner, survived a forced march of hundreds of miles in the Russian winter, and spent fifteen months in a Russian prison. Poncelet made good use of his confinement, reconstructing and extending ideas of Monge and Lazare Carnot about the properties of projections of geometric figures. Upon his repatriation in 1814, Poncelet brought back mathematical manuscripts that he would develop into his *Traité des propriétés projectives des figures* (1822). According to Chasles, he also returned with a Russian abacus and suggested that it might be used to teach small children in the schools of Metz. From there the instrument spread throughout France.[3]

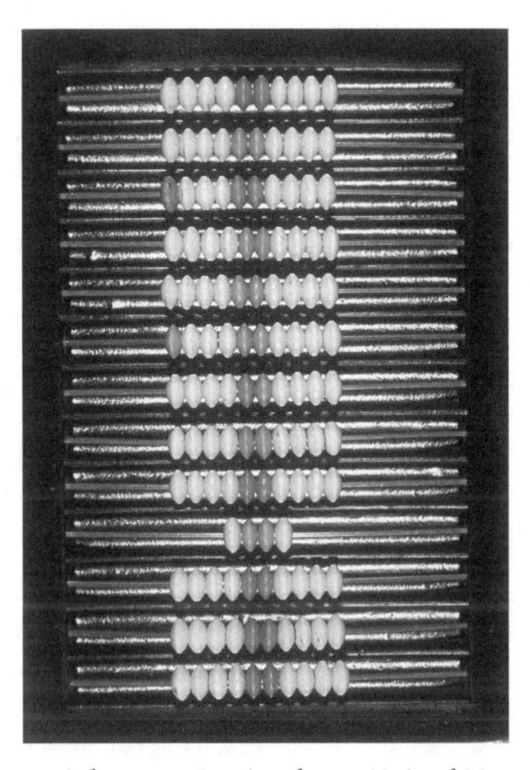

Figure 6.1. A twentieth-century Russian abacus. National Museum of American History collections, gift of Edith R. Meggers. Smithsonian Negative no. 79-10143.

At about the same time, the industrialist Robert Owen opened an "Institution for the Formation of Character" at his model Scottish industrial village of New Lanark. One part of this establishment was an infant school for young children. The school not only offered day care for those whose parents and older siblings worked in the mills but also provided an introduction to academic subjects, music, dance, and military drill.[4] Owen's school proved sufficiently successful that similar infant schools soon opened around London. The first of them had as its master James Buchanan, who had taught at New Lanark. Buchanan, in turn, trained the tradesman Samuel Wilderspin and his wife, Sarah Anne Wilderspin, who opened a school at Spitalfields in London in 1819. These early infant school teachers firmly believed that young children would learn better if they could associate numbers with tangible objects. In addition to providing simple objects for counting, Samuel Wilderspin experimented with using buttons moved along a string. He soon substituted wooden balls moving along wires for the buttons and string. Wilderspin initially placed one ball on the first wire, two on the wire to the right of it, three on the third row, and so forth up to twelve balls on the rightmost wire. Moving the balls up and down would provide a simple introduction to counting and to basic operations of arithmetic. The beads could also be arranged to represent various geometric figures.

Wilderspin was very proud of the device, describing it in an 1825 book as "one of the best instruments that was ever introduced into an infant school." He also noted there that "as persons may not know how to give directions for making this instrument, I have had a number made on purpose to obviate this difficulty."[5] Wilderspin soon modified his instrument so that the beads moved crosswise (Figure 6.2). Upon further reflection he concluded that it would be more useful to have twelve beads on each wire, with the beads painted alternately black and white. He dubbed the instrument the "arithmeticon" and suggested that it should be used to represent numbers. The lowest row represented units, the next one up tens, the next hundreds, and so forth. As others soon pointed out, this representation of numbers was much like that on a Russian abacus, although the latter had only ten beads per row. Wilderspin later reported that he had never seen an abacus at the time of his invention. He had first published his ideas about infant school teaching in 1823 and mentioned the abacus for the first time in the third, 1825, edition of his book.[6] Selections from his work and that of other infant school advocates were published in New York in 1827.[7]

Americans could and did read about infant schools from other sources as well.[8] In 1824 Wilderspin persuaded William Wilson, the vicar in the town of Walthamstow near London, to open an infant school there. Wilson not only opened a school but also wrote a general account of the education he provided. Wilson's *System of Infants' Schools* had reached its second London edition by 1825 and was published again there the following year. Ex-

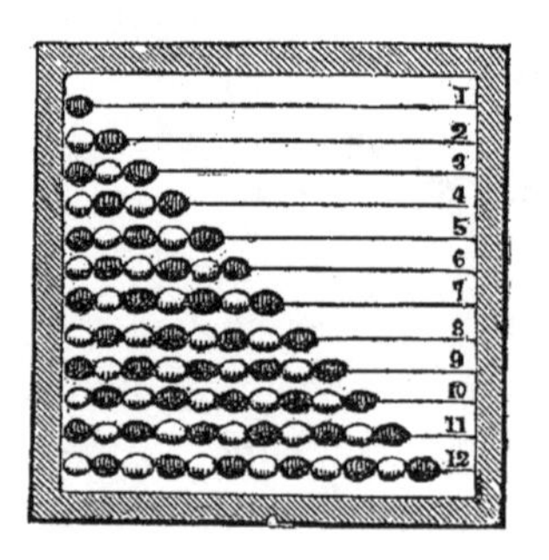

Figure 6.2. Samuel Wilderspin's early numeral frame, 1829. After *Infant Education: Or Remarks on the Importance of Educating the Infant Poor* (London: Simpkin & Marshall, 1829), 188.

tensive extracts from it were reprinted in the *American Sunday School Magazine* in 1826, and a revised version of the book appeared in New York in 1830.[9]

Wilson recommended that infant school teachers train children to clap their hands or beat their feet in unison and suggested that students should learn to count and to repeat the multiplication tables to this rhythm. He then used a large abacus, or numeral frame, to "place before the eyes of the children a representation of the combinations of numbers which they have already committed to memory."[10] More specifically, Wilson used a wooden frame with twelve horizontal wires, each wire carrying twelve balls (Figure 6.3). Combining balls on a single wire illustrated the addition and subtraction of small whole numbers. Assuming that every ball represented a unit and taking balls from two adjacent wires in pairs, Wilson obtained rectangles that represented multiples of two; moving balls from three wires represented multiples of three, and so forth. This use of a 12 × 12 abacus may well reflect the importance of duodecimals in the currency, weights, and measures of contemporary Britain.

Wilson thought that the balls on his abacus could be used to illustrate diverse principles. Like Wilderspin in his later works—and like Russian users of the abacus—he suggested that balls on different wires could sometimes be assumed to have different place value. Those on the top row, he said, might represent units, on the second row tens, on the third row hundreds, and so on (this is the opposite of the place value order on both the Russian abacus and that of Wilderspin). Wilson also thought that the numeral frame might be used to teach more than arithmetic. By arranging the beads appropriately, one could also illustrate the shape of various polygons. Mindful of art and the physical sciences, Wilson also went on to suggest that rows of balls should be painted in different colors in order to illustrate interrelations among various hues.

By 1830 Wilson had developed a more consistent way of using the abacus. In *A Manual of Instruction for Infants' Schools* he proposed using an abacus with a 10 × 10 array of balls, with one ball representing a unit at all times. Children should first be taught to clap their hands in measured time and then, gradually, to count up to 100 while clapping. Next, they should learn to count the 100 balls of the abacus, both from 1 up to 100 and from 100 down to 1. Simple addition and subtraction followed and then multiplication using beads from as many rows as needed. Wilson retained his belief that different rows of balls should be differently colored, using a somewhat

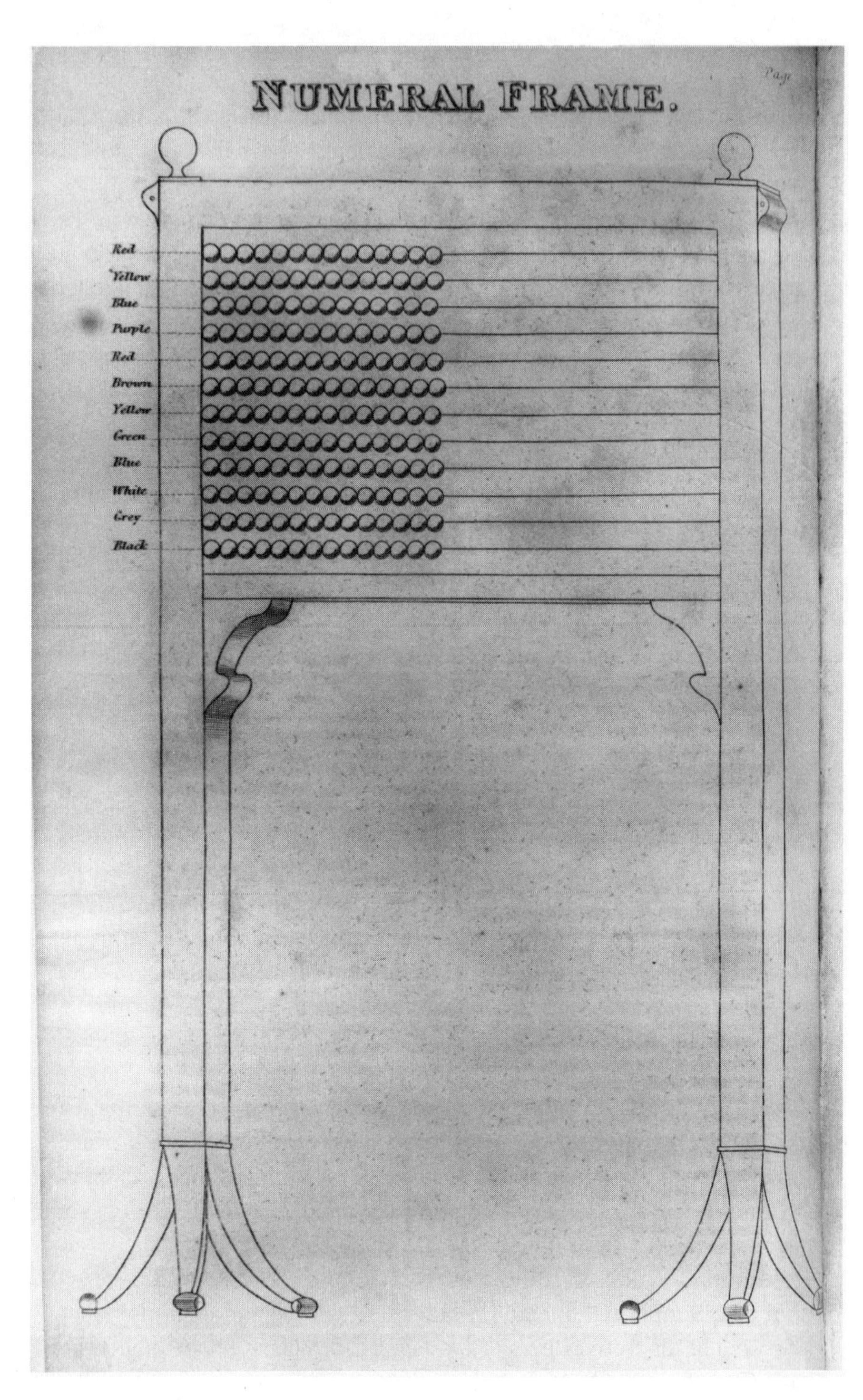

Figure 6.3. William Wilson's numeral frame, September 1826. From *The American Sunday School Magazine,* 3:257. Courtesy of American Antiquarian Society collections.

different arrangement of colors. He made no mention of using the abacus to teach about place value.[11]

The abacus also attracted the attention of the French Pestalozzian teacher William S. Phiquepal. Phiquepal, like Poncelet, was a veteran of the Napoleonic Wars but had fought on the side of the Royalists. He thus found it expedient to leave France and spent some time in Pennsylvania, where he worked as an assistant at a school run by one of Pestalozzi's students, Joseph Neef.[12] Following the Restoration, Phiquepal returned to France and attended medical school in Paris. He remained much interested in education, however, and, in about 1820, opened a Pestalozzian school for boys at the Paris home of the Scottish-born philanthropist William Maclure. His school not only taught basic reading, writing, and arithmetic but also offered practical instruction for carpenters, coopers, blacksmiths, and similar tradesmen. For this work he employed a range of teaching devices that went considerably beyond what Pestalozzi and Neef had used.[13]

By 1824 the restrictions of the Bourbon monarchy persuaded both Maclure and Phiquepal that the United States offered greater opportunity for educational reform than France. Phiquepal moved to Philadelphia, where he planned to open a school for boys. This institution complemented a school for girls, run there under Maclure's sponsorship as well and headed by the Frenchwoman Marie Duclos Fretageot. With Phiquepal came four of his students and considerable apparatus.

In a letter published in 1824, Maclure described to his fellow geologist Benjamin Silliman of Yale University the advantages of the techniques that Phiquepal had developed. Pestalozzi, as well as his disciples Joseph Neef and the Bostonian Warren Colburn, had represented numbers by lines arranged in written tables.[14] Phiquepal introduced "tangible substances and machinery."[15] Whereas the earlier authors had relied on description, he offered prints and representations, which made the entire process of learning easier, more direct, and more exact. By 1826 Maclure had prepared a somewhat more extensive article entitled "An Epitome of the Improved Pestalozzian System . . . ," which Silliman also published in the *American Journal of Science*. According to this summary, Phiquepal and Fretageot first and foremost followed Pestalozzi and advocated the direct examination of objects as a foundation for knowledge. They envisioned a much wider range of aids to instruction, however, than Pestalozzi had used. Mechanism was taught from machines or direct models of them, arithmetic by an instrument called the "arithmometer" (the abacus), geometry by instruments called the trigonometer and the mathemometer, and mathematical forms by using solid figures. The slate and pencil were used to teach writing; skeletons, preparations, and wax figures to teach anatomy; and specially built instruments to teach music. Thus, whereas Pestalozzi and Neef had made do with simple charts and familiar objects, Maclure and his associates proposed an array of classroom equipment. The article made no attempt to describe individual instruments in detail.[16]

Not long after Phiquepal arrived in the United States, both he and Fretageot became intrigued with the ideas of another recent immigrant, Robert Owen. Owen had left New Lanark behind but remained convinced that it was important to educate all members of society. He envisioned an egalitarian community freed from traditional notions of private property, religion, and conventional marriage. Maclure had met Owen in Britain and heard about his ideas from Fretageot and from other friends in Philadelphia. He feared that Owen's plan was impractical "because the materials he has to work upon are stubborn, crooked and too often bent in an opposite direction from their owne [*sic*] most evident interests."[17] Despite these doubts, Maclure agreed to underwrite part of the venture. In December 1825 Maclure, Phiquepal, Fretageot, Owen, and several others sailed from Pittsburgh to the Utopian community of New Harmony, Indiana. Neef would join them there in March 1826.

The teachers soon had rough classes under way. Sarah Cox Thrall, a pupil in the girls' department, later recalled some of the apparatus: "I remember that there were blackboards covering one side of the schoolroom, and that we had wires with balls on them by which we learned to count." These new educational tools were part of a daily round of simple food, barnyard chores, primitive accommodations, and uniform clothes.[18] Soon, however, both the community of New Harmony and its schools were rent by divisions. Owen's challenges to conventional mores were widely publicized and made it even more difficult to attract students. The teachers quarreled among themselves, and by October the instructors were running three separate schools. These rifts were minor compared to the larger problems encountered by New Harmony. In June 1827, after losing a lawsuit filed by Maclure, Owen abandoned the enterprise, and the community collapsed.

Maclure did not give up his educational endeavors quite so quickly. He founded a School of Industry in New Harmony in May 1827, leaving it largely under the direction of Fretageot when he retreated to warmer climates. He also subsidized the School Press for several years and, in 1838,

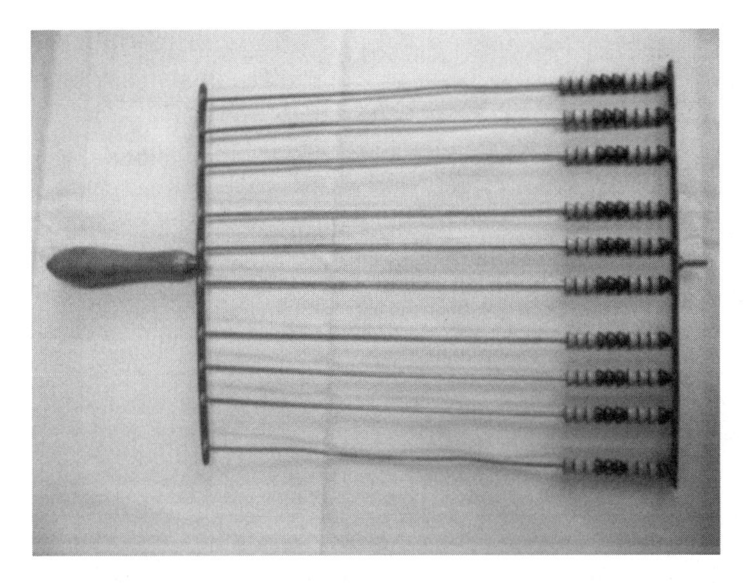

Figure 6.4. William S. Phiquepal's abacus from New Harmony, Ind., ca. 1825. Courtesy of Working Men's Institute collections.

established the Working Men's Institute Library. The latter institution survives in New Harmony to this day. Nonetheless, New Harmony did not remain a center of Pestalozzian endeavors. Neef left for Cincinnati in 1828, Phiquepal departed with the British reformer and heiress Fanny Wright the next year, and Fretageot returned to Paris in 1831.[19]

Phiquepal and his coworkers left apparatus behind, however, including two abaci, which remain in New Harmony as mute testimony to the founders' views on teaching. The numeral frames are quite different from those of infant school teachers such as Wilderspin and Wilson (Figure 6.4). They both have ten rows of ten wooden beads, not twelve rows of twelve, and the beads are not spherical but conical in shape.[20] In each row the leftmost bead is black, the next three are unpainted, the next three black, and the final three unpainted. The instruments have a wooden handle. These differences from infant school numeral frames confirm Maclure's statements that Phiquepal had brought the idea of the schoolroom abacus with him from France, rather than adopting the tool from Owen or other infant school advocates. No detailed description of how Phiquepal used the abacus seems to have survived.[21]

The Numeral Frame as Product

Most educators in Western Europe and the United States associated the numeral frame with basic education in arithmetic for young children. Curiously, the form of the instrument that sold most successfully in nineteenth-century America was that of Josiah Holbrook, who centered much of his attention on education for adults. Holbrook was a Yale graduate who shared the view of Pestalozzi and Maclure that manual labor could usefully be combined with education. In 1824 he opened an Agricultural Seminary on his farm in Derby, Connecticut, that was to combine instruction in chemistry, mechanics, and land surveying with actual practice. The school failed the following year but not before Holbrook had attended several of Benjamin Silliman's lectures on chemistry, mineralogy, and geology.[22]

This experience may have led Holbrook to a new appreciation of both laboratory demonstrations and education beyond the conventional limits of schooling. In any event, by 1826 he had come up with a plan for "associations of adults for mutual schooling," or, as they came to be called, lyceums. Holbrook organized several lyceums in Worcester County, Massachusetts, and soon moved to Boston to extend the movement on a statewide and even a national basis. He published a brief account of his ideas in William Russell's *American Journal of Education,* revising it in 1827 into a proposed constitution for a lyceum.[23] At the close of this short pamphlet Holbrook extolled the value of lyceums. Among other things, he said, they might improve the common schools by establishing greater uniformity in books and instruction and even by creating institutions for teacher training. Holbrook also emphasized the importance of learning from objects and apparatus, suggesting that lyceums should be repositories for books, instruments, and natural history collections.[24]

Collections of the sort that Holbrook envisioned had been expensive. In 1824, when William Maclure had approached instrument makers in London about building apparatus for his Pestalozzian schools in Philadelphia, he found that they were attached "to old habits of making highly finished and ornamented instruments for gentlemen's cabinets." This approach might produce a good appearance but was not what Maclure needed. As he put it, "The instruments for the schools must be fashioned with only the useful, rough, and strong in mind so that they will require little labor and cost cheap."[25] Maclure relied on Phiquepal and his students to produce such devices.

Holbrook had no established clientele demanding elegant instruments and was greatly aware of a demand for inexpensive apparatus. In 1828 he copyrighted a small book entitled *Easy Lessons in Geometry for Use in Infant and Primary Schools, as Well as Academies, Lyceums and Families,* which was a manual for a set of three sheets of diagrams and several geometrical solids that was manufactured under his direction. By 1832 the book had reached its fifth printing.[26] Holbrook soon expanded his offerings to include the numeral frame. In 1829 he published two further editions of his proposed constitution for a lyceum. As in 1827, he extolled the importance of education for adults and called for regular meetings at which they would share knowledge. At the close of the pamphlets he included a letter that specifically recommended lyceums as forums for teacher education. Finally, Holbrook noted that apparatus suited for families, schools, and lyceums was readily available for sale. According to the January version of his pamphlet, remittances were to be sent directly to Holbrook. The only arithmetic teaching apparatus available was a chart. By the end of the year Holbrook's apparatus was available for sale from his publisher, Perkins & Marvin of Boston. For arithmetic teaching there was a chart, a set of cubes, and a "numeral frame or Arithmometer."[27] Notices of the new apparatus soon appeared in major cities.[28]

An 1833 description of Holbrook's arithmometer shows a wooden frame with a handle at the bottom and twelve cross-wires, each with thirteen beads. The text accompanying the drawing indicates, however, that the instrument in fact had twelve rows of twelve beads, like the numeral frames of Wilderspin and Wilson. The beads were spherical and colored alternately dark and light, as on Wilderspin's arithmeticon.[29] A second publication from the same year used the same drawing but described the use of Holbrook's instrument in more detail (Figure 6.5). Following Wilson's 1830 approach, Holbrook assigned unit value to each of the balls. He did urge that the numeral frame should be used from the very beginning in teaching children to count so that they acquired a clear idea of the meaning of individual numbers, rather than simply memorizing names. Once children grasped the meaning of numbers, teachers also could use the numeral frame in teaching addition, multiplication, proportion, fractions, and the like. Wilson had advocated using the numeral frame to teach about simple geometry and col-

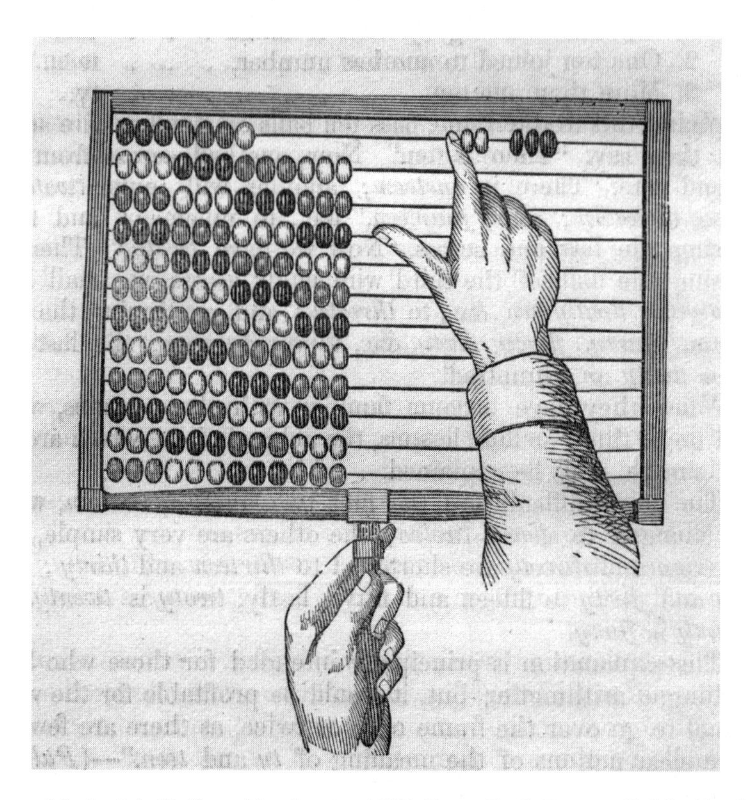

Figure 6.5. Josiah Holbrook's abacus, 1857. From F. C. Brownell, *The Teacher's Guide to Illustration: A Manual to Accompany Holbrook's School Apparatus* (Hartford, Conn.: Holbrook School Apparatus Co., 1857), 21. Smithsonian Negative no. 2006-21928.

ors as well as arithmetic; Holbrook did not suggest going beyond arithmetic.[30]

When Holbrook took up the cause of improving education through lyceums and teacher education, he did not abandon his earlier idea that formal education could usefully be combined with manual training. By 1837 he had persuaded several investors to establish a Lyceum Village in Berea, Ohio. The village combined a school, a grindstone factory, a printing press, and facilities for making Holbrook's school apparatus. Students were expected to study and recite for four hours a day and to work for six hours. The children's effort included "preparing for sale, cabinets of nature and art, scientific apparatus, and various simple instruments of knowledge fitted to their age and skill."[31] Thus, Holbrook's numeral frame represents not only the spread of arithmetic teaching in the United States but, at least for part of its history, the fruits of child labor.

The Numeral Frame in the Boston Public Schools

Holbrook's views on the use of the abacus as a tool for introducing basic arithmetic fit well with contemporary arithmetic teaching, particularly as practiced in the Boston public schools. In 1820 Warren Colburn, a mechanic who had gone on to attend Harvard University, opened a school in Boston. His classroom experience and reading convinced him that entirely new arithmetic textbooks were needed in the United States. Colburn wrote and published *An Arithmetic on the Plan of Pestalozzi, with Some Improvements* (1821), a book that would remain in print for more than six decades and inspire a host of imitators.

In the first edition of his book Colburn included three plates showing charts to be used in arithmetic teaching that were modeled after charts used by Pestalozzi and described by Neef.[32] All three authors carefully distinguished between calculating, which required knowing the names and properties of numbers, and ciphering, which involved actually writing numerals down. They argued that young children should learn to calculate before worrying about how to write numbers. Thus, in their charts numbers were represented by vertical lines (|, | |, | | |, | | | |, and so on), rather than by digits (1, 2, 3, 4, and so on). Fractions appeared as subdivisions of a square, rather than being written ⅓, 4/7, and the like.[33]

Textbooks recommended for use in the Boston schools were examined by the School Committee's Committee on Books. For many years the approved arithmetic had been Nathan Daboll's *Schoolmaster's Assistant*, which followed the form of contemporary English arithmetics. By 1823 Colburn's *First Lessons in Arithmetic* had replaced Daboll in some Boston classes.[34] In 1826 the older text finally gave way to a combination of Colburn's *First Lessons* and his introduction to written arithmetic, *Arithmetic upon the Inductive Method of Instruction Being a Sequel to Intellectual Arithmetic*. Early editions of Colburn's *Sequel* resolutely excluded many of the topics from commercial arithmetic that had encumbered earlier arithmetics. Indeed, the 1826 committee that examined the book found it "more simple, & at the same time more scientific & perfect, than any other with which the Com^tee are [*sic*] acquainted."[35]

With the introduction of the new books, the writing masters at the city's eight grammar schools, who officially were charged with teaching writing and arithmetic, developed new ideas about how they should teach arithmetic. In a quarterly report dated January 1827, Peter MacKintosh Jr., writing master at the Hancock School, requested permission to paint Colburn's diagrams on the walls of his classroom.[36] In June 1829 all seven writing masters petitioned H. T. Otis, mayor of Boston and chairman of the School Committee, to have elementary arithmetic added to the curriculum of the primary schools, which had not been possible when the first subject taught had been writing numbers. Given that mental arithmetic did not require the use of the pen, it was quite appropriate for young students. Indeed, one of the masters, Frederick Emerson, had written a small book of lessons suited to primary school use.[37] Prominently displayed on the cover of the 1829 version of this book was an image of a numeral frame.[38]

In suggesting that arithmetic was an easier subject than had previously been thought, the writing masters may have threatened their own livelihood. As enrollments burgeoned and masters sought to improve their classrooms, the Boston School Committee struggled to contain costs. In May 1830 the committee voted to place grammar schools that enrolled white students under the direction of a single master, the reading master, who would continue at a salary of $1,200 per year. He would be assisted by a submaster charged especially with writing who received the former salary of a master's assistant—$600 per year. In addition, each school would have two teachers, paid $100 per year, and four young monitors, paid $50 per year. In other words, at each school a staff of four paid a total of $3,600 per year was to be replaced by a staff of eight paid a total of $2,200 per year. The grammar schools were also to be segregated by sex, with four for boys and three for girls; the sex of the subordinate staff was to match the sex of the students.[39]

This reorganization did not last long. By 1833 submasters had been restored to their former salaries, and schools for boys had male assistants, while those for girls had more numerous but very poorly paid female adjunct teachers.[40] A feminine teaching force, and associated low salaries, would become an enduring feature of the schools. More immediately, Boston appropriated $500 of the money it saved in salaries under the new organization for the purchase of the blackboards and other school apparatus for the eight grammar schools. The final order placed by the board included seven abaci at a cost of $11 each. For comparison's sake, blackboards purchased at the same time cost $12 to $15. These prices were somewhat higher than those listed in Holbrook's *Easy Lessons in Geometry* from about that time.[41]

It seems likely that a less expensive form of abacus may have been acquired for use in primary schools in conjunction with a later edition of Emerson's *Primary Lessons in Arithmetic.* An account of the history of the Boston School Committee notes that "numerical calculators" were purchased for the primary schools during the 1830s, and it is quite possible that these instruments were abaci.[42] Emerson designed and published a set of large cards for use in primary schools that illustrated aspects of reading, writing, and arithmetic. The charts that concerned arithmetic were also specifically designed for use with an abacus.[43]

From the Numeral Frame to the Arithmetical Frame

Reactions to the new arithmetic—like responses to other reforms in mathematics teaching—ranged from enthusiasm such as that expressed by the Boston writing masters to indifference to outright opposition. Those who felt indifferent simply continued to teach as they had been taught. Daboll's *Schoolmaster's Assistant* went through at least forty-five printings between 1799 and 1851, with twenty-seven of them appearing after 1820.

Opponents questioned both the value of object teaching generally and its use in mathematics. In August 1831, for example, Boston minister Hubbard Winslow read an address to the American Institute of Education. His talk, entitled "On the Dangerous Tendency to Innovation and Extremes in Education," was published as a pamphlet in 1835. According to Winslow, teachers who adopted Pestalozzian ideas had thrown aside the experience of thousands of years far too hastily. He feared that "the mind of this generation is restive, feverish, impassioned, and consequently prone to a reckless radicalism."[44] His talk outlined recent trends in manual education, intellectual education, and moral discipline and found them all wanting. His comments on the second topic are particularly relevant here. Winslow acknowledged that innovations in teaching for young children might produce a "precocious exhibition of large and splendid acquisitions of popular knowledge." Yet the world did not need a "luxurious growth of mushroom scholars."[45] A proper education produced learning that students could wisely use. This benefit, Winslow argued, came from the contact of the mind of a student with the mind of a scholar. Winslow dismissed recent textbooks as being "adapted to please rather than to profit."[46] He thought that visible signs, figures, and machines might have some role in teaching. Although such objects might give children a more distinct view of certain topics, their use failed to develop the imagination. Students who relied too much on visible illustrations, Winslow said, would have difficulty when they reached more abstract topics. More generally, the new teaching methods failed to promote mental discipline.

Winslow did not specifically mention mathematics teaching in his pamphlet. There were, however, occasions when the Boston Primary School Committee voiced its objections to the introduction of subjects it considered too advanced for the young students. In 1830, for example, it passed a resolution directing instructors to follow the program of studies laid down in its regulations. Committee members feared that the schools were not emphasizing their proper subjects, "correct reading and thorough spelling."[47] On another occasion that same year the committee expressed doubts about the training that some of the infants coming to the primary schools had received. According to the *Annals* of the committee, "The attempt to learn [*sic*] children of this tender age spelling lessons, and even lessons in arithmetic and geography, by singing them in concert, while marching to time, gave them a restless habit and a sing-song style which it was subsequently found almost impossible to eradicate, and caused more trouble to the teachers of the Primary Schools than all the advantage the pupils had derived from the instruction they had received."[48] Clearly, infant schools were not universally admired.

Most authors who wrote on arithmetic seemed more interested in improving on the abacus than in objecting to its use. In 1831 William C. Woodbridge, editor of the *American Annals of Education,* prepared an article for his journal entitled "Palpable Arithmetic," in which he described two new devices for arithmetic instruction that had been demonstrated at a recent meeting of the American Lyceum in New York. Both of them illustrated the idea of the place value of a digit. One, the Visible Numerator, patented by Oliver A. Shaw of Richmond, Virginia, represented digits of different place value by blocks of differing size (units were small cubes, tens 1 × 1 × 10 prisms, hundreds 1 × 10 × 10 blocks, and thousands 10 × 10 × 10 cubes). These blocks were demonstrated to the Boston School Committee but were not adopted.[49] A second device, designed by Woodbridge himself, combined a conventional numeral frame with one in which the balls moved vertically. As on a Chinese or Japanese abacus, the balls that moved vertically had different place values. Woodbridge's device worked well on a flat surface, but the balls on the vertical wire all fell toward the bottom when the frame was held up to a class. Needless to say, this instrument also did not prove to be popular.[50]

Of greater influence were Thomas Palmer's comments about the numeral frame in a prize essay on education that he wrote for the American Institute of Instruction in 1838. Palmer's paper was published in 1840 and would be widely quoted and reprinted. He argued that teachers need not actually purchase a numeral frame; they could simply use the frame from a broken slate, wire, and a few beads. He strongly recommended that each wire of the abacus have only ten beads, rather than the twelve used by Holbrook. This arrangement would greatly assist in teaching place values, without requiring the purchase of additional apparatus like that of Woodbridge or Shaw. As noted earlier, Holbrook was soon producing numeral frames with ten rather than twelve beads per wire. It is less clear that teachers busied themselves with making their own numeral frames.[51]

Meanwhile, Frederick Emerson prepared a series of three arithmetics, which he believed should replace Colburn's two books. The subcommittee of the Boston School Committee that reviewed textbooks demurred at first, but in 1834, after Colburn had died and they had received a petition from eight grammar school masters requesting the change, they adopted all three parts of Emerson's *North American Arithmetic* for use in the Boston primary and grammar schools.[52] Emerson reduced the portion of the course devoted exclusively to mental arithmetic and added such commercial topics as the weights, measures, and currency of foreign countries. It became common later in the century for arithmetics to have a short introduction to mental arithmetic, followed by extensive written problems.

This combination of mental and written arithmetic eventually led to a new form of numeral frame. The instrument, patented in 1864, had a series of rotating slats, with several digits painted on each side of each slat. By rotating the slats and marking off problems with a thread, teachers could provide a wide range of arithmetic problems. The inventor of this "arithmetical frame," a teacher named Henry K. Bugbee, even provided a written *Key* that gave answers to the questions.[53] Thus, teaching arithmetic operations was once again united with written arithmetic. Mental arithmetic, along with palpable tools for teaching it, were out of fashion. In the twentieth century the abacus would increasingly be seen largely as an exploratory toy for children, not a centerpiece of instruction. In this way the instrument became divorced not only from its historical origins as an aid to calculation but also from its nineteenth-century place as a classroom tool.

CHAPTER SEVEN

The Slide Rule
Useful Instruction for Practical People

> I amused myself during the passage, at such times as my health would permit, in the study of mathematics. I had previously made myself acquainted with the improved engineers' sliding rule, and decimal fractions; and committed to memory several useful factors . . . Before I left the ship, I had prepared several tables for various purposes.

The classroom abacus was a form of demonstration apparatus, not a tool for lifetime use. Once students understood counting and arithmetic, they were expected to put the abacus aside. Similarly, the cube root block offered a tangible representation of a binomial equation. When students had mastered the metaphor—or even if they did not understand it—routine algorithms and mathematical tables were available for finding specific cube roots. Of course, instruments could be used for more than demonstration. In Russia and much of Asia, and among some Asian Americans, the abacus was a practical as well as a classroom tool. Another object with strong ties to the workaday world was the slide rule. The principle of the instrument was discovered in seventeenth-century Britain, and specific forms of rule were developed for several occupations. These slide rules were not aids to understanding but means of replacing mental effort with routine physical manipulation.

The first slide rules used in the United States followed British forms, with specific applications to carpentry and to engineering. Although none of these instruments was ever widely used in this country, a few engineers made use of them, and several compendia on practical mathematics published in the antebellum years mentioned them.[1] In the 1880s a growing demand for computations, improvements in the design of slide rules, and careful marketing combined to increase demand for the instrument. Slide rules were especially important to professional engineers, and much of the instruction in their use took place in courses in engineering, not mathematics. In the early twentieth century those seeking to reform secondary education in the United States, along with slide rule manufacturers, encouraged instruction in the slide rule for high school and even junior high students as well as for those studying technical subjects in college. Thus, the slide rule came to have a place in the mathematics classroom that it would retain until the introduction of the handheld electronic calculator in the 1970s.

Origins

In the early seventeenth century the Scottish nobleman and mathematician John Napier discovered a function that he named the logarithm. It has the remarkable property that the sum of the logarithms of two numbers equals the logarithm of the product of those two numbers. In other words, instead of multiplying two numbers together, one can add their logarithms; the product is the anti-logarithm (or inverse logarithm) of this sum. Similarly, division can be reduced to subtraction. Mathematicians, astronomers, and other table makers had long known that people doing calculations add and subtract much faster and more accurately than they multiply and divide. Hence, logarithms were quickly adopted by astronomers such as Johannes Kepler and other calculating people.

Several mathematically minded Englishmen pointed out that logarithmically divided scales could also be of great use in computations done by people who had little knowledge of logarithms or indeed arithmetic. Edmund Gunter, professor of astronomy at Gresham College, London, noted that with a logarithmic scale and a pair of dividers, one could find the sums and differences of the logarithms of numbers and thereby multiply and divide. In the 1620s the English clergyman and mathematician William Oughtred described an instrument with a rotating dial that had logarithmic scales on both the outer edge of the dial and the adjacent circle of the frame. One could multiply or divide two numbers by rotating the dial, dispensing with the dividers. This was the first slide rule.[2]

Other slide rules, with linear rather than circular scales, were proposed over the course of the seventeenth century. In 1677, for example, Henry Coggeshall published a pamphlet describing a linear rule designed especially for calculations relating to timber.[3] A revised version of his small book appeared in 1682, with further editions by later authors published as late as 1767. Because one of the major early exports of the British colonies in North America and later the United States was timber, it is not surprising that a form of Coggeshall's rule called the "carpenter's rule" would come to be one of the first slide rules sold in this country.[4] The carpenter's rule was made from two wooden one-foot rules that were held together at one end by a metal joint. Unfolded, one side became a simple two-foot measuring rule. The upper part of the other side of the twofold rule contained a groove that held a brass slide, with logarithmic scales on the upper and lower edges of both the slide and the adjacent parts of the groove. The inner edges and lower part of the carpenter's rule commonly were marked with other scales that were useful to carpenters and spar makers (Figure 7.1).

A second British slide rule that would prove of some importance in the United States was developed in the late eighteenth century by James Watt and Matthew Boulton, working with their associate James Southern. They designed a 10-inch wooden rule specially suited to engineering purposes

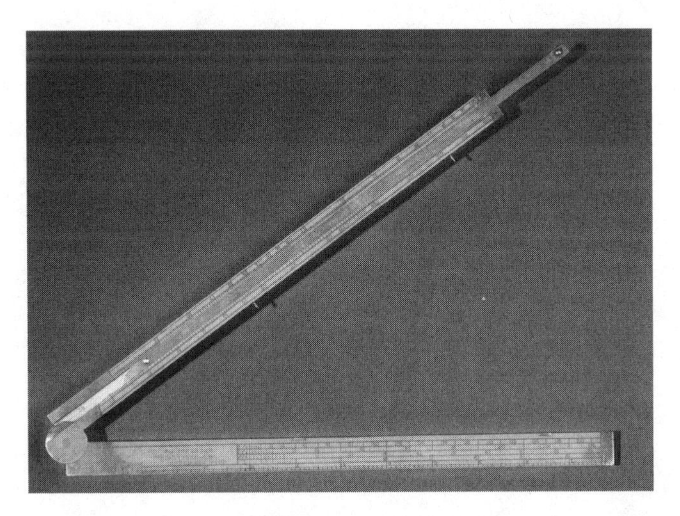

Figure 7.1. Carpenter's rule, ca. 1842. Sold by S. A. Jones & Co. of Hartford, Conn. The scales were labeled, going from top to bottom, *A, B, C,* and *D*. The *A, B,* and *C* scales were identical and divided logarithmically from 1 to 10 to 100 (i.e., from 1 to 10 twice). Divisions on the *D* scale, or "girt line," were twice as large as on the other scales for use in approximating the volume of solids of known height and circumference. National Museum of American History collections. Smithsonian Negative no. 2004-40575.

Règle à Calculer.

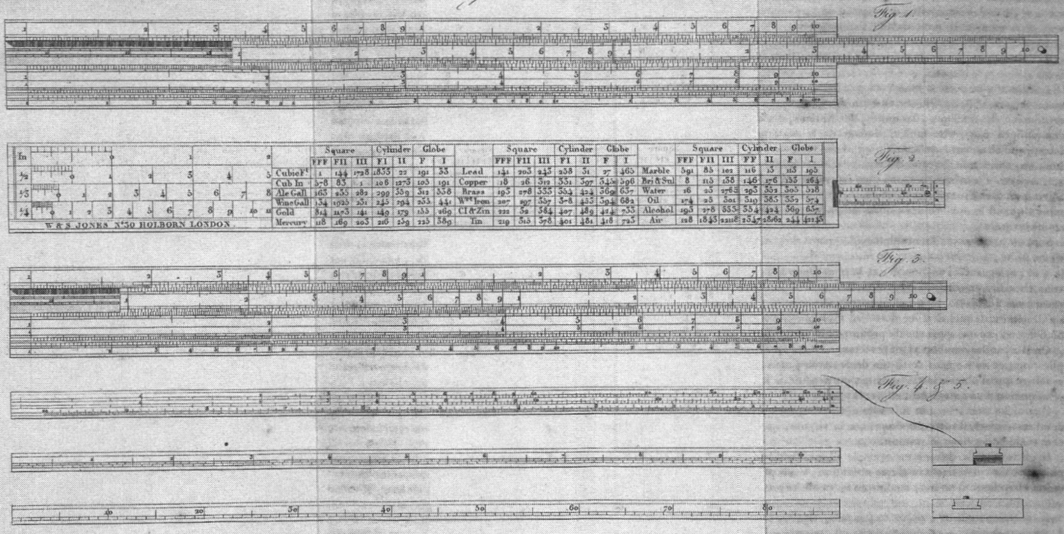

Fig. 1.

Fig. 2.

	Square			Cylinder		Globe			Square			Cylinder		Globe			Square			Cylinder		Globe	
	FFF	FII	III	FI	II	F	I		FFF	FII	III	FI	II	F	I		FFF	FII	III	FF	II	F	I
Cubic F^t	1	144	1728	1833	22	191	33	Lead	141	203	243	238	31	27	465	Marble	591	85	102	116	13	113	193
Cub In	578	83	1	108	1273	103	191	Copper	18	26	312	331	397	343	596	Bri & Sul	8	115	138	146	176	133	264
Ale Gall	163	233	282	299	359	312	338	Brass	193	278	333	354	424	369	637	Water	16	23	2763	293	352	303	528
Wine Gall	134	1925	231	243	294	253	441	W^t Iron	207	297	357	378	453	394	682	Oil	174	25	301	319	383	332	574
Gold	814	1173	141	149	179	155	269	CI & Zin	222	32	384	407	489	424	733	Alcohol	193	278	333	354	424	369	637
Mercury	118	169	203	216	259	225	389	Tin	219	315	378	401	481	418	723	Air	128	1843	22118	2347	28162	244	42245

W & S JONES N° 30 HOLBORN LONDON.

Fig. 3.

Fig. 4. & 5.

[illegible] Del. et Sculp. [illegible]

(Figure 7.2). A few were built at Boulton and Watt's manufactory in Soho, and more were produced by the firm of W. & S. Jones of London. Like a carpenter's rule, the so-called Soho rule had A, B, and C scales that were divided logarithmically from 1 to 10 twice. The D scale had divisions that were twice as large as on the other scales and ran only from 1 to 10. The A and B scales were used for simple multiplication and division, the C and D scales for finding squares of numbers and square roots. The back of the slide often had scales of logarithmic sines and tangents as well as a scale for finding logarithms and anti-logarithms of numbers. The reverse side of Soho rules sometimes had scales useful in drawing and also tables for calculating the volume and weight of solids of differing shape and materials.[5]

In 1814, during Napoléon's exile in Elba, the French engineer and Egyptologist Edmé-François Jomard visited England to examine both antiquities and the latest in industrial practice. He purchased a Soho rule from W. & S. Jones and prepared an account of the instrument for the French Société d'Encouragement pour l'Industrie Nationale.[6] It may well be that one or more of the English engineers who immigrated to the United States in the early nineteenth century brought a Soho slide rule along.[7] Soho rules would be sold in the United States in the late nineteenth century.

Of somewhat more immediate influence in the United States was an "improved" engineer's slide rule devised by Joshua Routledge of Bolton in about 1810. Like the carpenter's rule, Routledge's rule had two one-foot wooden parts held together by a metal joint at one end. The upper part was grooved with a metal slide. Routledge's rule had the A, B, C, and D scales of a Soho rule, with similar tables for calculating the volume and weights of solids of differing shapes (Figure 7.3). The back of the slide on an English Routledge rule in the Smithsonian collections has a scale of inches. Routledge also included tables that were specifically intended for use in calculations relating to steam engines.[8]

Knowledge and use of the slide rule diffused slowly in the British colonies of America and the early United States. In the antebellum years perhaps two dozen books published in the country included brief discussions of some form of slide rule.[9] These were practical treatises for workingmen

Figure 7.2. (Opposite) Soho-type slide rule, 1815. From Edmé-François Jomard, "Description d'une règle à calculer, employée en Angleterre et appelée sliding rule; précédée de quelques réflexions sur l'état de l'industrie anglaise, en avril 1815," ca. 172. Smithsonian Negative no. 2004-40663.

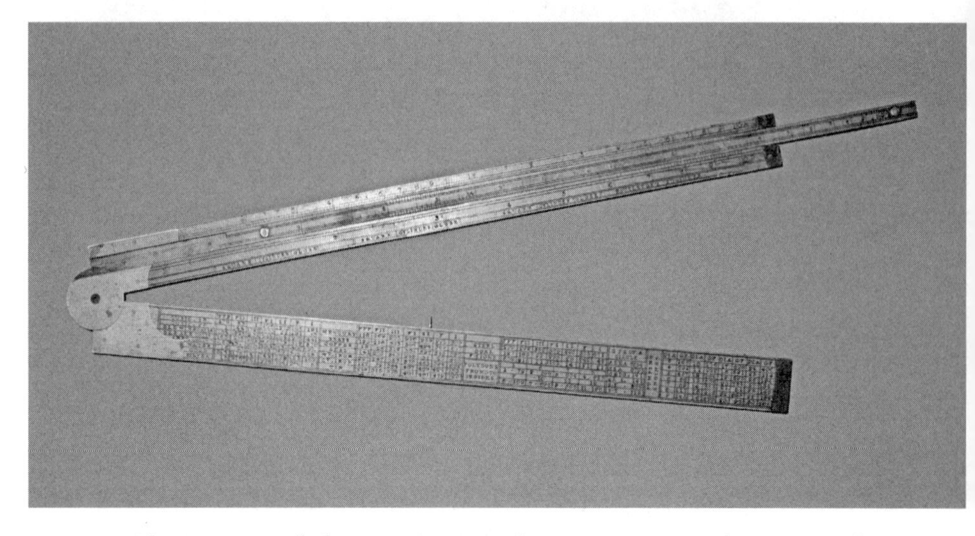

Figure 7.3. Routledge's engineer's rule, ca. 1850. National Museum of American History collections, gift of Mrs. Leila D. Lovelace. Smithsonian Negative no. 82-6212.

(none seems to have included women as possible readers), not texts for classroom study. Some were pirated editions of English books; others were by American authors who drew heavily on English sources. They emphasized procedures that readers could follow to get the answers they sought, not mathematical principles. Thus, early American users of the slide rule probably had no better sense of why it worked than most present-day users of electronic calculators understand the operating principles of those devices.

Slide Rules for Professional Engineers

In the late nineteenth century the number of engineers in the United States grew rapidly. Technical education, which once had occurred largely on the shop floor and in informal mechanics institutes, increasingly took place in technical institutes, engineering schools at private universities, and state technical colleges. The Morrill Act of 1862 provided funds for agricultural and mechanics institutes in every state of the Union. Successful entrepreneurs also devoted some of their wealth to engineering education.

The first American slide rule designed for engineering work had been patented in 1851 by the Swedish-born and educated civil engineer John W. Nystrom of Philadelphia. Nystrom designed a circular instrument with scales that were arranged on arcs of spirals (Figure 7.4). It not only included scales for ordinary multiplication and division but made provision for multiplying sines and cosines of angles. Nystrom used his "calculator" in the design of steamboat propellers.[10] He also relied on it to compute the tables in a small book entitled *Pocket-book of Mechanics and Engineering*. This volume went through eighteen editions by the time of Nystrom's death in 1885. Nystrom's rule was advertised here. It also was listed in catalogs of the instrument maker William J. Young of Philadelphia from 1874 to 1882. At a price variously given as $15 to $40, it found few buyers.[11]

In France from the 1790s and in Germany from the 1870s, advanced technical education was directed as part of a national system. This was not the case in the United States. Uniformity in engineering instruction emerged through the establishment of regular programs of engineering education, the diffusion of graduates of these programs to other schools, the publication of textbooks, and the activities of national engineering societies. At the same time, the activities of instrument makers such as Keuffel & Esser (K&E) of New York encouraged the use of various tools, including slide rules. Several events from around 1880 hint at the changing market for engineer's slide rules. One sign came from Britain. In 1879 George Fuller, a

Figure 7.4. U.S. patent model for Nystrom's calculator, 1851. National Museum of American History collections. Smithsonian Negative no. 87-8042.

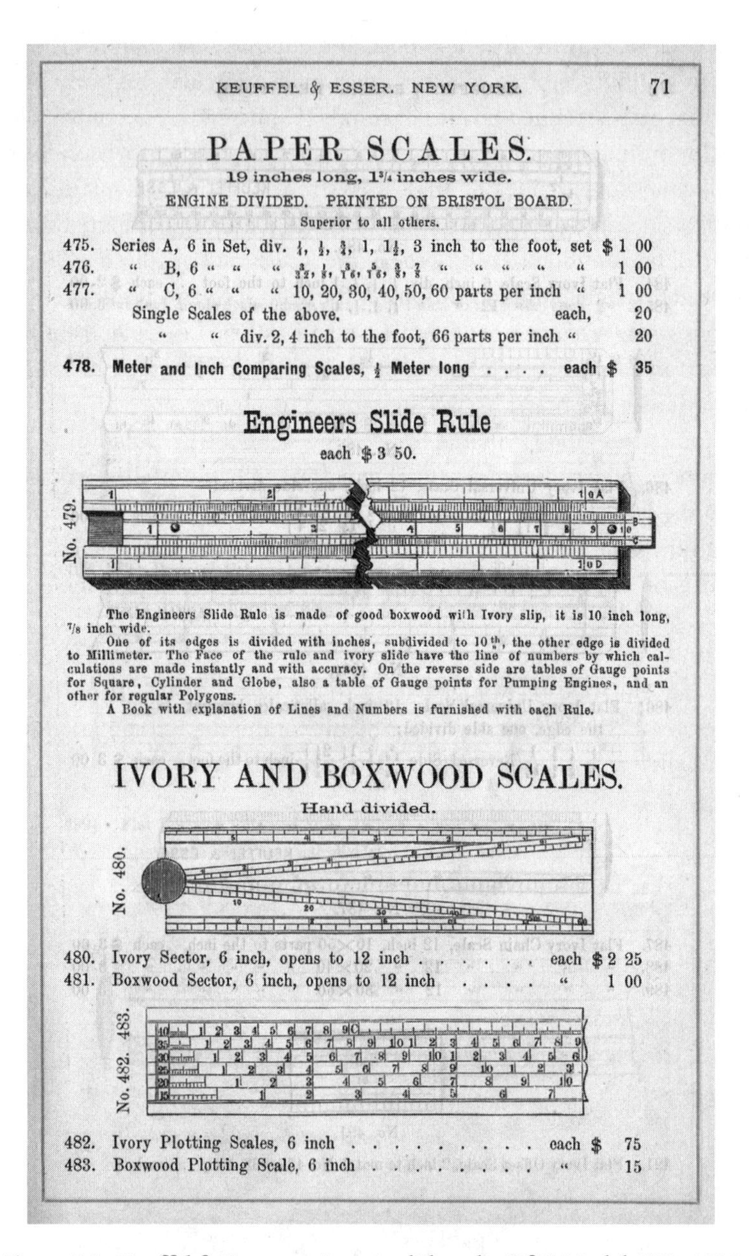

KEUFFEL & ESSER. NEW YORK. 71

PAPER SCALES.

19 inches long, $1^3/_4$ inches wide.

ENGINE DIVIDED. PRINTED ON BRISTOL BOARD.

Superior to all others.

475. Series A, 6 in Set, div. $\frac{1}{4}$, $\frac{1}{2}$, $\frac{3}{4}$, 1, $1\frac{1}{2}$, 3 inch to the foot, set $ 1 00
476. " B, 6 " " " $\frac{3}{32}$, $\frac{1}{8}$, $\frac{3}{16}$, $\frac{5}{16}$, $\frac{3}{8}$, $\frac{7}{8}$ " " " " " 1 00
477. " C, 6 " " " 10, 20, 30, 40, 50, 60 parts per inch " 1 00
Single Scales of the above. each, 20
" " div. 2, 4 inch to the foot, 66 parts per inch " 20

478. Meter and Inch Comparing Scales, $\frac{1}{2}$ Meter long . . . . each $ 35

Engineers Slide Rule

each $ 3 50.

No. 479.

The Engineers Slide Rule is made of good boxwood with Ivory slip, it is 10 inch long, $^7/_8$ inch wide.

One of its edges is divided with inches, subdivided to 10th, the other edge is divided to Millimeter. The Face of the rule and ivory slide have the line of numbers by which calculations are made instantly and with accuracy. On the reverse side are tables of Gauge points for Square, Cylinder and Globe, also a table of Gauge points for Pumping Engines, and another for regular Polygons.

A Book with explanation of Lines and Numbers is furnished with each Rule.

IVORY AND BOXWOOD SCALES.

Hand divided.

No. 480.

480. Ivory Sector, 6 inch, opens to 12 inch . . . . . each $ 2 25
481. Boxwood Sector, 6 inch, opens to 12 inch . . . . . " 1 00

No. 482. 483.

482. Ivory Plotting Scales, 6 inch . . . . each $ 75
483. Boxwood Plotting Scale, 6 inch . . . . " 15

Figure 7.5. Keuffel & Esser engineer's slide rule (K&E Model 479), 1880. The image is from the company's catalog. Courtesy of Smithsonian Institution Libraries.

railway engineer and professor of engineering in Queen's College, Belfast, received a U.S. patent for a new form of slide rule, which had scales arranged in spirals on a cylinder. Fuller's instrument would be produced by W. F. Stanley of London and eventually sold in the United States.[12] A second development occurred in 1880, when Keuffel & Esser Company published its thirteenth catalog, listing as item no. 479 an "engineer's slide rule." This instrument was a simple wooden linear slide rule 10 inches long with an ivory slide (Figure 7.5). Its form and scales were those of the front of a Soho rule (there was a small knob at one end for moving the ivory slide, which would preclude useful sliding scales on the reverse of the slide).[13] According to a company history, William Keuffel made a trip to Europe that year to select instruments and materials from makers in Switzerland, Germany, and England.[14] Precisely where this rule was purchased is not clear.

A similar rule, apparently of American manufacture, was shown the next year in a book by Robert Riddell, teacher of the artisan class in the Central High School of Philadelphia. Riddell wrote several books on practical topics relating to carpentry. He had advocated the use of the slide rule in a book entitled *The Carpenter and Joiner Modernized* (3rd ed., 1880) but apparently showed no image of the instrument. In 1881 Riddell published *The Slide Rule Simplified, Explained, and Illustrated.* Bound with the copy of the book in the Smithsonian library is a broadside showing and describing a slide rule made by Stephens & Company of Riverton, Connecticut. Although Riddell indicated that his book was intended particularly for young carpenters and joiners, the featured slide rule is not a carpenter's rule or even a Routledge engineer's rule but a wooden linear rule with brass slide and the scales of a Soho rule. Riddell makes no reference to any scales on the back of the slide for trigonometric functions or for finding logarithms, and none are evident from the plate on the broadside. We have as yet found no specific documents that illuminate the connection between Riddell and Stephens & Company.[15]

Also in 1881, Edwin Thacher, a "computing engineer" for the Keystone Bridge Company in Pittsburgh, Pennsylvania, received a patent for an improvement in slide rules. Thacher was a graduate of Rensselaer Polytechnic Institute (RPI) who spent much of his career designing railway bridges. To assist in his calculations, he designed a cylindrical slide rule. Thacher's rule, though it fit neatly on a desk, was equivalent to a linear slide rule over 59 feet long (Figure 7.6). It had scales for multiplication and division and another with divisions twice as large for use in finding squares and square

Figure 7.6. Classes of 1894 and 1895 using instruments at the Thayer School of Engineering, Dartmouth College. A Thacher cylindrical slide rule is on the table at the *center* toward the *right*. Courtesy of Dartmouth College Library.

roots. There were no trigonometric scales. To produce his "calculating instrument," Thacher, like Fuller, turned to Stanley of London. The firm even designed a special dividing engine for preparing the scales for the instrument, which were printed on paper sheets that were pasted to the drum and the slats.[16]

Finally, in 1883 Keuffel and Esser began offering in its catalog a much cheaper slide rule that would find widespread classroom use. This model was a modified form of the Soho rule that had long been used in France. As noted earlier, Jomard had brought the Soho rule to Paris in 1814. The French instrument maker Etienne Lenoir and his son Paul-Etienne manufactured wooden slide rules like the Soho rule and passed along the business to a successor with the surname of Gravet (later the firm of Tavernier and Gravet).[17] In 1851 a young artillery officer named Amadée Mannheim proposed a new arrangement of the scales on a Soho rule. On Mannheim's rule the top

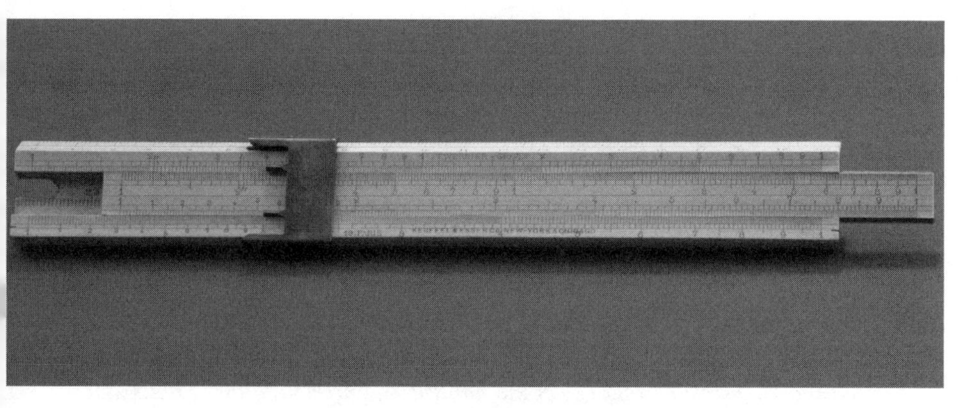

Figure 7.7. Mannheim slide rule made by Tavernier-Gravet for Keuffel & Esser, ca. 1890. National Museum of American History collections, gift of Keuffel & Esser. Smithsonian Negative no. 2003-26092.

scales on the base and slide remained identical, running from 1 to 10 twice. The lower two scales, one on the slide and one on the base, also were identical, running from 1 to 10 only once. This allowed for more precise computations. Mannheim's rule also had a cursor, which made it possible to use the top and bottom scales together and hence to take squares and square roots (Figure 7.7). As on the Soho rule, the back of the slide had a scale of logarithms of numbers and scales for logarithms of trigonometric functions.[18] Knowledge of the Mannheim slide rule was soon required for admission to the École Polytechnique in Paris and the École Militaire of Saint-Cyr, and instruction in it was even offered at certain French lycées.[19]

From the mid-nineteenth century international expositions had offered an opportunity for juries and ordinary visitors to compare inventions from around the world. Several of the slide rules just described were exhibited in 1885 at the International Inventions Exhibition held in London. Stanley displayed both Fuller and Thacher rules, while Gravet showed the Mannheim rule. The English physicist and instrument inventor Charles V. Boys attended the exhibition and took the occasion to write an article on the slide rule that was published in *Nature* and reprinted in New York in Van Nostrand's *Engineering Magazine.*

Boys firmly believed that scientists and engineers should use slide rules. As he wrote: "It is a perpetual source of amazement to those who are familiar with this instrument that its use is not almost universal. People of every class have to make simple calculations, while those engaged in scientific work, in designing apparatus, or in invention, perpetually cover sheets of paper with figures, all of which trouble, and the loss of time which it involves, might be saved by the intelligent use of a good slide rule, and yet, for reasons difficult to find out, the habitual use of this instrument is limited to a very small proportion of the calculating community."[20] Boys went on to describe several of the slide rules exhibited in 1885, placing special emphasis on the advantages of the Mannheim slide rule.

This diverse activity attracted the attention of American engineers. In 1886, as part of its "Topical Discussions and Interchange of Data," the *Transactions of the American Society of Mechanical Engineers* included a discussion of the use of computing devices and tables. The eminent consulting steam engineer Charles E. Emery expressed the opinion that "the most efficient form of slide rule is probably that formerly made by Mr. John W. Nystrom." He hoped that the rule would still be produced even though Nystrom had died. This wish proved to be in vain. Other engineers such as Frederick A. Halsey spoke more generally about the advantages that either they or workmen they knew derived from having slide rules but expressed the view that to take full advantage of the instrument one needed to use it efficiently. Olin Scott was more skeptical. He had found from hard experience as a builder of machinery for measuring apparatus that it was very difficult "to induce any man, having any mathematical ability whatever, to substitute a machine for mental operations." He noted several instances in which he had tried to introduce scales onto measuring instruments that would reduce calculations required, only to find that purchasers were not interested.[21]

Instrument dealers were quick to encourage the new interest in the slide rule. Keuffel & Esser imported both the Fuller and the Thacher rules from London and Mannheim rules from Tavernier-Gravet in Paris. Moreover, the firm hired the Englishman William Cox to help publicize its products. From 1891 through 1894 Cox edited a magazine known as the *Compass,* which described a variety of mathematical instruments sold by K&E, including the slide rule. He also wrote a manual on the Mannheim slide rule and published articles on the instrument in the journal *Engineering News.* Cox designed a two-sided, or duplex, slide rule, patented in 1891, that would become a standard product of the company (Figure 7.8). Other American dealers, including Eugene Dietzgen in Chicago and L. E. Knott in Boston, began to sell slide rules.

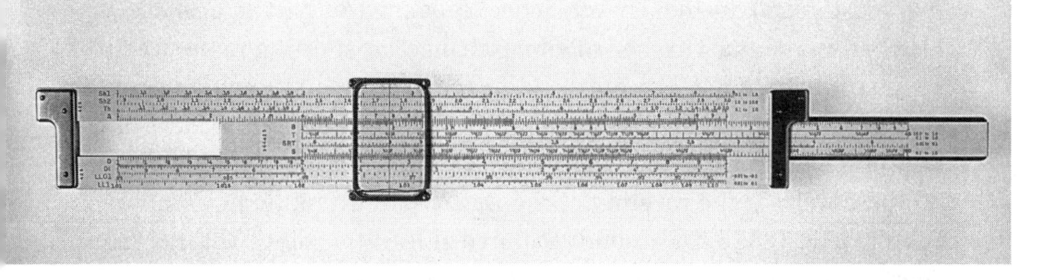

Figure 7.8. Keuffel & Esser duplex slide rule, ca. 1900. National Museum of American History collections, gift of Keuffel & Esser. Smithsonian Negative no. 62754.

The Slide Rule Enters the Engineering School

In 1900 John Butler Johnson introduced a discussion of the slide rule into the fifteenth edition of his standard textbook, *The Theory and Practice of Surveying*. Johnson had long taught civil engineering at Washington University and had recently become dean of engineering at the University of Wisconsin. He was well aware that some readers might not consider the slide rule to be a usual topic for a course in surveying and commented in the preface: "While the author has long been a constant user of slide-rules of all kinds, he had not thought to include it in a description of surveying instruments. It is now introduced here because its use is not taught elsewhere in our engineering schools."[22] Johnson may have chosen to add a reference to the slide rule to his book because he had just been working with the Keuffel & Esser Company on developing an inexpensive form of the Mannheim slide rule for student use; he explicitly mentions this product in the text. Other professors who were advocates of the slide rule came from different fields and introduced it into the courses they happened to teach or direct. A few of these courses were in mathematics; others were in such disparate areas as civil engineering, mechanical engineering, and physics. It was only in the 1920s, as engineers sought more uniform curricula, that instruction in the slide rule generally became a task of mathematics departments.[23]

Throughout the 1880s and 1890s Keuffel & Esser took pains to publicize its new slide rules. Documentation of this process as it took place at the Thayer School of Engineering at Dartmouth is particularly rich and will serve as an example. In 1887, the first year that Keuffel & Esser sold the Thacher cylindrical slide rule, Robert Fletcher, professor of civil engineering and director of the Thayer School, obtained one. It is possible that Fletcher was sent a Thacher rule in exchange for attesting to the utility of the instrument. A letter recommending the instrument to a member of the staff of the chief engineer's office for the aqueduct commissioners of the state of New York survives in his correspondence.[24] Fletcher proudly showed off the novelty at the meeting of the Dartmouth Scientific Association on February 16, 1887.[25] Dartmouth acquired at least one other Thacher calculator, as two examples of the instrument survive in its collections.[26] In December 1889, just as K&E was beginning its publicity campaign for the Mannheim slide rule, Fletcher spoke to the Dartmouth Scientific Association on the "common slide rule."[27] Soon all engineering students at Dartmouth were expected to use Mannheim slide rules as part of their laboratory work.[28] An 1895 photograph in the Dartmouth archives shows a roomful of students with diverse apparatus. A Thacher cylindrical slide rule and a linear slide rule are on a table in the foreground, along with a planimeter and drawing instruments (Figure 7.6).[29] Fletcher made sure that his students used the most economical slide rules available. In 1897 K&E introduced the "student's slide rule" so admired by Johnson. The instrument cost only a dollar, compared to $5 for a regular Mannheim slide rule offered in the same catalog.[30] At this time the total annual expenses of a frugal student at the Thayer School were estimated as running from $350 to $400, "including tuition, books and stationery, board, fuel, light, and drawing instruments."[31] The less expensive slide rule was definitely appealing. By 1898 Fletcher was ordering the instrument for each of his students.[32]

Some engineering students learned about slide rules in mathematics classes. The manual that Dartmouth students used was written by Charles W. Crockett of Rensselaer Polytechnic.[33] Crockett had attended RPI, graduating in 1884. He stayed on as an instructor in mathematics and astronomy and in 1893 was appointed the professor in this area and head of the department. That very year the RPI *Handbook of Instruction* indicated for the first time in its description of mathematics department offerings that "lectures on the theory and the various forms of the slide rule are also delivered."[34] No instructor was named, but Crockett's influence looms large. In a similar vein Arthur E. Haynes, professor of engineering mathematics at the University of Minnesota, seems to have been responsible for introducing instruction on the use of the slide rule there. Haynes taught mathematics in the engineering school at Minnesota from 1896, and the instruction he offered to sophomore engineers on the slide rule is explicitly mentioned in university catalogs from at least 1906–7.[35]

Some slide rule enthusiasts came from the ranks of mechanical engineers. They tended to introduce the slide rule as part of an advanced laboratory course in experimental mechanics. Such instruction is first mentioned in the *Annual Announcement* of the Stevens Institute of Technology for 1895–96 and was continued there through at least 1901–2. By 1916 teaching of the slide rule had become a general requirement for engineering majors and was covered in the freshman year math course in plane and analytic geometry.[36] Similarly, beginning in 1896–97, all midshipmen at the United States Naval Academy received practical instruction in steam engineering that included work with the slide rule.[37] Official teaching of the slide rule remained part of this course until the 1920s, when responsibility for teaching the topic shifted to the mathematics department. Following the same pattern, in the early twentieth century, when mechanical engineering was introduced into the curriculum of the Maryland Agricultural College, the slide rule was taught in a mechanical engineering course for seniors on "testing."[38]

At some schools instruction in the slide rule was the responsibility of the physics department. Catalogs of the Georgia School of Technology (later the Georgia Institute of Technology) indicate that by 1905–6 the slide rule was taught in the plane trigonometry course. A year later, however, Charles J. Payne had introduced a "Physical Laboratory" that met three hours a week throughout one year. "A systematic study of the slide rule" was part of the course and remained so at least through 1911.[39] At the U.S. Military Academy at West Point instruction in the slide rule was the province of the physics department until the topic was taken over by the mathematics department in 1943.[40]

The Diffusion of the Slide Rule

During the first half of the twentieth century use of the slide rule became common among numerate Americans. Rules were made from materials that ranged from paper to wood to metal to plastic. In addition to instruments for general calculations such as those we have described, inventors copyrighted and patented instruments to carry out specific computations needed in fields ranging from surveying to hydraulics to chemistry to industrial engineering to statistics. Manufacturers also distributed inexpensive paper rules as advertisements for products that ranged from electrical resistors to tractors to porcelain enamel coatings.[41]

Some mathematics professors tried to incorporate instruction in the slide rule into more general courses on computation. Such work was of particular importance at a time when the interpretation of vast amounts of quantitative data challenged not only scientists and engineers but also social scientists, businessmen, and government bureaucrats. The need for data reduction had not only encouraged the improvement and diffusion of the slide rule but created a market for such other relatively new instruments as the commercial calculating machine, integrators, and harmonic analyzers. In a 1911 Ph.D. dissertation written in the Department of Mathematics of the University of Chicago, Theodore Lindquist examined in detail the mathematics taught to freshmen at 130 engineering schools. He paid no attention to instruction in the slide rule alone but commended 4 (unspecified) schools that had introduced courses in computation that included instruction on the use of calculating instruments. Lindquist also polled some 650 mathematics teachers and engineers, eliciting opinions about what should be taught to freshmen engineers. Somewhat over a third of those who replied favored a course in techniques of computation, while about a fifth of them were opposed. Both those who favored the course and those who opposed it were concerned about fitting more material into the crowded academic schedule of engineering students.[42]

One of the most ambitious courses in computation offered for engineers was introduced not long after Lindquist completed his dissertation. In 1914 Joseph Lipka of the MIT mathematics department began offering a "Mathematical Laboratory" that met two hours a week. The MIT catalog described it as:

> a course for practical instruction in numerical, graphical and mechanical calculation and analysis as required in the engineering or applied mathematical sciences. The course will include: methods for checking the accuracy of arithmetic and logarithmic computations; numerical solution of algebraic, transcendental and differential equations; graphical methods in the processes of arithmetic, algebra and the calculus; curve fitting to empirical data; the use and principles of construction of instruments employed in calculation, such as slide-

rules, arithmometers, planimeters and integraphs; and many kindred topics.[43]

In later years Lipka added to the course description specific mention of such "kindred topics" as nomography and the construction of graphical charts as well as approximate methods of differentiation and interpolation. Perhaps tellingly, this course in numerical methods was not required by engineering departments. At a much more elementary level, in the academic year 1917–18 Lipka began offering an elective course of four exercises for first-semester freshmen that provided them with a basic introduction to the slide rule.[44]

While a few professors hoped to provide students with a general introduction to numerical analysis and related instruments, most colleges and engineering schools found little room for the topic in their curricula. Indeed, there was a considerable move to encourage high schools and even junior high schools to offer the instruction in computing devices that was provided, namely an introduction to the slide rule.[45] As early as 1903, George W. Myers of the College of Education at the University of Chicago recommended slide rules for use in the secondary high school "mathematics laboratory." Myers described two ways of equipping such a laboratory—one "fairly complete" and the other "less pretentious." The more expensive mathematics laboratory would boast both logarithmic slide rules and computing machines. The less expensive version had a dollar logarithmic slide rule and books of mathematical tables.[46] C. E. Comstock, instructor in mathematics at the Bradley Institute in Peoria, Illinois, published his comments on the mathematical laboratory for the secondary school that same year. He, too, recommended diverse modern apparatus, including the protractor and other drawing instruments, geometric models, physical apparatus, and the slide rule.[47] By 1923 a report of the National Committee on Mathematical Requirements (NCMR) of the Mathematical Association of America could describe several schools that covered the subject. Writing on "optional topics" that might be included in grades 7, 8, and 9 of secondary school, the committee reported that it looked with favor on efforts "to introduce earlier than is now customary certain topics and processes which are closely related to modern needs, such as the meaning and use of fractional and negative exponents, the use of the slide-rule, the use of logarithms and of other simple tables."[48]

Many of the recommendations of the NCMR were ignored by those responsible for school mathematics teaching. In the years following its report, however, Keuffel & Esser regularly advertised the slide rule as a device for teaching trigonometry in both the *American Mathematical Monthly,* a journal aimed largely at college teachers, and in the *Mathematics Teacher,* a magazine for secondary school teachers and mathematics educators.[49] Instruction in use of the instrument came to be given to students planning advanced scientific and technical studies, including apprenticeship programs for electricians and mechanics. Both K&E and rival firms made and sold oversized slide rules for classroom use, although these instruments were not necessarily listed in company catalogs.[50] Thus, the slide rule, like the abacus long before, could become a piece of demonstration apparatus. It would remain a part of school and college mathematics teaching until the introduction of inexpensive handheld electronic calculators in the mid-1970s.

CHAPTER EIGHT

The Cube Root Block
Teaching "Evolution" in the Schools

> The extraction of the cube root can be explained most easily by the use of the Cube Root Block. In fact, no person who is unacquainted with Algebra or Geometry can know the reason for this rule without the aid of some such illustration.

Throughout most of the nineteenth century it was standard for American arithmetic textbooks to include a section on the extraction of roots as a culminating topic late in the course. Often this lesson was presented under the title "evolution," referring to the notion that the root must be "evolved," or "unfolded," from the number in which it had been "involved," or "infolded," by the inverse process of raising to a power (e.g., squaring or cubing). In the case of the square root the method used was almost always that described in an arithmetic text published by James Porter in 1845:

> Rule. Separate the given number into periods of two figures each, beginning at the units' place.
>
> Subtract from the first period at the left hand the greatest square it contains, setting the root of that square as a quotient figure, and doubling said root for a divisor, and bring down the second period to the remainder for a dividend.
>
> Try how often the said divisor (with the figure used in the trial thereto annexed) is contained in the dividend, and set this figure in both the divisor and root; then multiply and subtract, as in division, and bring down the next period. Double the ascertained root for a new divisor, and repeat the process to the end.[1]

Porter supplied only one justification of his method:

> Proof. Square the root, adding in the remainder, if any, and the result will equal the given number.[2]

But many of Porter's contemporaries supported the method with a geometric argument involving a dissected square. A corresponding geometric argument for a method of extracting cube roots also became popular, relying on dissection of a cube. For many decades, even into the twentieth century, actual three-dimensional models made of wood were employed by instructors for this purpose.[3] In this chapter these square and cube root methods will be referred to collectively as the "binomial method," for they fundamentally depend on the binomial expansion $(a + b)^n$, for $n = 2$ and $n = 3$—that is, on squaring or cubing the two-term expression $(a + b)$. The binomial method has serious deficiencies from the viewpoint of modern mathematics, either pure or applied. It is isolated from other mathematical techniques, and it has not been a source of fruitful generalizations to other problems; even pursuing the most obvious extension, to extraction of fourth roots, has usually been found to be a painful and unrewarding venture. Further, for those seriously engaged in extracting square and cube roots for practical purposes, it is a laborious method; other alternatives, known well before the nineteenth century, accomplish the task with far less effort.

Nevertheless, the binomial method reigned supreme in mathematics education in the United States for a long time and continued to be taught past the middle of the twentieth century. Cube root extraction and the cube root block were the first elements to disappear, but even here the decline was not straightforward. The history of the cube root block and related methods of arithmetic teaching illuminates the changing relationship between elementary and advanced mathematics, the professionalization of mathematics, and the role of visualization in mathematical instruction.

The Binomial Method and Its Geometric Interpretation

If one wishes, for example, to extract the square root of 5,329 by the binomial method, one first groups the digits in pairs:

$$\overline{53}\ \overline{29}$$

Taking the leftmost pair, one seeks the smallest digit whose square is less than 53, namely 7. One places this 7 above the 53, places the square of 7 beneath the 53, subtracts, and then brings down the next pair of digits:

```
 7   _
53  29
49
 4  29
```

Next, one doubles the 7, obtaining 14, placing this off to the side:

```
       7   _
14_   53  29
      49
       4  29
```

One then seeks a digit *b*, which can be placed both to the right of the 14 and above the 29, so that 14*b* times *b* is as close as possible to 429 without exceeding it. In this example *b* should be chosen to be 3 because 143 times 3 is exactly 429:

```
       7   3
143   53  29
      49
       4  29
       4  29
           0
```

Given that there is no remainder, we have found the exact root, namely 73, a result that can easily be checked.

Why does this method work? The first step of grouping the digits is simply a means of determining the number of digits in the root. In the case of the square root the groups consist of two digits because the square of a single digit will result in at most two digits. Units when squared give at most tens, tens when squared give at most thousands, and so on. Therefore, working in reverse, the root of a number will have a units digit derived from the tens and units of the given number, a tens digit derived from the thousands and hundreds, and so on. Similar reasoning leads to grouping the digits by three in the case of cube roots.

Proceeding with the square root case, if the already determined portion of the root is a, the object is to choose the next portion, b, so that $(a + b)^2$ is as close as possible to the given number. Now $(a + b)^2 = a^2 + 2ab + b^2$, and because a^2 is already determined, we subtract it away, meaning we must choose b so that the remainder is approximated by $2ab + b^2 = (2a + b)b$—hence, the admonition in the 1845 text about "doubling said root for a divisor," that is, producing $2a$. The text's further instructions to choose a "trial" number (our b) and then to "set this figure in both divisor and root; then multiply and subtract" is a verbal description of adding b to $2a$ and multiplying the result by b, hence $(2a + b)b$, and then subtracting to find the remainder. The process can then be repeated as desired. The cube root method is more involved but can be explained in similar fashion, with reference to the fact that $(a + b)^3 = a^3 + 3a^2b + 3ab^2 + b^3$.

The expansion of $(a + b)^2$ can be depicted geometrically by appropriately dissecting a square of side $a + b$ into four sections: one of area a^2, two of area ab, and one of area b^2. Figure 8.1 shows such a square used in a 1928 textbook treatment of square root extraction. Similarly, the expansion of $(a + b)^3$ can be depicted geometrically by dissecting a cube into eight solid blocks: one block of volume a^3, three of volume a^2b, three of volume ab^2, and one of volume b^3. In both the square root and cube root cases the geometry serves to emphasize the idea of starting with a lower approximation to the root, represented by the side of a square (or cube, respectively), and then improving this approximation by attaching to the original figure appropriate rectangles (respectively, rectangular blocks), so as to maintain the square (respectively, cubic) form.

The steps in the geometric interpretation of the cube root method are shown especially clearly in Figure 8.2, from an 1871 textbook. The pictures in this figure strongly suggest the use of an actual three-dimensional model, one built out of wood, for instance. Figures 8.3 through 8.5 show some ex-

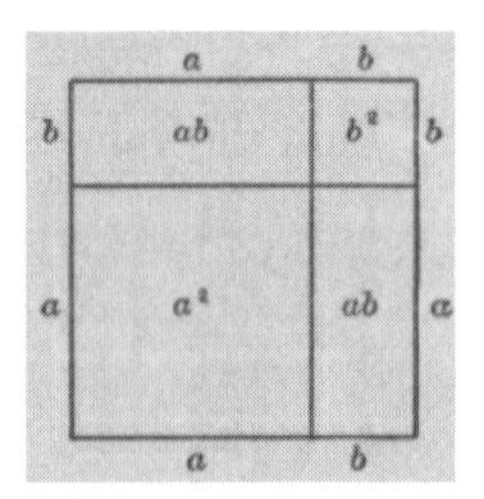

Figure 8.1. Dissected square, 1928. From Ernst R. Breslich, *Senior Mathematics* (Chicago: University of Chicago Press, 1928), 1:165.

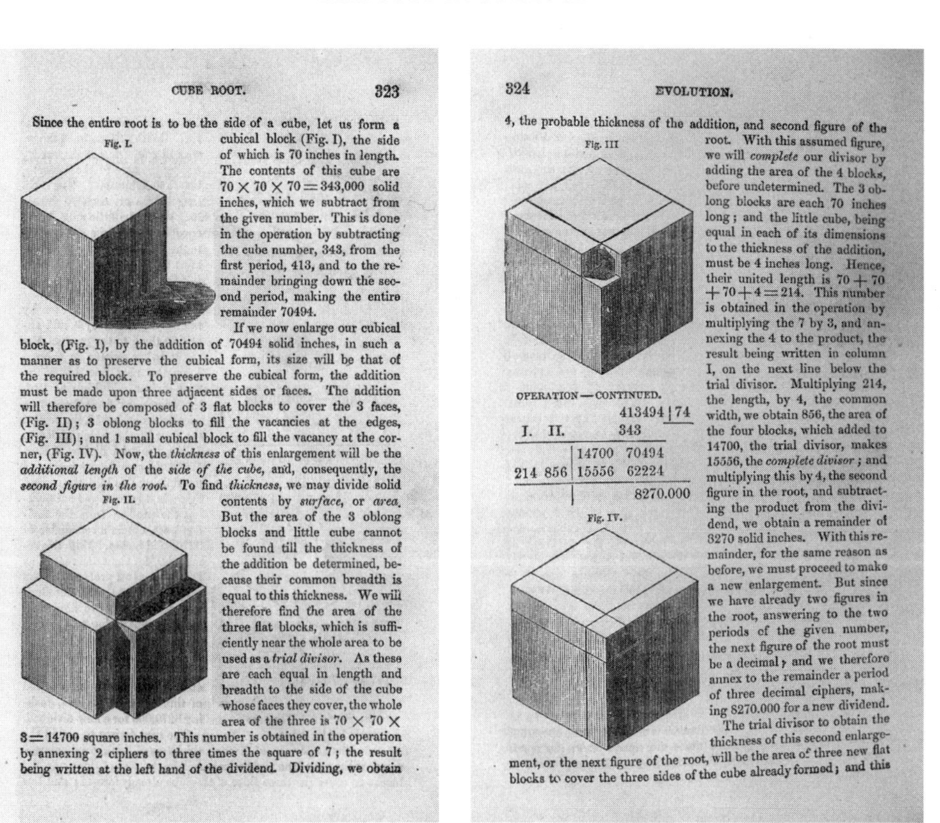

CUBE ROOT. 323

Since the entire root is to be the side of a cube, let us form a cubical block (Fig. I), the side of which is 70 inches in length. The contents of this cube are 70 × 70 × 70 = 343,000 solid inches, which we subtract from the given number. This is done in the operation by subtracting the cube number, 343, from the first period, 413, and to the remainder bringing down the second period, making the entire remainder 70494.

If we now enlarge our cubical block, (Fig. I), by the addition of 70494 solid inches, in such a manner as to preserve the cubical form, its size will be that of the required block. To preserve the cubical form, the addition must be made upon three adjacent sides or faces. The addition will therefore be composed of 3 flat blocks to cover the 3 faces, (Fig. II); 3 oblong blocks to fill the vacancies at the edges, (Fig. III); and 1 small cubical block to fill the vacancy at the corner, (Fig. IV). Now, the *thickness* of this enlargement will be the *additional length* of the *side of the cube*, and, consequently, the *second figure in the root.* To find *thickness*, we may divide solid contents by *surface*, or *area*. But the area of the 3 oblong blocks and little cube cannot be found till the thickness of the addition be determined, because their common breadth is equal to this thickness. We will therefore find the area of the three flat blocks, which is sufficiently near the whole area to be used as a *trial divisor*. As these are each equal in length and breadth to the side of the cube whose faces they cover, the whole area of the three is 70 × 70 × 3 = 14700 square inches. This number is obtained in the operation by annexing 2 ciphers to three times the square of 7; the result being written at the left hand of the dividend. Dividing, we obtain

324 EVOLUTION.

4, the probable thickness of the addition, and second figure of the root. With this assumed figure, we will *complete* our divisor by adding the area of the 4 blocks, before undetermined. The 3 oblong blocks are each 70 inches long; and the little cube, being equal in each of its dimensions to the thickness of the addition, must be 4 inches long. Hence, their united length is 70 + 70 + 70 + 4 = 214. This number is obtained in the operation by multiplying the 7 by 3, and annexing the 4 to the product, the result being written in column I, on the next line below the trial divisor. Multiplying 214, the length, by 4, the common width, we obtain 856, the area of the four blocks, which added to 14700, the trial divisor, makes 15556, the *complete divisor*; and multiplying this by 4, the second figure in the root, and subtracting the product from the dividend, we obtain a remainder of 8270 solid inches. With this remainder, for the same reason as before, we must proceed to make a new enlargement. But since we have already two figures in the root, answering to the two periods of the given number, the next figure of the root must be a decimal; and we therefore annex to the remainder a period of three decimal ciphers, making 8270.000 for a new dividend.

OPERATION—CONTINUED.

I.	II.			
			413494	74
			343	
		14700	70494	
214	856	15556	62224	
			8270.000	

The trial divisor to obtain the thickness of this second enlargement, or the next figure of the root, will be the area of three new flat blocks to cover the three sides of the cube already formed; and this

Figure 8.2. Cube root algorithm, 1871. From Daniel W. Fish, *Robinson's Progressive Arithmetic* (New York: Ivison, Blakeman, Taylor & Co., 1871), 323–24.

amples of such models (in two-dimensional rendering, of course) that have been offered for sale in the United States over a period of more than one hundred years.

The History of Root Extraction Methods

The binomial method has a long and richly multicultural history, sketched by Martin Nordgaard in 1924.[4] Nordgaard found evidence of a very similar method as far back as 375 C.E., with Theon of Alexandria. Hindu mathematicians over the next few centuries, such as Aryabhata and Brahmagupta, provided crucial simplifications by using place value notation and the con-

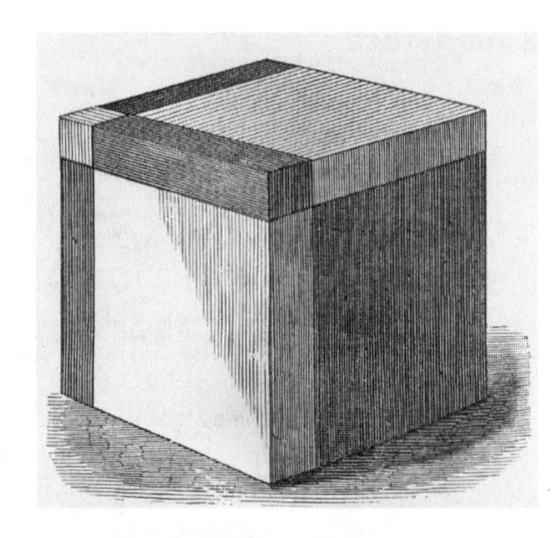

Figure 8.3. Cube root block, 1856. From F. C. Brownell, *The Teacher's Guide to Illustration: A Manual to Accompany Holbrook's School Apparatus* (Hartford, Conn.: Holbrook School Apparatus Co., 1856), 34. Smithsonian Negative no. 2006-21929.

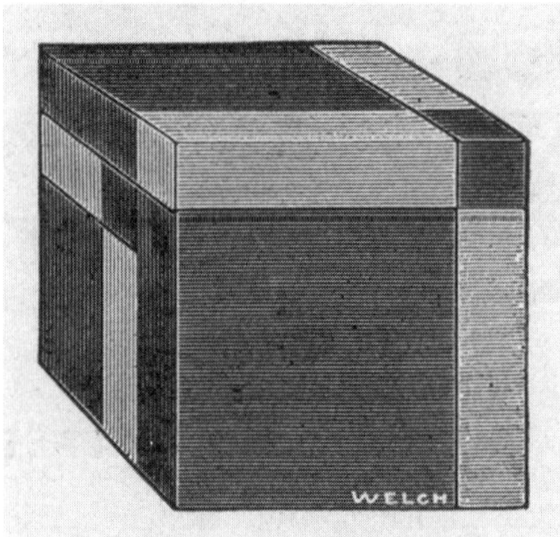

Figure 8.4. Two-place cube root block, 1922. From *Catalog "G" Laboratory Apparatus and Supplies for Physics and Chemistry* (Chicago: W. M. Welch Scientific Co., 1922), 52. Courtesy of Smithsonian Institution Libraries.

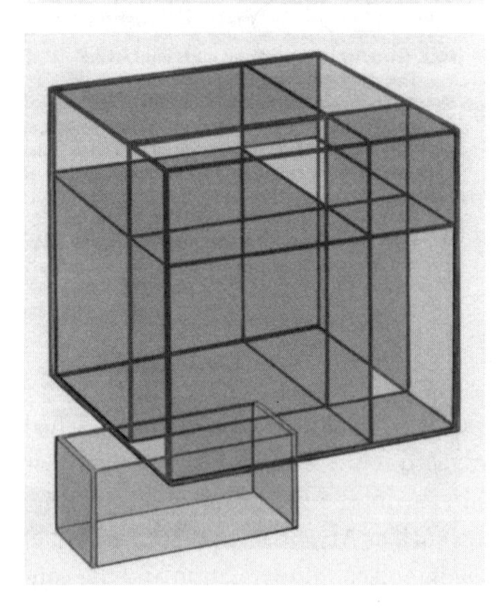

Figure 8.5. Dissected cube, 1965. From *Welch Mathematics Catalog* (Skokie, Ill.: Welch Scientific Co., 1965), 7. Courtesy of Smithsonian Institution Libraries.

cept of zero. These Hindu methods came to Europe in the thirteenth century, via Arab intermediaries, and Europeans such as Leonardo of Pisa (aka Fibonacci), Georg Peurbach, and Hieronimo Cardano subsequently contributed to their enhancement. Writing about the same time as Nordgaard, mathematical educator and historian David Eugene Smith noted that "early writers gave the rules without any explanation, or at most with merely a reference to the fact that $(a + b)^2 = a^2 + 2ab + b^2$," but that "a conviction of the value of the reasoning involved in the subject led various writers in the 16th century to give clear explanations based on the geometric diagram. The use of the blocks for explaining the cube root was found somewhat later, and became fairly common in the 17th century."[5]

The seventeenth century also saw the development of important new mathematical techniques capable of being used for extracting roots. Before 1620 the Scotsman John Napier and his English friend Henry Briggs had introduced the basic idea of logarithms.[6] Using this tool, one can reduce the complexity of arithmetic problems: multiplication reduces to addition, division reduces to subtraction, raising to a power reduces to multiplication, and extracting a root reduces to division. Of course, one can only obtain these benefits if one has a table of logarithms, the production of which requires a substantial amount of initial arithmetic effort. In computing his tables, Briggs developed a scheme based on the successive extraction of square roots. Computing the logarithm of 2, for example, required 47 square root extractions. It is significant that when faced with such massive root extraction problems, Briggs did not use the traditional binomial method but instead was motivated to search for shortcuts. In effect he invented part of what today would be called the "calculus of finite differences."[7]

The example of Briggs suggests that the cumbersome nature of the binomial method was recognized at an early date. By the middle of the nineteenth century those doing calculations for practical purposes were quite explicit on this point. After dutifully explaining the binomial method for square roots, an engineering handbook published in 1851 advised: "The rules for the cube and higher roots are very tedious in practice: on which account it is advisable to work by means of logarithms."[8]

Another alternative to the binomial method, the iterative technique known today as the Newton-Raphson method, or Newton's method, was originally suggested in the latter half of the seventeenth century by Isaac Newton and Joseph Raphson, although it was not expressed in modern form until 1740 by fellow Englishman Thomas Simpson.[9] It had been perfected and extensively analyzed by the early nineteenth century, notably by the French mathematicians J. B. J. Fourier and A. L. Cauchy.[10] Like logarithms, the Newton-Raphson method's utility extends beyond root extraction; it can be applied to compute approximate solutions to a huge variety of equations, of which two of the simplest cases are $x^2 = c$ and $x^3 = c$, where c is any fixed number. The Newton-Raphson method can be expressed purely as a formula, to be applied repeatedly, but it also has a revealing geometric interpretation.[11] The prerequisites for fully appreciating this geometric interpretation, however, are more formidable than those for the binomial method: the rudiments of differential calculus, or at least an understanding of analytic geometry up to the notion of finding a tangent to a curve. In the case of the square root the Newton-Raphson formula simplifies to $x_{k+1} = (½)[x_k + c/x_k]$, where c is the number whose square root is sought. This formula can be readily presented in a way that is suitable for students of arithmetic who have advanced no further than long division, as in these instructions from a 1928 textbook:

1. Make as good a guess as you can of the square root sought.
2. Divide the given number by the guessed root and get the number of decimal places wanted.
3. Average the guessed root and the quotient obtained.
4. Use the average as a new divisor and divide again.[12]

These steps can easily be justified in terms of the meaning of division and the definition of square root.

The Binomial Method in the United States

Despite the simplicity of the Newton-Raphson method for square roots, we have found no instance of its inclusion in an elementary American text earlier than in the text cited from 1928. The binomial method dominated American education through the nineteenth century and well into the twentieth. In the 1829 preface to *Practical and Mental Arithmetic on a New Plan* Roswell C. Smith asserts that "as this mode of teaching recognises no authority but that of reason, it was found necessary to illustrate the rule of the extraction of the cube root, by means of blocks, which accompany this work."[13] Likewise, Oliver Shaw, writing in 1832, gives an extensive explanation of the binomial method for both square and cube roots, with diagrams, and emphasizes his belief that the geometrical interpretation is the key to truly understanding the method.[14] Not everyone agreed with Smith and Shaw; Porter's 1845 text makes no appeal to geometry at all. But by the mid-1850s the geometric interpretation was clearly becoming well established as a teaching method. The Holbrook Company was marketing cube root blocks by 1856, and Greenleaf's text of the same year uses diagrams to illustrate both the square and cube root rules.[15] From this time through the end of the century the binomial method appears to have been a standard feature of advanced arithmetic instruction, frequently accompanied by geometrical explanation.[16] Along with these texts we find steady marketing of cube root blocks by sellers of scientific and educational apparatus. The text accompanying the 1856 advertisement for Holbrook's cube root block (shown in Figure 8.3) is especially interesting: "The extraction of the cube root can be explained most easily by the use of the Cube Root Block. In fact, no person who is unacquainted with Algebra or Geometry can know the reason for this rule without the aid of some such illustration."[17] This statement suggests that the cube root block was seen not necessarily as a device to engage the student with the more abstract realms of mathematics but, rather, as a means by which the student might avoid such engagement.

Cube root blocks were usually advertised along with other objects to aid in elementary mathematics instruction—for example, numeral frames, sets of geometric forms, and dissected cones. The initial popularity of the cube root block thus appears to have been part of the growing movement toward "object teaching" noted in earlier chapters, as pioneered in Europe by J. H. Pestalozzi and his student Friedrich Froebel.[18] This connection was made explicitly in the 1871 advertising supplement published by J. W. Schermerhorn & Company of New York, in which is found a section entitled "Object Teaching Aids," including all the objects just named, plus "Kindergarten Blocks au Froebel."[19]

The use of objects, pictures, and manual activity in American education continued to receive promotion in the last decades of the nineteenth century. General educators who looked favorably on such approaches included Calvin Woodward, Francis Parker, and John Dewey.[20] Only slightly later, more specialized mathematical educators introduced the notion of "laboratory" methods of teaching mathematics, with strong emphasis on graphs, models, and other "concrete" means of imparting abstract concepts.[21] Additionally, these same mathematical educators and others were calling for a "unification" of the mathematical curriculum, including an effort to break down the separation between the teaching of arithmetic, algebra, and geometry. One might think that educators in such an environment would continue to appreciate the binomial method, whose steps can be carried out with simple arithmetic, which is based on an algebraic relationship and which can be modeled geometrically. But in fact it was precisely the turn of the century that marked the beginning of the decline in popularity of the binomial method in American education. The geometric appurtenances of the method disappeared first, beginning with the cube root block. James W. Queen & Company of Philadelphia, which in 1867 was selling both a "cube root block" and a closely related object, a "dissected trinomial cube," by 1882 was selling only the trinomial cube and sixteen years later was offering no cubes at all.[22] Other companies dropped the item over the next couple of decades: the L. E. Knott Apparatus Company had a cube root block in its 1916 catalog, but it was gone by 1925; the Welch Scientific Company still offered the item as late as 1922 but had discontinued it by 1929.[23] Plastic cubes similar to the old wooden blocks made an appearance much later but with an important difference. The text that accompanies the ad from 1965 shown on Figure 8.5, unlike earlier advertisements, made no claim that the block was useful for teaching cube root extraction; it was presented merely as an aid to understanding the expansion of $(a + b)^3$. This change in purpose can be seen in the structure of the block. In the block from 1965 the value of a is not as large in comparison with b as in the earlier blocks; no longer is the viewer to think of a as the biggest whole number whose cube is less than $(a + b)^3$.

On the textbook side, 1913 is the latest year for which we have found a text published with a fully geometric treatment of both the square and cube root binomial method. All other texts examined from that decade leave out either the geometry or cube root or both.[24] It appears that few or no American textbooks after 1920 explain the binomial method for cube root in any form at all.

Decline of the Binomial Method

It is not hard to find a proximate cause for the decline of the binomial method: mathematical educators decided that its usefulness was limited and pressed to have it de-emphasized in the curriculum. This sentiment was already marked in the famous "Committee of Ten" report of 1893, in which the cube root was classed among those topics that should be "curtailed, or entirely omitted" because they "perplex and exhaust the student without affording any really valuable mental discipline."[25] Other committees of mathematical educators continued to make similar statements over the next few decades.[26] The 1928 textbook in which the binomial method was abandoned in favor of Newton-Raphson forthrightly declared: "Eighth-grade pupils are not asked to learn so much about 'square root' as pupils once were, because it is now thought best to spend more time on other topics."[27] Explaining why this sentiment developed when it did and in the way it did is more difficult, but at least five related factors contributed to the decline of the binomial method:

- Competition for time within the curriculum.
- The rise of algebra as a subject for secondary education.
- The decline in popularity of the mental discipline thesis.
- The increasing availability of alternative techniques for root extraction.
- The move to unify mathematical instruction.

Competition for Time within the Curriculum

The question of how much time would be devoted to different school subjects was pervasive at all levels of education in the United States from the middle of the nineteenth century. The burgeoning of subjects thought appropriate for formal education gave rise to the elective system, which in turn put even more pressure on the classical subjects (mathematics, Latin, Greek) to justify themselves in competition with the new subjects (English, history, the sciences, modern languages).[28] Mathematics was able to justify itself anew by appealing to its utility in science and engineering, but this meant incorporating new (or at least seventeenth-century) mathematics into the curriculum, resulting in additional time pressure. All efforts at curriculum reform, from the Committee of Ten report onward, reflect pressures to eject topics from an overly crowded curriculum. Root extraction did not fare well under the scrutiny of educators trying to compare the usefulness of different mathematical topics. In a survey of the employment of arithmetic by parents of grade-school students conducted in 1916–17, it was found that calculating square roots was among those processes used by the parents "so few times as to suggest their omission from the arithmetic work of the elementary grades." The case for omitting lessons about computing cube roots was judged to be even stronger, as not a single parent was observed in the act of extracting such a root.[29]

The Rise of Algebra as a Subject for Secondary Education

One of the mathematical subjects for which room was sought in the curriculum was algebra. Algebra offered the enticement of being at once more "modern" than arithmetic and geometry and at the same time being a means to condense these older subjects into a more compact form for teaching, especially arithmetic.[30] A single algebraic formula could incorporate numerous rules of arithmetic. The binomial method of root extraction had been very appropriate as the capstone of an arithmetic course. The technique required only a knowledge of multiplication, addition, and subtraction (division could be avoided, except in the disguised form of trial and error multiplication), giving the student practice in applying these basic operations. The steps in the method could easily be put in tabular form, comfortably similar to long division. But these factors were much less compelling in the context of an algebra course. Algebra texts of the nineteenth century included the binomial method, but here it was a minor subtopic in the manipulation of exponents, no longer a culminating topic that allowed one to demonstrate mastery of all the preceding material. Moreover, from Warren Colburn's algebra of 1825 to the books published in the 1910s, there appear to be few, if any, American algebra texts that used geometry to explain the binomial method.[31] The writers of arithmetic textbooks had embraced geometry as a tool to explain a process that was difficult to justify with mere words, but algebra could achieve this result, more compactly and with fewer limitations. The geometric interpretation of the binomial method provided no new insight that could not already be obtained by a good command of algebra. Indeed, as David Eugene Smith put it in 1902:

> The formula is to be preferred to the diagram, as a basis for work, because
>
> 1. The geometric notion limits the idea of evolution to the square and cube roots;
> 2. The formula method makes the cube and higher roots very simple after square root is understood;
> 3. We are working with numbers, not with geometric concepts;

4. The formula lends itself more easily to a clear explanation of the process.[32]

The Decline in Popularity of the Mental Discipline Thesis

Elementary arithmetic had been considered primarily as a tool of commercial activity in England and its colonies since the seventeenth century, but arithmetic in the early United States began to be justified as a way to train a person's general reasoning powers as well.[33] The preface to Benjamin Greenleaf's 1856 arithmetic text makes this point: "The end to be sought in the study of Arithmetic [the author] regards as twofold,—a practical knowledge of numbers and the art of calculation, and the discipline of the mental powers."[34] Successfully using the binomial method of root extraction required careful organization and attention to detail, desirable characteristics with which to imbue the mind. That square and cube root extraction were looked upon as sources of mental discipline can hardly be doubted in the face of such otherwise absurd problems as "extract the cube root of 2205 to 19 places."[35] Such extravagant examples of root extraction become harder to find later in the nineteenth century and into the twentieth, although echoes of the disciplinary use of root extraction can be seen in the following rhyme from the 1950s:

> Little Jack Horner
> Sits in a corner
> Extracting cube roots to infinity,
> An assignment for boys
> That will minimize noise
> And Produce a more peaceful vicinity.[36]

The general mental discipline thesis, that certain specific areas of study prepared the mind for a broad range of challenges, came under increasing attack late in the nineteenth century.[37] The influential psychological studies of Edward L. Thorndike and Robert S. Woodworth reported in 1901 were widely taken as having disproved the "transfer of training," although the investigators themselves were more cautious in their interpretation.[38] Thereafter, the claims for mathematics as a source of mental discipline became more muted. To quote David Eugene Smith once again, this time from 1909: "Cube root may well be delayed until the pupil studies algebra, because it has so few practical applications. Even square root is valuable more as a bit of logic than as a practical subject, since those who use it most employ tables."[39] Smith's comments suggest the increasingly tenuous position of the binomial method in the early twentieth century, with doubts being cast on its practical value, while at the same time its disciplinary value was being defended, with no great enthusiasm, as a mere "bit of logic."

The Increasing Availability of Alternative Techniques for Root Extraction

Given that a person desires to extract a root, there arises the technical question of how to do so most efficiently. As noted earlier, the cumbersome nature of the binomial method was already the subject of commentary in the mid-nineteenth century and very likely many years earlier. As D. E. Smith observed, by the early twentieth century tables had become widely available to assist practical problem solvers in root extraction, as had slide rules, often with special scales for square and cube root.[40] These developments surely did not help the popularity of the binomial method in the schools, but it is doubtful that they would have proved decisive on their own. As long as the mental discipline thesis was not strongly challenged, there could be no great objection to giving students inefficient methods with which to work. As some would have contended, the more inefficient the better.

It was years before teaching efficient methods of computation was explicitly acknowledged as being relevant to the school curriculum. After the advent of the electronic calculator, the 1978 Yearbook of the National Council of Teachers of Mathematics, *Developing Computational Skills,* discusses square root calculations purely in terms of the Newton-Raphson method and comments on "the remarkable speed with which the algorithm converges."[41] Such a notion was alien to the period when the binomial method was first losing favor, a period that definitely predated calculators, although these devices may well have administered the coup de grace.

The Move to Unify Mathematical Instruction

Ultimately, the effort to tie together different levels of mathematics education may have been the most important force for determining the fate of the binomial method in the schools, but it is also the most difficult to express and document. Put briefly, as long as there was a certain degree of isolation between elementary mathematical instruction and other uses of mathematics, none of the factors already discussed was capable of toppling the binomial method.

At the end of the nineteenth century some American mathematicians began to see the gap between elementary and advanced mathematics as a problem requiring a solution. The most notable expression of this concern was the address given by E. H. Moore when he was retiring as president of the American Mathematical Society in 1902.[42] One of the most significant features of this address was Moore's stress on developing an instructional program in mathematics that would build steadily and naturally to its culmination in mathematical research. Everyone would receive the same initial education, but those with different needs would step off the ladder of instruction at different rungs. There were surely concerns for the professional and economic survival of mathematics at the base of this attitude. Moore explicitly raised an alarm over the prospect of mathematicians losing control over the teaching of service courses to engineers.[43] In an environment in which each level of the mathematical community depended for much of its livelihood on teaching the level immediately beneath it, if mathematical researchers were to thrive, reproducing themselves and even adding to their numbers every generation, then they needed to exercise more control over less advanced instruction.

Moore's choice for achieving this control and unification was to emphasize models, graphs, and other means of visualization. This unifying impulse apparently resulted in a brief resurgence of the geometrical treatment of square roots in some textbooks. The textbooks promulgated by Moore's colleagues at Chicago follow his advice to break down the barriers between algebra and geometry and are the first known to us that use both the formula and the picture in explaining the binomial method (for square root only).[44]

Yet in the longer run Moore's attitude was not favorable to a topic such as the binomial method, which in the context of modern mathematics was exceedingly restricted. The geometry of the binomial method was seen as an interesting curiosity, in contrast to the geometry of the Newton-Raphson method, which was seen as richly suggestive of new problems and methods. The real interest of Moore and other like-minded mathematicians was the use of broad unifying concepts such as "function." Many functions can be usefully visualized as graphs, but many cannot; the concept transcends visualization. Moore was thus not at all an "object teacher" in the same sense as the general educators in the nineteenth century, and it should not be surprising that his influence was very different.

Moore and other mathematicians of the turn of the twentieth century were not successful in their larger program to reform school mathematics.[45] Nevertheless, they planted the seed of the idea that elementary mathematics instruction should not be allowed to proceed in splendid isolation from other mathematical endeavors. This idea, even applied erratically, was ultimately fatal to the binomial method. The cube root block was now seen as merely a cumbersome way of representing a special case of a general algebraic formula. The geometry dropped away, except as an occasional historical or cultural footnote. The old rules persisted for a time, until the hand calculator era.

CHAPTER NINE

Blocks, Beads, and Bars
Learning Numbers through Manipulation

The beauty of the Cuisenaire rods is not only that they enable the child to discover, by himself, how to carry out certain operations, but also that they enable him to satisfy himself that these operations really work, really describe what happens.

We want the rods to turn the mumbo-jumbo of arithmetic into sense. The danger is that the mumbo-jumbo may engulf the rods instead.

By the early twentieth century teaching tools were increasingly made for individual students. Numerous engineering, science, and mathematics students owned and used their own slide rules. Elementary school students had rulers and protractors. Tools for group demonstration such as the teaching abacus gave way to devices for a single child. Specially designed blocks, beads, and bars were sold for very young children to learn counting, simple arithmetic, and basic geometry.

The idea that children should learn basic mathematics by handling objects was not new. Parents undoubtedly had taught children to count familiar things for centuries. Pestalozzi and his disciples had urged teachers and mothers to provide children with dried beans or other simple objects as they learned to count. Samuel Wilderspin not only championed the abacus but encouraged teachers at infant schools to use one-inch cubes to represent numbers.[1] These objects, however, do not seem to have been widely adopted; they were not part of any general educational theory and did not easily lend themselves to use in large classes.

From the late nineteenth century Americans used a wide array of apparatus specifically designed to teach children basic mathematical concepts through manipulation. These tools were sold in conjunction with diverse pedagogical reforms, five of which merit mention here. The first was the kindergarten movement, which had its roots in the mid-nineteenth-century work of the German Friedrich Froebel, and reached the United States after the Civil War. Froebel introduced several objects for the education of very young children, including a set of wooden cubes that were designed, in part, to introduce numbers and elementary arithmetic. A second set of objects, developed by the Italian physician Maria Montessori, was first introduced into the United States just before World War I. Montessori's system, grounded in nineteenth-century social sciences, included rods to illustrate basic mathematical ideas. In later years she extended her techniques to primary schools.

In the wake of Montessori's work three people developed apparatus intended specifically for mathematics teaching. The first, Catherine Stern, was a physicist who had run a Montessori school in Breslau. She and her family found it expedient to immigrate to the United States in 1938 and settled in New York City. Stern published a book on arithmetic teaching in 1949. Her methods, which were based on structural psychology, used blocks to introduce numbers and arithmetic processes; they were first used in preschools and then extended to primary schools. During the same period the Belgian primary school teacher Emile-Georges Cuisenaire designed his own set of colored rods for arithmetic teaching. The British-based educator Caleb Gattegno became an enthusiastic advocate of Cuisenaire's methods, tying them to ideas of modern mathematics and manufacturing the rods in England. They were introduced in the United States in the mid-1950s, becoming associated with the "New Math" movement of the 1960s. Finally, Charles Tacey, an associate of the English firm that distributed Montessori materials, designed a set of interlocking cubes that represented an improvement on her apparatus. Since the 1960s these blocks have sold in the United States under the name of Unifix cubes.

These five systems were introduced into the United States in the order in which they developed, yet over the years their popularity has waxed and waned. Froebel's materials for teaching kindergarten were largely replaced by other objects by the 1920s, but they still have some advocates today. Montessori's methods were not generally discussed in the United States between 1920 and the 1960s, although her apparatus has since become quite popular. Stern's materials were very modestly successful when they were first introduced, receiving a small boost at the time of the New Math. They are also still being made. Cuisenaire rods were popular in the 1960s and came back into vogue again in the 1990s. Unifix cubes were put on the market in the 1960s but became popular much more recently.

Those interested in the education of very young children have long been on the fringes of the academic community. Froebel, Montessori, Stern, and Gattegno all took some time to settle into the role of early childhood educators and had complicated careers on the boundaries of conventional academe. Tacey worked within a family of school supply dealers. He not only was outside of scholarly communities but placed little emphasis on the philosophical underpinnings of his work.

The changes recounted here have little to do with technical improvement. The alterations in the cubes and blocks used in arithmetic teaching, when they occurred, were quite small. They reflect factors such as the development of new materials, pricing, marketing, and educational ideas, rather than improved technical capabilities of the objects themselves. Thus, the story of such simple apparatus offers a useful contrast to accounts of contemporary printing, computing, and electronic technologies. It also points up once again the international character of educational reform. Froebel, Montessori, Stern, and Cuisenaire's advocate Caleb Gattegno all spent much of their lives outside their native countries and found important disciples in disparate lands. Finally, the history of manipulative devices demonstrates once more the importance of marketing and of government reform to classroom practice. Instrument dealers and government agencies both encouraged and discouraged the use of all of these materials. The role of these external factors looms especially large in the historical record, as the children who used these beads and bars often were too young to record their reactions to them and the teachers who used them left relatively scant accounts.

Froebel's Kindergarten Apparatus

Friedrich Froebel, a native of a small town in central Germany, was the son of a Lutheran pastor. He absorbed much of his father's piety but had no strong sense of vocation. After an apprenticeship as a forester, he studied mathematics and botany at the University of Jena and then considered careers in farming, surveying, and architecture. He also taught at a Pestalozzian school in Frankfurt and, from 1808 until 1810, in Switzerland under Pestalozzi himself. After further university studies, a stint in the Prussian army, and work in the mineralogical museum at the University of Berlin,

Froebel opened his own school in 1816 and spent the rest of his life as an educator. It was only in the 1830s, while he was in charge of a Swiss orphanage, that he came to think especially about the teaching of very young children. Froebel started his first school for them in 1837. Two years later he gave such a place for the cultivation of children the name "kindergarten." In the course of the 1840s he vigorously promoted kindergartens through writings, training schools, travel, and sale of apparatus. Germany boasted seven kindergartens in 1847, and another forty-four opened in the revolutionary year of 1848. Then, in 1851, Prussian authorities banned kindergartens as socialistic and atheistic. Froebel died the following year, and his disciples scattered across Europe, to Great Britain, and eventually to the United States.[2]

Kindergarten was intended to complement a child's home life, meeting only a few hours each day. Activities included singing, dancing, and games as well as quieter moments in which each child kept busy with a copy of one of Froebel's gifts, or "occupations." Froebel believed that the same objects could teach in several different ways. His third gift, for example, was a set of eight identical cubes, each one inch long on a side, which stacked to form a two-inch cube. One way to use them was as building blocks, representing abstractly such objects as a chair, a wall, a church, or stairs. In this guise the cubes were what Froebel called a "gift of experience." The blocks could also be grouped to represent simple ideas about addition, subtraction, and proportion (Figure 9.1). Used in this manner, the blocks became what Froebel called a "gift of knowledge."[3] Kindergartens also had tables marked with a grid of lines, much like graph paper; each square in the grid was the size of the face of one of the small cubes. Arranging the cubes symmetrically on the grid produced pleasing patterns, making the blocks into a "gift of beauty."[4] In addition to combinations of solids, Froebel's gifts included objects designed to teach properties of planes, lines, and points. One early American account of Froebel's gifts even suggested, for the benefit of older students, how some blocks might be placed to illustrate a special case of the Pythagorean theorem.[5]

In the years following the Civil War Froebel's ideas attracted considerable attention in the United States, particularly among German immigrants and New England transcendentalists. The first kindergarten in the country had opened in Watertown, Wisconsin, in 1856. Through the determined efforts of women such as the Bostonian educator Elizabeth Palmer Peabody, the idea of the kindergarten spread. In the summer of 1869 Edward Wiebé, a

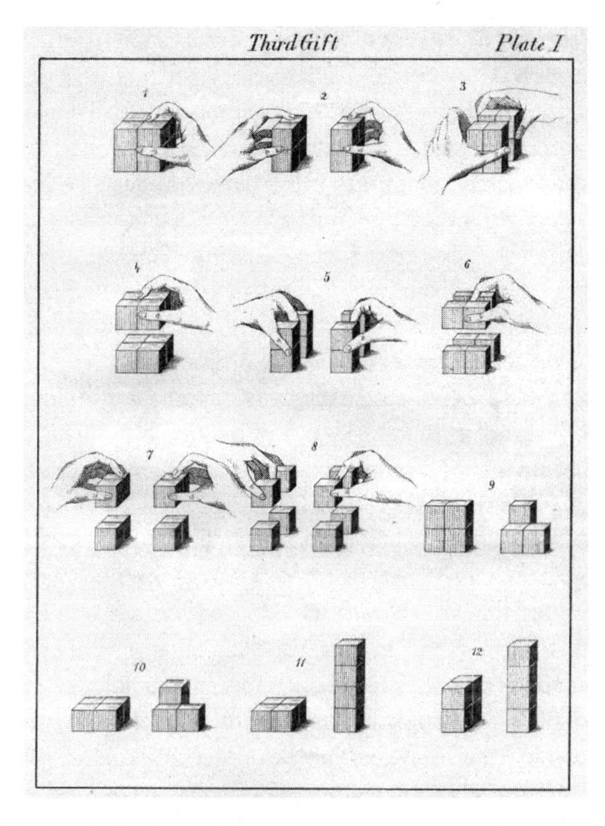

Figure 9.1. Froebel's Third Gift, used to represent simple fractions, 1869. From Edward Wiebé, *The Paradise of Childhood* (Springfield, Mass.: Milton Bradley & Co., 1869), pl. 1. Courtesy of Smithsonian Institution Libraries.

German-born music teacher in Springfield, Massachusetts, persuaded his neighbor, the board game maker Milton Bradley, to attend one of Peabody's lectures. Wiebé had studied kindergarten methods in Germany with those close to Froebel and had written a book on the subject, which he hoped Bradley would publish. Much impressed by Peabody's talk, Bradley agreed and also began to manufacture teaching apparatus based on Froebel's design. By the 1870s two New York publishers, J. W. Schermerhorn and Ernst Steiger, were also distributing kindergarten materials.[6]

Kindergartens received extensive attention as a result of the World's Fairs held in Vienna in 1873, in Philadelphia in 1876, and in Chicago in 1893. As one might expect, some observers thought they could improve on Froebel's gifts and occupations. The French-born American physician Edouard Seguin, who was very interested in child development, served as a judge at the Vienna exposition and criticized Froebel's choice of small cubical blocks for his gifts. Seguin suggested that if kindergarten advocates had chosen a block "with their senses, as it must speak to the senses of the child, instead of with their mind, they would certainly never have selected the cube, a form in which similarity is everywhere, difference nowhere, a barren type, incapable, by itself, of instigating the child to comparison and action." He thought that a larger object would be easier for a child to handle, and a rectangular block offered a greater variety of properties.[7]

Even those who adopted Froebel's general approach were quite willing to sell alternative materials. Bradley, for example, sold blocks painted with letters of the alphabet, not just plain cubes. On a more mathematical level he also sold a "curvilinear gift." Froebel had talked of introducing more curved surfaces among his gifts, and Minnie Maud Glidden took out a U.S. patent in 1897 for an "educational appliance" that consisted of a cube, a sphere, a cylinder, and a cone, each dissected in several ways. Glidden, who taught at the kindergarten at the Pratt Institute in New York City, hoped that this apparatus would lend itself to a wider appreciation of mathematical designs. Bradley greatly simplified her proposed gift, offering only a dissected cylinder. Further suggestions about how dissected geometrical forms might be used as gifts came in Glidden's later work as well as from kindergarten teachers in Chicago and Los Angeles.[8]

Such deviations from Froebelian orthodoxy greatly disturbed Elizabeth Peabody. When an educational journal advertised kindergarten materials developed by Mt. Holyoke graduate Emily Coe, Peabody severed her connection with it.[9] At the same time, a serious rift arose between kindergarten traditionalists and educators attracted by the promise of contemporary psychological theories. Many of the latter believed that it was more important to cultivate young children's social skills than to develop their intellectual and aesthetic prowess. These teachers replaced Froebel's abstract forms with blocks, which groups of children could build into large structures resembling everyday objects. They thought such play was more suitable for kindergarten children than learning arithmetic. As kindergartens increasingly became part of the public school system in the United States, inclinations toward teaching subject matter in a distinctive manner with special materials declined.[10] Hence, the abstract spirit of Froebel's gifts had little direct impact in American mathematics teaching.

Maria Montessori's Didactic Apparatus

In the early twentieth century the Italian physician Maria Montessori proposed a set of "didactic apparatus" for young children that built on Froebel's work as well as that of more recent psychologists, anthropologists, and physicians. Montessori had been raised in Rome, the daughter of a civil servant. She attended a local primary school and then, encouraged by her mother, a technical school and a technical institute. She enjoyed her mathematical studies and briefly considered a career in engineering. She was also greatly interested in natural history, however, and in 1890 enrolled in a two-year premedical program at the University of Rome. She was admitted as the first female medical student at the University of Rome, receiving her M.D. degree in 1896.[11]

After graduation Montessori worked as an assistant at a Roman hospital and established a small private practice. She also became a voluntary assistant in the psychiatric clinic of the University of Rome. Her observations of children at insane asylums led her to read the earlier work of Jean-Marc-Gaspard Itard and Edouard Seguin and to become familiar with devices they had designed for special education. Montessori also audited courses on pedagogy and physical anthropology, learning about the work of Pestalozzi, Froebel, and the Italian measurer of skulls Cesare Lombroso. In the late 1890s she began to lecture on teaching, and in 1900 she was named codirector of a demonstration school of a Roman teachers' college. Montessori abruptly left this position in 1901 and soon gave birth to a child fathered by the codirector of the school. Leaving her child to be raised by others, she returned to the University of Rome to study pedagogy, hygiene, and experimental psychology. From 1904 she taught a course on anthropology and education for students in natural science and medicine at the Pedagogic School of the University of Rome.

Montessori's schools for young children emerged outside the Italian system of public instruction. Late-nineteenth-century industrialization had encouraged Italians to move to overcrowded urban areas, creating vast slums. When a group of Roman bankers renovated some slum buildings, they arranged to have centers, or *casa dei bambini,* built in them for the care of resident children. The first of these centers opened in 1907, with several dozen youngsters. Eager to try the materials she had developed for special education with normal children, Montessori agreed to oversee the unit. Using and refining her instructional devices, she developed a system of what she called "auto-education."

Like Froebel, Montessori firmly believed that children should learn by manipulating carefully selected objects. To emphasize the importance of children learning on their own, she dubbed the women who tended to the day-to-day activities at her schools "directresses," not teachers. The rhetoric underlying Montessori's system was not a romantic philosophy of nature, as with Froebel, but nineteenth-century psychology and anthropology. She placed great emphasis on measurements of children's weight and height. Moreover, she was much less concerned about instilling a sense of beauty and cosmic unity than with establishing motor skills and delicate sense perception. To give children a sense of relative size, for example, Montessori designed three sets of blocks. The first, used very early in her program, was a group of ten nested wooden cubes, painted pink, and decreasing in edge length from 10 centimeters to 1 centimeter. Children used the blocks to build a tower. To make the highest possible tower, they had to pick the largest block and to judge the order in which the blocks decreased in size. In a second set of ten wooden blocks, the blocks were of uniform depth but decreased regularly in height and width. They were arranged to form what Montessori called "the broad stair." A third set of ten wooden blocks consisted of square prisms that varied only in length. The shortest was 10 centimeters long, the next 20 centimeters, the next 30 centimeters, and so forth, until one reached the largest block of 100 centimeters in length. These rods were divided into sections 10 centimeters long, and the sections were painted alternately red and blue. The "long stair," as these blocks were called, was first used to teach children to compare lengths (Figure 9.2).[12]

When the *casa dei bambini* proved successful, Montessori and those she had trained not only opened similar schools in other Roman slums but adopted their methods for wealthier children. Montessori schools soon opened in Milan and the Italian part of Switzerland. Montessori offered her first training course in the summer of 1909 and also wrote an exposition of her methods entitled, in the English translation published in New York in 1912, *The Montessori Method: Scientific Pedagogy as Applied to Child Education in "The Children's Houses" with Additions and Revisions by the Author.* In this book Montessori described how apparatus might be used with more advanced students. The blocks of the long stair, for example, could be used to introduce addition, subtraction, and the meaning of written numerals. Montessori also proposed using counting boxes divided into sections. A

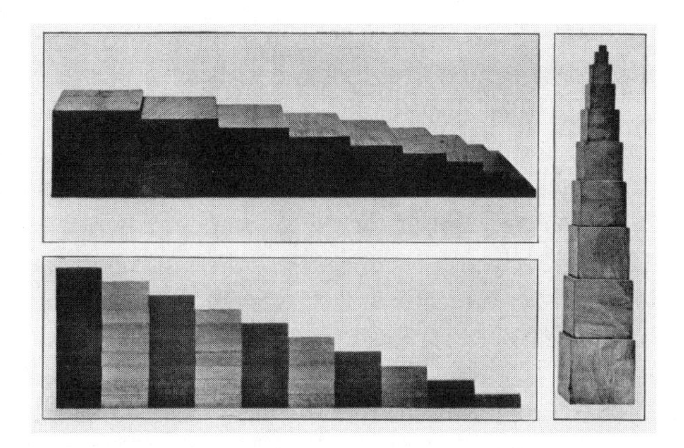

Figure 9.2. Montessori's Broad Stair, Long Stair, and Pink Tower, 1912. From Maria Montessori, *The Montessori Method* (New York: Frederick A. Stokes Co., 1912), 191. Courtesy of Smithsonian Institution Libraries.

digit cut out of paper and mounted on a card was put at the back of each section. Children received a bunch of counting sticks and placed an appropriate number of sticks in front of each card. Other cards were used to teach two-digit numbers.

In addition to teaching about relative size and basic arithmetic, Montessori believed that children should learn about simple flat shapes. Following the ideas of Seguin, she had boards made with holes cut in them to hold simple forms such as triangles, quadrilaterals, other polygons, and circles that decreased in size. Children using the forms learned coordination, which, as Montessori explained in a U.S. patent application in 1909, was a first step toward writing.[13] Montessori also had metal forms made, which slightly more advanced students traced on paper and colored. She also believed that forms inset into matching holes in boards could be used to introduce plane geometric figures that differed in shape but had equal area. She applied for a U.S. patent for a few simple tessellations of this sort in 1913 and received it in 1916.[14] This apparatus was not as widely used as her blocks or form boards.

Montessori's work attracted attention in the United States much more quickly than Froebel's ideas had. She applied for her first U.S. patent only two years after the opening of the first *casa dei bambini.* Jennie B. Merrill, the supervisor of kindergartens for part of New York City, published several articles about her work in *The Kindergarten-Primary Magazine* in 1909 and

1910. Merrill found much of possible interest in the accounts of Montessori's methods that she had read or had translated, although she herself preferred traditional Froebelian apparatus. Wider publicity came from articles published in 1911 and 1912 in *McClure's Magazine*. The first three were by Josephine Tozier, who had actually visited Montessori and her schools; the fourth was by Montessori herself. Carl R. Byoir, an Iowa-born law student at Columbia University, chanced upon Montessori's article and was much impressed. He visited her in Italy and acquired rights to distribute her teaching apparatus in the United States. By late 1911 Byoir had opened the House of Childhood, Inc., in New York City to sell the materials. He reportedly earned $63,000 in profit from the venture before turning his full attention to a career in public relations in about 1916.[15] Anne George, an elementary school teacher from Chicago, became Montessori's first American student in the winter of 1910 and opened the first Montessori school in the United States in the fall of 1911, working at a private home in New York state. George also prepared the English translation of *The Montessori Method*.

Montessori's name became even better known in the United States in 1913, when she spent three weeks in the country on a lecture tour. Her methods received further attention in 1915, when she spent several months in California under the auspices of the National Education Association. There she not only trained teachers but oversaw the operation of a Montessori school at the Panama-Pacific International Exposition, a World's Fair held in San Francisco. The school attracted numerous visitors and reportedly won two gold medals.[16]

Despite this welcome, Montessori's influence soon faded. She sought to control and to profit from her methods, recognizing only those directresses that she herself had trained. While seeking to remain in charge, Montessori broke her ties with many of her followers in the United States. More generally, many Americans felt no need to send very young children away from home for schooling, and those who did tended to be more worried about inculcating common social skills and assumptions than training the senses. William Heard Kilpatrick, an influential philosopher of education at Columbia University's Teachers College, visited a *casa dei bambini* in Rome and was not impressed. In 1914 he published *The Montessori System Examined,* in which he dismissed both Montessori's methods and the psychological theory on which they were based.

Undeterred by such criticisms, Montessori prepared an account of how her methods might be extended to primary school. This work was published in Rome in 1916 as *L'autoeducazione nelle Scole Elementari,*[17] a book that appeared in New York the following year as *The Advanced Montessori Method.* In it Montessori described in detail her ideas about teaching grammar, reading, arithmetic, geometry, music, and poetic meter. Of particular relevance here are the sets of beads she proposed for arithmetic teaching. One set was a smaller form of the blocks of the long stair. Montessori represented a digit by stringing the corresponding number of beads onto a wire, or "bead-bar." For each digit the beads were a distinctive color—there were ten orange beads in the ten bead-bar, nine dark-blue beads in the nine bead-bar, and so forth. The bead-bars would be recast by Stern and by Cuisenaire as colored rods. With Tacey they became interlocking cubes.

In addition to these materials, Montessori taught children about decimal relations by combining ten groups of 10-bars into a "100-chain" and ten 100-chains into a "1000-chain." She argued that actually counting the number of beads in these chains would give children a sense of the relative size and interrelationships of numbers. To teach elementary multiplication, Montessori returned to the bead-bars, arranging them in rectangles (e.g., both three adjacent 2-bars and two adjacent 3-bars gave a product of six beads). The beads could also be arranged to form squares and cubes, introducing these concepts. Once students understood the idea of place value, Montessori believed directresses should use a numeral frame to present numbers, first using four wires to represent numbers up to 9,999, then using seven wires and displaying numbers as high as 9,999,999.[18]

Montessori returned to the United States in 1916–17, in part to promote her materials and in part to attend the wedding of her son. She received much less attention than previously, and there is little indication that her new materials attracted interest. She would not return again but carefully retained control over the certification of teachers using her method. After Montessori's death in 1952, the English-trained Montessori teacher Nancy McCormick Rambusch wrote a series of articles about her ideas in the Catholic magazine *Jubilee.* Rambusch went on to found the Whitby School in Greenwich, Connecticut, in 1958 and to serve as the first president of the American Montessori Society from 1960 to 1963.[19] By the fall of the next year there were about one hundred Montessori schools in the United States, at least a third of them newly opened.[20] The first graduate program offering Montessori teacher training opened at Xavier University in Cincinnati, Ohio, in 1965.[21] As federal funding for preschools increased and a larger portion of mothers worked outside the home, the number of Montessori schools grew. By 2002 there were about five thousand of these schools across the country.[22]

Stern Blocks and Structural Arithmetic

During the years between the world wars Maria Montessori lived in Barcelona, traveling from Spain to lecture and give training courses in Austria, England, Germany, Holland, and Italy. Her pupils included the mathematical physicist Catherine Stern, who would refine Montessori's teaching apparatus and recast her ideas into a more modern psychological framework. Stern would also immigrate to the United States and participate in debate surrounding the New Math. Her work is not as widely known as that of her predecessors such as Froebel and Montessori or successors such as Georges Cuisenaire, yet it signals the transition from apparatus designed for exploring a wide range of experiences, such as that of Froebel and Montessori, to material designed to teach a single academic subject. At the same time, Stern worked not only at her own preschools but in conjunction with university scholars, suggesting a change in context for innovations in elementary mathematics teaching apparatus.

Catherine Stern, born Käthe Brieger, was the daughter of a physician and a devoted kindergarten volunteer. She studied physics at the university in her native Breslau but left school after her father's death in 1914, first to teach and then to undertake wartime service in a hospital. Despite these distractions, in 1918 she completed a doctoral dissertation on crystallography at the University of Breslau. She was one of the first generation of German women to be able to attend graduate school in physics in their native country.[23]

The following year Brieger married the physician Rudolf Stern. She gave birth to a daughter in 1920 and a son in 1926. Married women and mothers were not encouraged to teach in German universities or gymnasia. Instead, Stern changed careers, studying the Montessori teaching method, conducting a preschool in her home, and then, in 1924, opening Breslau's first Montessori school. Stern was not willing to confine herself either to preschool teaching or to Montessori's methods. She obtained certification to teach grade school and, from 1929, ran an after-school club for older children. By 1933 Montessori had been burned in effigy in Berlin, and all German Montessori schools were closed. And although Stern was a practicing Lutheran, both she and her husband were of Jewish descent and soon faced difficulties from the Nazis. The family decided to immigrate to New York City, where Stern resumed her work with children. Not long after she arrived in 1938, she published a booklet describing the play materials that she had developed for training the senses of children from age five to fourteen.[24]

In 1940 Stern met the gestalt psychologist Max Wertheimer, who had fled from his position in Berlin and was teaching at the New School of Social Research in New York City. She became his research assistant, gave lecture-demonstrations in his classes, and began an account of her methods for teaching arithmetic that would be published in 1949 as *Children Discover Arithmetic*. After Wertheimer's death in 1943, Stern, her daughter, and her future daughter-in-law opened the Castle School, an experimental preschool in New York City that taught children from two to four years old. She also continued to write and, from 1950, published her mathematics teaching materials for classroom use with Houghton Mifflin.

Stern's apparatus was designed not for the general training of the senses but for the specific task of guiding students to understand patterns and solve problems in arithmetic. *Children Discover Arithmetic* describes blocks that filled many of the functions of Montessori's bead-bars and bead-chains in a manner that Stern found more intellectually satisfying. Stern thought that Montessori placed too much emphasis on counting and not enough on making numerical relations visually apparent. When two of Montessori's bead-bars were put end to end, for example, the chain was longer than the bead-bar representing the sum of the numbers. To avoid this difficulty, Stern replaced the bead-bars with colored square prisms of differing length. Then if two five-rods were placed end to end, the length of the combined rods equaled that of a 10-rod. Stern hoped that these rods would not only represent numbers more precisely but also encourage children to see patterns in arithmetic combinations. Adding one was like stepping up a single step in a stairway of digit rods, adding two like mounting a steeper stairway, and so forth. Stern believed that having an understanding of such patterns would be much more useful to children than rote memorization of sums.

To teach numeration and basic arithmetic, Stern designed several different cases that held her rods. The first, called a counting board, had ten parallel grooves, each the length of one of the rods. Above each groove was a place for a cube with a number written on it. Very young children discovered which rod fit into each groove, learned the names of the corresponding numbers, and in time figured out which written numeral should be placed at the head of each column. A second set of cases, called "pattern

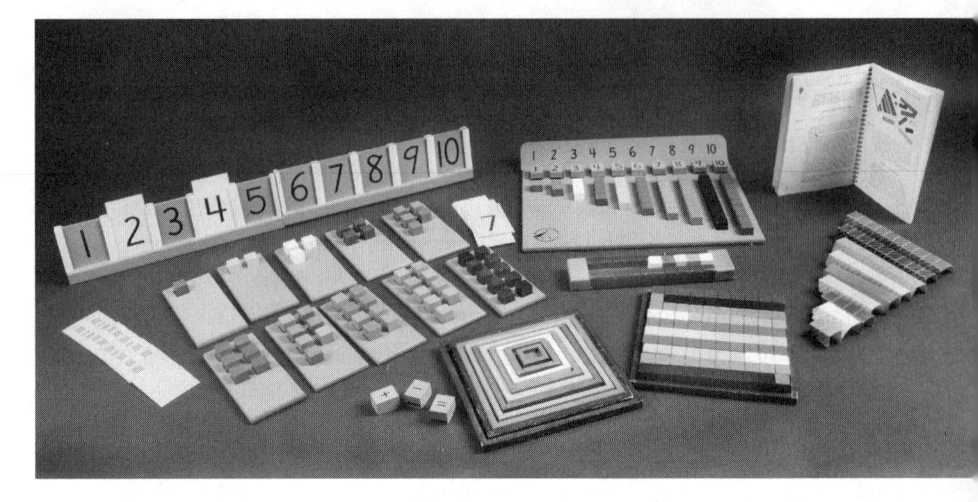

Figure 9.3. Stern's teaching apparatus, ca. 1960. Included are a number stand and pattern boards on the *left,* a pile of number cases and a unit block at the *center front,* a number track and a counting board behind them, and a set of rods and instructions on the *right.* National Museum of American History collections, gift of Peggy S. Mabie. Smithsonian Negative no. 2006-2056.

boards," associated visual patterns with digits and made manifest the distinction between even and odd numbers. All the pattern boards had indentations in them the size of unit rods. The first held one of these cubes, the second two adjacent cubes, the third two cubes in one column and one in the adjacent column, and so forth, up to two columns with five cubes each (Figure 9.3). This apparatus is a variation on a method that Montessori had proposed for exercising the number memory of very young children.[25] Stern used these patterns to represent digits in later exercises as well.

Stern then presented children with a third kind of case, called the "Unit Block." It held ten rods, each of them 10 units long. She encouraged the youngsters to find other rods that were, when combined, 10 units long, and in this way she introduced simple arithmetic. Taking blocks away from the columns in the unit block introduced subtraction. Further square cases that held shorter rods were designed to teach about other combinations of numbers. To represent larger numbers, Stern used a long grooved block, or "number track," with the digits from 1 to 100 written along the edge. The number blocks fit into the groove, indicating combinations of greater size. Stern broke with Montessori in her discussion of place value, suggesting that the numeral frame was a poor way of explaining the concept. She much preferred to distinguish ones, tens, hundreds, and thousands by representing them distinctly with unit rods, 10-rods, 10 × 10 squares, and so forth (Figure 9.4).[26] She was neither the first nor the last to do this. Oliver A. Shaw of Richmond, Virginia, had made a similar suggestion with his "Visible Numerator" in the 1830s. In the 1960s the Hungarian Zoltan P. Dienes would attempt to extend the apparatus to represent digits in different bases. Recently, apparatus like that Stern advocated for teaching place value has sold under the name of "base ten blocks."[27]

Stern's book was favorably received in a number of quarters.[28] She happily adopted the suggestion of reviewers that her materials might also be useful in primary schools. In 1950 Houghton Mifflin published a manual for use with her materials entitled *Experimenting with Numbers: Teacher's Manual for Use with Beginners.*[29] Teachers and schools could purchase the manual, with the required apparatus, for $24. By 1956 Stern and her associates had prepared an entire series of books for children from kindergarten

Figure 9.4. Stern's place value apparatus, ca. 1960. National Museum of American History collections, gift of Peggy S. Mabie. Smithsonian Negative no. 2006-2050.

through grade 2 that sold under the title *Structural Arithmetic.*[30] She would go on to extend her ideas to teaching multiplication and to reading. According to one biographical account, Stern "was a very private person, indifferent to public acclaim or conventional honors and devoted to her close friends and family."[31] Not overwhelmingly interested in promoting her ideas as products, she would find her contributions obscured by the efforts of others.[32]

Indeed, several firms introduced somewhat similar blocks for arithmetic teaching at about the same time. Several of these were described in the columns of *Mathematics Teacher.*[33] Ideas of Charles Tacey and his associates at the English firm of Philip & Tacey proved particularly influential in the long run. This company, one of England's oldest educational supply companies, traced its roots to the early nineteenth century. It was a distributor of Froebel's kindergarten apparatus and Montessori's didactic apparatus. Tacey, like Stern, thought that Montessori's bead-chains should be replaced by continuous columns. Rather than using solid rods, however, he designed plastic cubes that linked together (Figure 9.5). He patented these cubes in

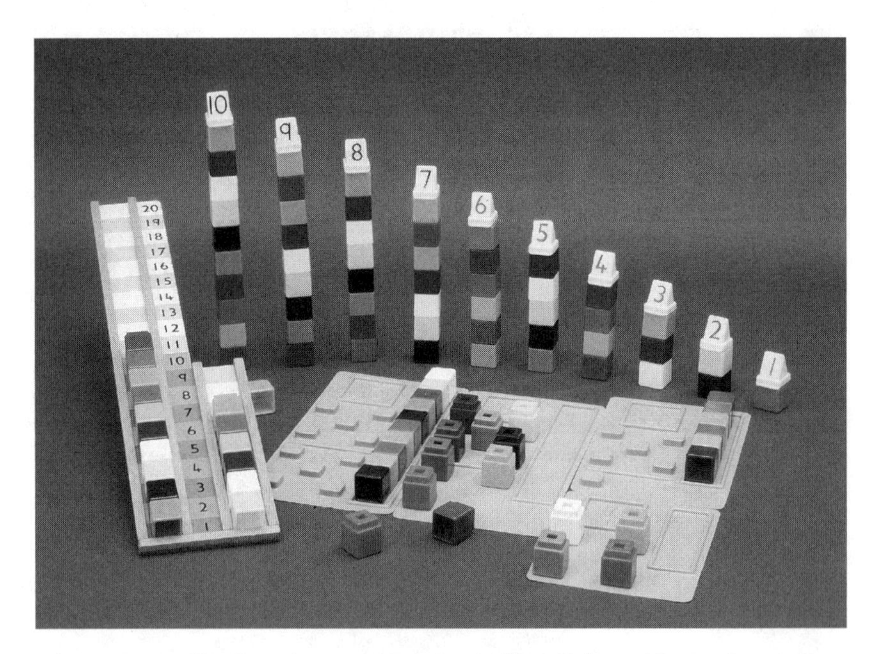

Figure 9.5. Unifix Counting and Notation Ladder, Cubes with Number Indicators, and Pattern Boards, ca. 1960. From the collections of Philip & Tacey Ltd. Smithsonian Negative no. 2005-37651.

the United Kingdom in 1955 and would sell them with trays and number tracks similar to Stern's equipment. These cubes sold in the United States beginning in about 1960. From 1976 they were distributed under the trademark Unifix by Didax, Inc., of Rowley, Massachusetts. A few enthusiastic teachers such as Californian Mary Baratta-Lorton and her husband, Bob Baratta-Lorton, and Kathy Richardson, now of Bellingham, Washington, have written and lectured extensively on the use of Unifix cubes. The colorful plastic materials were cheaper than wooden rods, quieter for groups to use, and extremely flexible. In 2001 Didax reported that over a billion of the small cubes had been sold for American classrooms.[34]

Cuisenaire Rods

There is a clear path from Montessori to Catherine Stern and Charles Tacey. Their contemporary, the Belgian teacher Emile-Georges Cuisenaire, also introduced objects for basic mathematics instruction that would come to be used in American mathematics teaching. Although Cuisenaire rods did in fact become widely known, the route to his invention is more obscure. Existing accounts of these developments, which were written largely by those who adopted and promoted Cuisenaire's ideas, are not entirely consistent.

In 1976 Louis Jeronnez of Waterloo, Belgium, prepared a tribute to Cuisenaire for the Belgian Society of French-Speaking Mathematics Teachers. According to his paper and a notice from about the same time by Cuisenaire's close associate Caleb Gattegno, Cuisenaire spent most of his life in the Walloon region of Belgium, in the southwestern province of Hainaut. He was born in the town of Quaregnon, obtained his teaching diploma at the École Normale of Mons in 1911, and carried out further studies in music at the conservatory there. He fought in the Belgian army from 1914 until 1918, surviving two gas attacks by the Germans. In 1919 he became head of the school in Thuin, later becoming director of the town's primary schools and kindergartens and then of its schools generally. In 1948 he left this administrative position to direct the local industrial school and nine years later retired completely. Cuisenaire was an honored member of the L'Association Cuisenaire de Belgique from its founding in 1970 and remained an active participant in the organization until not long before his death in 1976.[35]

Jeronnez reported that Cuisenaire responded evasively whenever he was asked about how he came to use rods in arithmetic teaching. His early publications had concerned teaching music and climate. Gattegno stated in 1976 that the first published account of the rods was in a booklet entitled *Les nombres en couleur,* published in December 1952. A slightly later account by the freelance writer Rosemary March dates the introduction of the rods to the late 1940s. According to March, the schoolmaster Cuisenaire was inspired by the plight of an eight-year-old who had been moved to tears by frustration with his mathematics books. This account is possible, although as the director of an industrial school, Cuisenaire would seem to have had few occasions to teach eight-year-olds.[36] A more recent publication by ETA/Cuisenaire dates Cuisenaire's first experiments with rods for teaching arithmetic to 1931.[37] In any event, in the early 1950s Cuisenaire published the brief work *Les nombres en couleur* and there described the use of colored rods of differing lengths in teaching arithmetic.

Cuisenaire, like Stern, focused his attention on arithmetic teaching, not on general systems of early childhood education. According to Gattegno and March, he initially used 30 centimeter rulers, cut to varying lengths and painted different colors. When they proved successful, Cuisenaire had more finished rods produced by a local carpenter. To avoid reference to counting altogether, there were none of the unit divisions on the rods that were found on Stern's apparatus.

Cuisenaire's work attracted sufficient attention among local teachers for one of them to invite Caleb Gattegno to see them demonstrated. Gattegno was an Egyptian-born educator who had obtained a Ph.D. degree in mathematics in Basel in 1937, an M.A. degree in mathematics education from London University in 1948, and a doctor of letters from the University of Lille in 1952. He had taught in Cairo from 1937 until 1945 and then moved to England, where he was associated with Liverpool University, London University, and then grammar schools around London. Gattegno was very interested in developing organizations for the improvement of mathematics education, both at a national and an international level. In 1952 he had founded the Association for Teaching Aids in Mathematics (ATAM), a British organization later renamed the Association of Teachers of Mathematics. That same year he participated in the founding of a Belgian society for French-speaking mathematics teachers, a connection that led to his meeting with Cuisenaire the next year.

Much impressed by what he saw in Thuin, Gattegno immediately wrote accounts of the Cuisenaire rods for the *Bulletin* of the ATAM and the (London) *Times Educational Supplement.*[38] He worked with Cuisenaire on an expanded introduction to Cuisenaire rods, which was published in 1954 as *Numbers in Colour: A New Method of Teaching the Processes of Arithmetic to All Levels of the Primary School.*[39] That same year he established the Cuisenaire Company in Reading, England, to distribute the rods. At the instigation of his secretary, Florence Parker, the company soon started to manufacture and not merely distribute the apparatus. The unit cube was 1 centimeter long, the rod representing 2 was 2 centimeters long, and so forth (Figure 9.6).[40]

According to Cuisenaire and Gattegno, the primary innovation in these rods was in the use of color. Cuisenaire chose the colors used to suggest in-

Figure 9.6. Cuisenaire Rods, ca. 1965. National Museum of American History collections, gift of Coralee C. Gillilland. Smithsonian Negative no. 96-4204-4.

terrelationships in the digits represented, with the colors becoming darker as the digits increased. The unit cube was uncolored. Powers of 2 formed a "red" family, 2 itself being red, 4 purple, and 8, darker still, brown. Multiples of 3 formed a blue family, with 3 light-green, 6 dark-green, and 9 blue. A third, yellow family included the yellow rod for 5 and the orange rod representing 10. Finally, the rod for 7 was black. Cuisenaire and Gattegno suggested that the rods be used to teach numeration, addition and subtraction, multiplication and division, reciprocal numbers, and proportion. They did not use grooved boards, cases, or tracks such as those proposed by Stern.[41]

In 1956 Caleb Gattegno prepared an introductory article on Cuisenaire rods for *Arithmetic Teacher,* giving it the title: "New Developments in Arithmetic Teaching in Britain: Introducing the Concept of 'Set.'" Here Gattegno argued that mathematical activity consists of "replacing actual actions by ones that are virtual and contemplating the structures therein contained."[42] A person thinking mathematically abstracted one or more relationships from a situation, transforming them into mental structures. Gattegno argued that Cuisenaire rods revealed the algebraic structure underlying arithmetic. The rods could easily be seen as divisible into equivalence classes, each consisting of rods of the same length. A single set of rods of length from 1 to 10 served as the quotient set, which was strictly ordered, with each rod either greater or less than any other. The rods, if extended indefinitely, would represent the integers; if considered in pairs, they corresponded to rational numbers. By moving rods about appropriately, students learned of the commutative property of addition and multiplication. Rearrangements of rods also could foster an understanding of division as decomposition into factors. Thus, Gattegno argued, the use of Cuisenaire rods would "bring modern mathematics into the primary stages of schooling." Instead of studying numbers, students would examine "sets and their decompositions, making apparent the dynamic operations that structure these sets."[43] Gattegno admitted that many primary school teachers might not see arithmetic in the terms he described. To introduce teachers to the rods, he had prepared an intensive course lasting one or two weeks, which had been taught successfully in several countries.

In the fall of 1956 Charles F. Howard of Sacramento State College in California went to England to observe Gattegno's demonstrations as well as classroom use of the Cuisenaire-Gattegno materials. He visited twenty-two classes at infant and junior schools around London and interviewed thirty-one teachers who were using the materials. In an article published in the

November 1957 issue of *Arithmetic Teacher* Howard reported that the teachers agreed that their students enjoyed using the rods, with brighter pupils making the best use of them. The insights the students reportedly gained corresponded quite closely to ideas emphasized in Gattegno's training materials. From the 1920s American arithmetic textbooks had stressed practical applications of the subject. Asked if students "experience difficulty in transferring their number ideas to social situations involving arithmetic," the British teachers reported that they did not know, as children received other instruction in such matters as using money. The Cuisenaire rods alone did not an arithmetic curriculum make. Both Howard and Ben A. Sueltz, the editor of *Arithmetic Teacher,* called for further studies.[44]

Cuisenaire and Gattegno actively promoted their products, Cuisenaire in French-speaking areas and Gattegno more generally. Cuisenaire rods were systematically introduced in Canada in 1957 and in the United States the following year. In March 1958 the Cuisenaire Company of America, based in New Rochelle, New York, and headed by educator Fritz Kunz, trademarked the term *Cuisenaire.* The mark covered not only the rods but a set of colored discs and Gattegno's geoboards. Gattegno and Fritz Kunz's son, John Kunz, toured the country. Reports on demonstrations of Cuisenaire rods appeared in newspapers such as the *Washington Post.*[45]

The Cuisenaire Company made strong claims for its products. A 1959 advertisement in *Arithmetic Teacher* boasted that "Numbers in Color" offered a new model that was "bringing fundamental changes in mathematics teaching in the U.S. and foreign countries." The ad claimed that the rods offered a concrete way to present not only arithmetic but topics in algebra, geometry, and set theory. Both educators and mathematicians recommended the device, and it had been used successfully in several European countries and in Canada. Moreover, the rods were "suitable for purchase under Title III, National Defense Education Act, 1958."[46] Cuisenaire's ideas also were popularized by William P. Hull, a teacher at Shady Hill School in Cambridge, Massachusetts, whose 1961 account of the devices was distributed by the National Council of Independent Schools.[47] Grade-school teacher John Holt observed Hull, Gattegno, and his own students using Cuisenaire rods from 1958 to 1961 and described the "real learning" that took place in his well-known book *How Children Fail.*[48]

Further accounts of Cuisenaire rods appeared in other popular discussions of new educational methods. In 1961 Martin P. Mayer published a general account of American education entitled *The Schools.* The section on mathematics teaching included mention of the use of Stern's blocks at the Friend's School in Locust Valley, New York, as well as a description of the use of Cuisenaire rods by David Page of the University of Illinois Arithmetic Project.[49] Photographs at the University of Illinois Archives also show Page and his students using Cuisenaire rods.[50] Cuisenaire rods were the subject of a 1961 article in *Life* magazine and a story in *PTA Magazine* that was reprinted in the *Reader's Digest*.[51] Indeed, the devices even were discussed on the editorial page of *Science,* a journal rarely concerned with elementary mathematics teaching. Associate editor Joseph Turner prepared a commentary entitled "Model Teaching" for the November 24, 1961, issue of the magazine. Turner described Cuisenaire rods in some detail, noting that models had an honorable place in mathematics teaching and the creation of mathematical ideas. He cautioned, however, against "becoming so zealous in the manipulation of the model as to loose [*sic*] sight of the mathematics the model is supposed to illuminate." He was particularly concerned about the use of the rods in teaching fractions.[52]

Turner's warning by no means discouraged enthusiasts for Cuisenaire rods. Advertisements and popular accounts continued to appear, and the rods were regularly demonstrated at meetings of mathematics teachers. Caleb Gattegno himself moved to the United States in 1966, though by the early 1970s he had broken with the Cuisenaire Company.[53] Amid the rhetoric of the "back to basics" movement, devices closely associated with the New Math movement were losing their appeal. Cuisenaire rods rarely were touted as the sole key to mathematical learning.

In the late twentieth century many educators argued that children should discover the principles of basic mathematics themselves, rather than learning them through rote drill. They used a variety of devices, including Cuisenaire rods and Unifix cubes. Such manipulative materials also acquired greater importance as a result of new practices in textbook selection. By the mid-1980s several states issued standards describing what should be included in textbooks purchased by schools using state funds. Meeting the requirements of states with a large school-age population, notably Texas and California, was particularly important to publishers. In 1983 the state of Texas mandated that the elementary mathematics textbooks to be adopted the following year include instructional information about "manipulatives." Three publishers developed texts in conjunction with kits designed by the Cuisenaire Company, and their textbooks were the only ones adopted by Texas for elementary school use in 1984. This decision had nationwide consequences. Precisely how this wider diffusion of objects influenced classroom practice remains a subject for study. Meanwhile, the Cuisenaire Company has been part of a more general consolidation of distributors of educational products. In 1990 it was acquired by Addison-Wesley. After various corporate mergers in the publishing field, it eventually was sold to Educational Teaching Aids (ETA).[54] The brightly colored blocks, cubes, beads, and bars used today to teach young children about numbers are among the simplest teaching tools. Their development has largely been the work of social reformers, teachers, and entrepreneurs outside the mathematical community. Occasionally, as with Cuisenaire rods and the New Math, objects were closely associated with specific reforms in elementary mathematics education.

PART THREE

Tools of Measurement and Representation

CHAPTER TEN

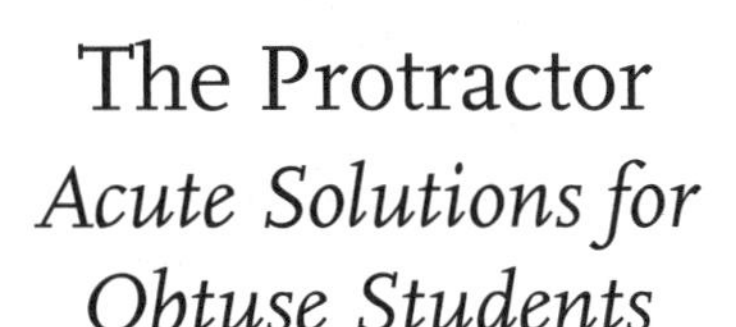

The Protractor
Acute Solutions for Obtuse Students

> Many facts about geometrical figures can be discovered by measurement and observation of carefully drawn figures. It is necessary to use the **protractor**, a tool for measuring angles . . . To get satisfactory results, the figures must be drawn and measured with greater accuracy than is usually possible. Conclusions reached from the study of one or two special figures may be incorrect.

In the late sixteenth century the need to record data generated by new surveying and navigation instruments inspired European mapmakers to develop drawing instruments such as the protractor. This device, graduated into degrees for measuring angles, was almost immediately manufactured in a standard shape (semicircular) from a standard material (hand-hammered sheets of brass) as a standard part of the engineer's toolkit. While the protractor was neither mathematically intricate nor visually glamorous, it was widely employed as a tool for applied mathematical work. Instruction in its use was probably passed down orally from master to apprentice from the seventeenth century through the early eighteenth century, when protractors began to appear in manuals for instrument makers and mathematical practitioners.[1]

Thus, like the slide rule and the abacus, the protractor's original purpose was practical. It was not needed to teach geometry formally, the standard approach in colonial colleges. Even in the early nineteenth century, the protractor played no role in college or secondary Euclidean geometry courses, in which all constructions were to be made only with compasses and an unmarked straightedge. Yet there were some students in other forms of education who learned "practical geometry" from textbooks based upon professional manuals and thus were required to master drawing instruments.

After 1850 geometry and trigonometry textbooks increasingly depicted protractors, and with the proliferation of primary and secondary schools, the educators who compiled these books addressed broader audiences. Students also advanced beyond the study of arithmetic at earlier ages than previous generations. To ease younger and more diverse pupils into subjects such as geometry and trigonometry, authors turned to concrete aids such as protractors. These efforts were unsystematic, and in general instructors only advocated angle measurement with protractors on an ad hoc basis until the turn of the twentieth century.

Then, between 1893 and 1923, educators came together in assigning protractors to a uniform, specific role in the secondary mathematics curriculum. Unlike the metric ruler, which was introduced in order to reform metrology, the adoption of the protractor reflected changes in teaching methodology and in the organization of the American educational system. Teachers encouraged all of their students to experiment with the fundamental concepts of geometry through protractor exercises. Once they had been introduced to the subject, the pupils would move on to analytical work and abstract geometrical principles—for which protractors were unnecessary and therefore set aside. This approach, emphasizing intuitive understanding as a preliminary to formal study, proved successful with the younger and less experienced Americans studying mathematics, including non-college-bound teenagers. Even after 1923, when educators turned back to mental discipline as the most suitable justification for teaching mathematics to everyone, this role for the protractor endured in written standards and classroom practice.

Protractors in Professional Manuals

In 1589 Thomas Blundeville noted that Gerhardus Mercator, Gemma Frisius, and other Continental nautical cartographers were using a "new instrument" that consisted of two quadrants of a circle and was marked both with 180 equal degrees and with 150 unequal degrees that corresponded to the meridian lines on what are now called Mercator maps.[2] Frisius had also invented the triangulation method for surveying around 1530. This was the process of dividing the earth's surface into a network of triangles by determining each location's distance or altitude relative to two known distances or altitudes. These measurements were taken with angular instruments bearing sights, such as the graphometer. As Philippe Danfrie explained in

1597, however, surveyors were transferring the measurements to paper with a protractor.[3] While use of this new drawing instrument would have spread at first through word of mouth, mathematical practitioners also needed to explain how to use the instrument in writing in order to see it accepted more widely in navigation and surveying. Indeed, Blundeville's 1589 treatise, which offers the first instance of the protractor in print, was pedagogical in nature, although it was targeted more toward history and geography students and travelers than at the mathematics students he tutored.[4]

While textual evidence of protractors is scarce for the seventeenth century, protractors began to appear in manuals for instrument makers and practitioners as well as in geometry textbooks in the eighteenth century. The most significant guide for instrument production was Nicolas Bion's *Traité de la construction et des principaux usages des instruments de mathematique* (1709). Bion began with an introduction to Euclidean plane geometry, covering the elementary constructions and proving those theorems needed for surveying and engineering drawing, such as dividing the line into parts and constructing similar polygons. When Bion turned in his third chapter to describing the design and applications of all of the mathematical instruments that existed at the turn of the eighteenth century, he first presented the protractor (Figure 10.1). He explained how to graduate a semicircular protractor and provided the helpful hint that protractors made of horn should be stored in books to prevent them from wrinkling. The protractor was used to draw and measure angles, of course, but also to inscribe a regular polygon in a circle and to construct a regular polygon on a line. Instrument makers throughout Europe referred to this treatise, and Edmond Stone translated it into English in 1758.[5]

In their geometry textbooks French authors also combined practical training in using instruments with geometrical propositions. Following the tradition of the Port Royal school, they generally valued real-world results over abstract, complex proofs, so protractors were frequently included in their presentations.[6] In *Éléments de géométrie* (1741), for example, Alexis Clairaut argued that it was most natural to learn geometry in historical order. Thus, like the ancient Egyptians, he began the textbook with length and distance measurements of land, finding solutions but not proving theorems. He suggested that an instrument was necessary to make one angle equal to another, so after going through the standard constructions and treating proportion algebraically, Clairaut explained how to use a graphometer and protractor.[7] He employed the protractor in drawing triangles and

Figure 10.1. Delure semicircular brass protractor, ca. 1720. This one resembles the protractor depicted in the plates for Nicolas Bion, *Traité de la construction* (1709), 28. National Museum of American History collections. Smithsonian Negative no. 64166S.

other plane figures, and he stressed the importance of measuring angles exactly. Once the student reader understood the principles behind land measurement, he could proceed to the later sections of the book to explore abstract relationships between rectilinear figures, circles, surfaces, and solids.

While approaches such as Clairaut's seemed appropriate for students at any stage of learning in France, mathematics teaching took a different direction in England and Scotland and, by extension, the American colonies. Most notably, geometry tutors and teachers preferred formal presentations of that subject, a trend that culminated with Robert Simson's edition of *The Elements of Euclid* (1756). Instructors used Simson's book and its successors to introduce geometry to college students, who memorized each proof in order. If learners reproduced the figures during their recitation of the proofs, they used straightedges and compasses only. As future ministers, lawyers, or doctors, they most needed to know how to prepare logical arguments, an ability that intellectuals believed was developed by an abstract approach to geometry. Skills in engineering drawing were less relevant.

After independence American college mathematics professors continued to rely on British versions of Euclid's *Elements of Geometry*. Yet other Americans followed British educational traditions that were less formal. As in Clairaut's textbook, these approaches emphasized angle measurement and brief, simple proofs prepared with the aid of symbolic language and instruments. Some middle-class children, for example, were learning subjects that required scientific drawing, such as geography or astronomy. Because they were not likely to attend college, these boys and girls might not need to appreciate fully the mental discipline aspects of geometry that were essential for learning the classics and embarking on vocations in ministry or the law. Textbooks of "practical geometry" were designed for these students to study with a tutor, in school, or on their own. After providing definitions, instruction in the use of compasses, and an explanation of how to inscribe figures in circles, for instance, British author T. Drummond showed "young ladies and gentlemen" that the protractor could be employed in working with scale drawings, and he used the instrument to prove seven theorems about lines and angles.[8]

Practical geometry textbooks printed in Great Britain and the United States were also read by American mechanics, construction workers, and draftsmen. Typically, the authors of these works assumed that protractors were required for dissecting lines, erecting perpendiculars, and constructing triangles as well as measuring angles.[9] Encyclopedias occasionally functioned as practical geometry texts as well. In the 1771 first edition of *Encyclopaedia Britannica* a protractor was depicted with other surveying instruments in the article on "Geometry" but was not discussed in the text. In George Gregory's *Dictionary of Arts and Sciences* (1816) protractors were illustrated and defined in the article on "Instruments."[10] On the other hand, early stand-alone surveying textbooks did not mention protractors, although later in the century authors and publishers did add illustrations of drawing instruments and discussions of angle measurement.[11]

In the eighteenth and nineteenth centuries British instrument makers continued to manufacture protractors from thin sheets of brass or, less frequently, copper and silver. They also found animal horn to be a popular material, as it was semitransparent and thus allowed a draftsman to place the protractor on top of his drawing. In the seventeenth century instrument makers had adopted Tycho Brahe's diagonal scale technique, combined with the scale of equal parts, to subdivide more accurately 180 degrees through bisection, trisection, and quinsection.[12] Although this method re-

Figure 10.2. Ramsden dividing engine, 1775. National Museum of American History collections, gift of Henry Morton. Smithsonian Negative no. 45722-B.

placed the process of engraving angles against a master pattern entirely by hand, patterns were still required, and the technique was not of the highest accuracy. In 1767 Jesse Ramsden built the first machine for marking degrees, the dividing engine. His second version of 1775, as well as versions by John and Edward Troughton and by William Simms, brought precision in angular division up to 5 seconds (Figure 10.2).[13] Protractors could be manufactured more consistently with the dividing engine.

Additionally, it became even easier and less expensive to make pocket-sized versions of drawing instruments, which engineers, draftsmen, and students could keep at hand. In all, the dividing engine, new materials, new production techniques, and larger markets made possible a scientific instrument industry that produced large quantities of small, basic objects such as protractors as well as more precise and complex devices. Americans imported and sold these items before an instrument industry was fully established in the United States at the end of the nineteenth century. In 1867, for example, William Y. McAllister, a Philadelphia optician and instrument dealer, offered stainless steel protractors for 75¢ to $6, horn protractors for 25¢ to $1.25, and brass protractors from 75¢ to $3.25. These semicircular protractors ranged from 4 to 8 inches in diameter.[14]

The Rise of Informal Geometry

While American workmen, engineers, and young students might have used protractors in the daily work of navigation, surveying, geography, engineering, and architecture, young men in college were still studying geometry by memorizing formal proofs. They used no instruments and made no connections to applications. They also learned trigonometry analytically. If and when they encountered practical mathematics in the uniform course of study, the purpose was mental exercise rather than vocational training. From Jeremiah Day's 1817 navigation and surveying textbook, for example, Yale students learned pencil-and-paper trigonometric manipulations for computing sailing routes in the absence of sextants and chronometers—a skill they were unlikely ever to require in daily life.[15] John Farrar of Harvard took his students outdoors to look through telescopes for astronomy class in the 1820s, but the group did not chart their observations on star maps or require drawing instruments for other tasks. At technical institutes modeled on the École Polytechnique and West Point, the protractor was not included in preliminary, formal mathematics courses. It appeared as a mechanical drawing tool for designing civil engineering projects such as railroads and public works.[16] As much as ever, educators justified mathematics instruction generally, and geometry specifically, as a means to mental discipline for college and elite academy students. Drawing instruments such as protractors were simply unnecessary.

Yet the audience for geometry instruction was beginning to change. Harvard made plane and solid geometry an entrance requirement in 1844, meaning that potential students would have to master the subject in the academy or public school they attended before college. Instructors in the high schools being established in American cities therefore adopted the mathematics textbooks used in colleges. They also appropriated the emphasis in these textbooks on putting young minds in order through the mental discipline imparted by mathematics. Most teachers embraced the teaching technique of recitation as well, but some authors decided to add drawing instruments to formal geometry in order to make the subject more appealing to its expanded audience.

In turning away from presentations styled after Euclid's *Elements of Geometry*, this subgroup of American educators looked to a subject now called "informal geometry." This field of study evolved in the first half of the nineteenth century from Johann Pestalozzi's educational ideas and from German efforts to create a self-contained elementary course for students who would not be attending university.[17] Some instructors of informal geometry relied nearly entirely on drawing and computation; W. Harnish in 1837 and Max Simon at the turn of the twentieth century each designed a systematic course based upon practical measurement with instruments. Although Americans appreciated the utilitarian bent of these German reformers, they also wanted to preserve the disciplinary value of geometry for developing powers of abstract reasoning even at elementary levels of instruction.

To unite utility with mental discipline, Thomas Hill, then president of Harvard, wrote *First Lessons in Geometry* (1855) to stimulate the creative thinking of six- to twelve-year-olds and *A Second Book in Geometry* (1863) to teach geometrical thought processes to teenagers. *First Lessons* asked students to experiment with real-life examples and to work out geometrical concepts on their own, in a manner typical of the informal geometry movement. In the first part of *Second Book* Hill provided a primer on writing proofs.[18] Protractors appeared in the second part of the text, as Hill advocated solving practical problems through constructions prepared with drawing instruments. Students began constructing proofs with a marked ruler and set of compasses as their tools. After explaining how to divide an angle into any number of equal parts and to draw an angle of a given number of degrees, Hill added a semicircular cardboard protractor to his readers' toolkit and provided instructions so they could make their own instruments.[19] He also wanted students to use a parallel ruler for drawing parallel lines. The second part of his textbook covered triangles, quadrilaterals, circles, areas, surveying, and heights and distances; the third part was devoted to solid geometry. Many of the student exercises placed throughout the text called for angle measurement and implied that a protractor was to be employed.

Despite its extensive reliance on the object, Hill's book did not herald any consistent role for the protractor in mathematics education. Rather, between 1850 and 1890 diverse approaches targeting a variety of audiences coexisted. A very few textbooks, most notably William George Spencer's *Inventional Geometry* (1877), furthered the development of informal geometry as a method for learning geometrical principles through experiment and discovery.[20] These works relied on numerous student exercises and often allowed additional tools such as protractors for preparing proofs. There were also some "elementary," or "practical," geometry textbooks designed for children, secondary school pupils, and self-study that allowed protractors in proofs and employed the instruments in applications.[21] Protractors frequently could be found in textbooks on drawing and in works sold to mechanics and self-trained engineers.[22] At the same time, older forms of geometry textbooks persisted, such as the Euclidean treatises memorized by college and high school students. Charles Davies's *Elements of Geometry* was slightly revised by J. Howard Van Amringe in 1885 and reprinted into the 1890s. Similarly, George Wentworth's *Elements of Plane and Solid Geometry* (1877) shortened proofs with symbolic language and added exercises but continued to emphasize abstract knowledge.[23] This type of textbook ignored protractors for a purely formal approach. Furthermore, the first works on teacher training and educational philosophy discussed only specific definitions and types of proof in their sections on geometry. The clear expectation was that instructors would transfer the content and methods employed in colleges directly into high schools.[24] Finally, the authors of some college and high school trigonometry textbooks added descriptions of the drawing instruments used by draftsmen and surveyors, including protractors, but protractors were not used to measure angles within the texts.[25]

Protractors as an Introduction to "Standardized" Geometry

Efforts began at the end of the nineteenth century to impose structure on American education in general and on geometry teaching in particular.[26] Instrument use and angle measurement figured in the committee and academic meetings, new professional organizations, and preparation of new mathematics textbooks through which frameworks for mathematics educational standards were proposed. The overall legacy of the standards movements was mixed, but education reformers did succeed in determining a specific role for the protractor for a clearly defined subset of American students.

By 1890 the fixed college curriculum had become overburdened with new subjects, while the classics appeared passé to citizens of an industrialized nation. Postsecondary educators replaced the uniform required course with disciplinary majors that could be pursued from freshman survey courses up through graduate degrees. They also finally completely phased out mandatory geometry courses in which students memorized Euclidean theorems and constructions, expecting that students would learn this material before college.[27] Yet there were still a number of colleges and normal schools that offered only the equivalent of high school education. Meanwhile, academies and high schools were distributed unevenly across the nation. Subjects were added on an ad hoc basis in these institutions, even though Americans showed increasing interest in public education for all young people. Similarly, a variety of forms of primary education existed across the nation. A sense that American education was in crisis became palpable.[28] Instructors believed that new teaching methods were needed to communicate essential mathematical knowledge to these transformed student bodies. Specifically, recitation of geometrical proofs in order to build mental discipline no longer seemed suitable for any educational level.

Thus, in 1893, under the auspices of the National Educational Association (NEA), the mathematics subcommittee of the Committee on Secondary School Studies (better known as the "Committee of Ten") advocated that inductive and experimental methods replace deduction in school mathematics. The members of the mathematics conference suggested revisions in arithmetic teaching, recommended the introduction of a course in "concrete" geometry, encouraged additional computation in algebra, and supported a standard curriculum in mathematics until high school.[29] Their report explicitly associated informal geometry with instruction in grades 5 through 8. From age ten American children were to learn about geometry systematically but in a hands-on fashion, in a course that should "include among other things the careful construction of plane figures, both by the unaided eye and by the aid of ruler, compasses and protractor; the indirect measurement of heights and distances by the aid of figures carefully drawn to scale; and elementary mensuration, plane and solid."[30] The report's authors expected students to lay aside their instruments when they reached high school because the teaching of "demonstrative" geometry was driven by axioms and postulates to prepare for college work and perhaps to enable students to learn some projective geometry at the secondary level. Meanwhile, protractors were not mentioned in the sections prepared by conferences on subjects that had served as applications for geometry in the past, such as geography.

Later discussions of mathematics education affirmed the desire to introduce instruction in geometry early in a student's education, with emphasis on the concrete aspects of the subject. The NEA's Committee on College Entrance Requirements, which met in 1899 and which shared at least one member (Henry B. Fine of Princeton) with the Mathematics Conference of the Committee of Ten, repeated the earlier committee's recommendation that informal geometry serve as the precursor to demonstrative geometry. The committee also called for a "correlated" approach to instruction, including some algebra and some geometry in mathematics instruction every year from seventh to twelfth grade rather than as distinct subjects for separate years of high school.[31] During the same period research mathematicians such as E. H. Moore began to try to influence school mathematics by pushing a "laboratory method" of teaching mathematics.[32] Also at the turn of the twentieth century, John Perry and the British Association for the Improvement of Geometry Teaching attempted to eliminate formal geometry in school entirely. This movement had limited impact in the United States on the whole, but its advocacy of experimental techniques with tools such as squared paper and the protractor reinforced similar American trends.[33]

An example of how protractors and informal geometry could be used to introduce a formal treatment of the subject can be found in *School Geometry: Inductive in Plan* (1897), by J. Fred Smith. Smith, the principal of the Iowa College Academy, was inspired by the Committee of Ten and an 1895 follow-up meeting, the Committee of Fifteen on Elementary Education, which both concluded that informal geometry ought to be taught to the very young. Smith suggested that original work that encouraged students to think was the best way to teach the basics of geometry to the 95 percent of students who did not finish high school. He recommended that teachers ensure that each pupil was provided "a pencil compass, a paper protractor, a short rule marked with common and metric scales, and a pad of good unruled paper."[34] He organized his textbook according to types of figures, such as lines and angles, triangles, and quadrilaterals, rather than by the books of Euclid's or Legendre's *Elements of Geometry*. Each chapter consisted of lessons, and each lesson introduced and described a concept that was tested with numerous pencil-and-paper exercises.

In other words, *School Geometry* looked much more like a late-twentieth-century middle school mathematics textbook than like nineteenth-century college geometry texts. Twenty pages into the chapter on lines and angles, Smith explained how to use a protractor, alongside the principle of opposite angles, and provided homework exercises on measuring and drawing angles with a protractor. Later in the chapter students compared the perpendicular lines they erected with compass and straightedge against perpendiculars they drew with a protractor.[35] After the experimental work in this first chapter, however, Smith used a formal approach—without drawing instruments—to cover the remainder of the course, which went as far as transformations and the division of figures.

At the same time, improved mass production techniques and new materials once again reduced the cost of protractors. By 1909 Keuffel & Esser (K&E), in New York City, sold brass, pocket-sized protractors at prices ranging from 9¢ to $4.50, depending on the diameter and production quality of the instrument. K&E also offered 5-inch cardboard protractors for 10¢ and Xylonite (celluloid) protractors of 5 to 10 inches for 45¢ to $5.50 (Figure 10.3).[36] As Smith had, J. C. Packard argued in 1903 that schools had material responsibilities to a class embarking on the study of geometry: "a small equipment of ruler, compasses, scissors, drawing board, protractor, parallel ruler, squared paper, T-square and a triangle of sixty and of thirty degrees with a right angle should be provided for each pupil."[37] Oversized, fiberboard protractors were first sold by Eugene Dietzgen and other companies in the 1920s, perhaps to help teachers at the blackboard demonstrate activities in informal geometry to an entire class (Figure 10.4).

By the 1910s it was common for high school geometry textbooks to use protractors to introduce students to the subject and then require that the instruments be set aside in order to master abstract proving techniques. In their 1916 *Plane Geometry*, for instance, Edith Long and W. C. Brenke laid an informal foundation for high school students that included measurement with a protractor. Then they warned against using intuition to jump to conclusions, changing course to a formal study of proofs for the remainder of the textbook.[38] Long was a high school teacher in Lincoln, Nebraska, while Brenke was a professor of mathematics at the University of Nebraska. Similarly, Webster Wells, a professional mathematics textbook author, and

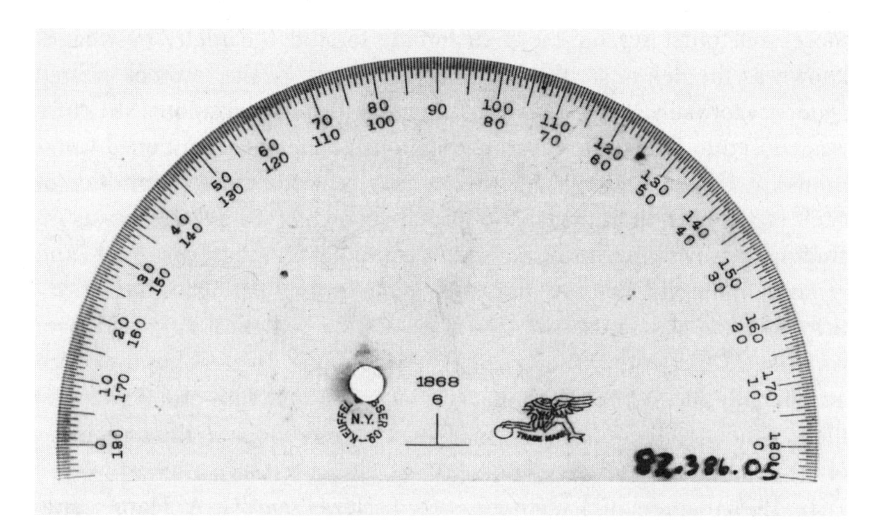

Figure 10.3. Xylonite semicircular protractor by Keuffel & Esser, ca. 1909–36. National Museum of American History collections, gift of John A. Betcher. Smithsonian Negative no. 83-13954.

Figure 10.4. Fiberboard Dietzgen protractor for blackboard use, ca. 1950. Its dimensions of 16 inches by 8 inches were typical for the form. National Museum of American History collections, gift of L. Thomas and Margaret G. Aldrich. Smithsonian Negative no. 2001-7657.

Walter W. Hart, a professor of mathematics at the University of Wisconsin, placed an explanation of the protractor at the end of the definitions in their 1915 high school textbook, *Plane and Solid Geometry.* They provided nine exercises in "experimental geometry" but then noted that "objections to studying geometry *only by the experimental method* may be given" and that for those and "other reasons, it is customary to study geometry by what is known as the demonstrative method."[39] The rest of the textbook treated geometry formally. Wells and Hart stated that their organizational structure was chosen to follow the recommendations of the aforementioned Committee of Fifteen Report, which essentially reinforced the Committee of Ten's statements on informal and formal geometry.[40] By 1926 Hart was including his own invention, a paper "combination protractor, rule, and square," in the endpapers of the textbook and congratulating himself on anticipating the standards (laid down by the Mathematical Association of America [MAA]) in 1923 (Figure 10.5).[41] Trigonometry textbook authors also occasionally added brief work on "graphic solutions" with protractors and then abandoned the tools in favor of analytical solutions in the remainder of the text.[42]

In 1921 mathematics instructors M. J. Newell and G. A. Harper suggested guidelines for teaching geometry to high school sophomores that again used protractors to prepare the students for abstract concepts: (1) explanation of the postulates and the straightedge and compasses as the only tools suitable for formal geometry; (2) construction exercises; (3) experiments with the protractor; (4) informal proofs of elementary theorems; and (5) formal proofs. Of protractors Newell and Harper said, "The use of the protractor need not be continued beyond the first few lessons. It is not, strictly speaking, a tool of geometry but is a valuable aid in developing a clear conception of the values of angles."[43] Yet, although educators and textbook authors assumed that this process was necessary for the many high school students whose earlier education might have been sporadic, they really wanted children to build a foundation in informal geometry before reaching high school. The students would then set their protractors aside when they took up formal plane and solid geometry in grades 10 through 12. In 1923 the MAA's National Committee on Mathematical Requirements recommended that seventh-graders learn informal geometry, consisting of "the direct measurement of distances and angles by means of a linear scale and protractor. The approximate character of measurement. An understanding of what is meant by the degree of precision as expressed by the number of 'significant' figures."[44] The report also urged teachers to cover areas, the technique of drawing to scale on squared paper, and constructions prepared with drawing instruments. Thus, in John Charles Stone's *New Mathematics* (1926) seventh-graders learned to measure and draw with a protractor so they could do experiments testing relationships between tri-

Figure 10.5. Paper protractor designed by Walter Hart, ca. 1936. It was enclosed with and illustrated in several editions of Hart's high school geometry textbooks, including Walter Wilson Hart, *Progressive Plane and Solid Geometry* (Boston: D. C. Heath & Co., 1936), 16–17. National Museum of American History collections, gift of Department of Mathematics, Brown University. Smithsonian Negative no. 88-11427.

angles—again with protractors—in eighth grade.[45] These activities, as well as construction of hexagons, pentagons, and parallel lines with a protractor, were reviewed in ninth grade with the aid of a 1931 textbook by Leo John Brueckner, Laura Farnham, and Edith Woolsey.[46]

The association between junior high educational standards and protractors proved to have remarkable staying power, even though reform movements waxed and waned throughout the twentieth century.[47] Emphasis on protractor experiments as a preliminary taste of geometry can be found in the renewed efforts after 1923 to justify mathematics in the school curriculum with the tenets of mental discipline, in the language of the "New Math" of the 1950s and 1960s, and in writings on the "Math Wars" of the last quarter of the twentieth century. As late as 2000, the National Council of Teachers of Mathematics wanted students in grades 6 to 8 to measure angles directly with a protractor after learning to recognize "benchmark" angles by sight.[48] Consistently, mathematics educators suggested that protractors and junior high mathematics were a natural match for three reasons. First, young minds needed several years of preparation before achieving the maturity necessary to begin demonstrative geometry. Second, experiments with protractors caught students' interest and introduced them to applications useful in the vocational work many would pursue after high school. Third, informal geometry correlated well with arithmetic, algebra, and trigonometry. Twentieth-century educators clearly viewed their universal student body as more “obtuse” and needing experimental learning to build the intellectual maturity required for an understanding of abstract proof. They solved this “acute” problem with protractors and then expected their students to set aside informal geometry with other childish things and proceed to formal geometry without instruments.

CHAPTER ELEVEN

Metric Teaching Apparatus
Making a Lasting Impression?

> The metric tables are given in nearly every series of Arithmetics. But when only tables are used, the impression is vague and the figures soon forgotten. The object itself always makes a clearer and more lasting impression than any verbal description of it.

The growing use of the protractor in the mathematics classroom in the late nineteenth century was closely tied to changes in the practice of teaching geometry, as the subject moved from the university into the secondary school. The same period also saw greater emphasis on the pedagogical possibilities of measures of length, volume, and weight. The topic of weights and measures had long been, and would remain, a part of introductions to commercial arithmetic. Increasing school use of apparatus in teaching the subject did not indicate a shift in basic course content. Rather, it reflected an attempt to reform metrology through the introduction of metric weights and measures.

Both cipher books and textbooks had included tables of weights and measures that students were expected to memorize. The information presumably was useful in commerce. The tables also offered myriad opportunities for students to review arithmetical processes by working problems that required converting one unit of measure into another. By the 1850s a few authors suggested that students might acquire a firmer sense of what these tables meant if they manipulated objects. Ira Mayhew, the superintendent of Public Instruction for the State of Michigan wrote in 1856:

> Weights and measures serve the same general purpose and may be rendered well-nigh as useful as slates and black-boards. Thousands of children recite every year the table, "Four gills make a pint, two pints

> make a quart, four quarts make a gallon," etc., month in and month out, without acquiring any distinct idea of what constitutes a gill or a quart and even without knowing which of the two is the greater. But let these measures be once introduced into the experimental play-room, and let the child, under the supervision of the Teacher or Monitor, actually see that four gills make a pint, etc., and he will learn the table with ten-fold greater pleasure than he otherwise would, and in one-tenth the time.[1]

Mayhew was not alone. By 1860 Henry Barnard listed a set of weights and measures as one of the "requisites for a primary school."[2] Barnard recommended buying this apparatus. Other authors suggested making simple length measures or borrowing them.[3]

Despite these suggestions, the systematic design and distribution of classroom weights and measures was not initially a result of a general desire to teach gills, gallons, and other traditional British units more effectively. Rather, it was part of a campaign waged by advocates of the metric system to introduce Americans to units of measure developed in France relatively recently. In the 1860s both the efforts of individuals and organizations and the legalization of metric units in Great Britain led to discussion about weights and measures generally and the metric system in particular. The U.S. Congress legalized the system in 1866 and encouraged efforts to popularize it. In the early 1870s the Prussian government made the use of metric measures mandatory and provided school apparatus to teach the system to children. Inspired by this example, and persuaded that the United States would soon also require the use of metric units, educators in Connecticut and Massachusetts considered how the system should be introduced in the classroom. The new measures were not part of the everyday experience of either students or teachers; they could not be borrowed from the kitchen, shop, or store. To encourage familiarity with the new units, metric advocates developed demonstration apparatus specifically for classroom use. They argued that decimal units would be simpler for children to learn and far more efficient to use than conventional units. Yet the rapid spread of metric measurement they envisaged did not come to pass. By the late 1880s those examining arithmetic teaching in the United States concluded that simplicity and efficiency would be gained not by teaching the metric system but by eliminating the topic from the arithmetic curriculum altogether. Charts and instruments for teaching about units of measure were still sold, but they tended to represent conventional units.

The Metric System in Antebellum America

The United States Constitution explicitly grants the federal government the power to regulate weights and measures, without specifying what units are to be used. In a 1790 report Secretary of State Thomas Jefferson proposed that the country either adopt weights and measures based on English use or introduce new units that were decimal multiples and submultiples of one another.[4] Congress ignored Jefferson's report.

In France units of measure differed from place to place far more than in England or the English colonies of America. In the wake of the Revolution of 1789 French citizens called for uniform weights and measures throughout the country. With the approval of the National Assembly and subsequent French governments, a commission of the Parisian Académie des Sciences and its successor, the Institut de France, developed entirely new units for measuring distance, volume, weight, angles, and even time. These units were interconnected; units of one quantity, such as length increased by powers of ten (centimeters, decimeters, meters). The fundamental unit of volume, the liter, was the volume of a cube 10 centimeters on a side. One kilogram was the mass of 1 liter of water. No such simple relations existed in English weights and measures between units of length (inches, feet, yards, and miles) or between units of length, volume, and mass. In short, the French introduced not only national standards but an entire system of standards. This system survives today, in modified form, as the metric system.[5]

As early as 1795, a French government decree about the metric system was republished in the United States,[6] and a few almanacs and encyclopedia articles described aspects of the new system.[7] The French government sent copies of its new standards to the United States in 1795.[8] Moreover, in 1805 Ferdinand R. Hassler came to the United States from Switzerland, bringing with him an iron meter bar and a brass kilogram weight that had been distributed by the French government in 1799. Hassler eventually was named head of the U.S. Coast and Geodetic Survey, and his kilogram and meter were used as its standards.[9]

Although the Coast Survey used metric units in its work, those who governed the early United States were primarily of English descent and used the older English units of measure. In 1821, at the request of the U.S. Congress, Secretary of State John Quincy Adams carried out a survey of the weights and measures that had been legally established by the states. Not surprisingly, these were generally modeled on English units. In the 1830s, not long after Britain had established the imperial system of weights and measures, U.S. Secretary of the Treasury Louis McLane ordered Hassler to provide customhouses with standard pounds, yards, and bushels. In 1836 Congress ordered that these standard weights and measures be supplied to the states as well. Fortunately, Hassler had copies of various British standards to use in this work. He died before completing the task, but his successor, Alexander D. Bache, continued the project.[10]

The Metric System Made Legal

By the time of the Civil War several American reformers believed that the metric system should be used in the United States. Some, like the Des Moines lawyer and politician John A. Kasson, thought that the adoption of metric units would ease international transactions. Appointed first assistant postmaster by President Lincoln, Kasson learned of the complex system of fees used to compute charges for international mail. Mindful that the U.S. Treasury was losing money on these transactions, Kasson called for an international conference of postal officials that would standardize rates and procedures. At this meeting, held in Paris in May 1863, the delegates recommended that international mail be weighed in metric units, a practice that was soon widely adopted.[11] Kasson attended the conference as the American representative but left the post office not long thereafter to serve in the U.S. House of Representatives. From this position he advocated legislation to make the metric system legal in the United States.

Further support for legalizing metric units came from New York congressman Samuel B. Ruggles. In 1863 he had attended an international statistical conference in Berlin, where delegates recommended that the metric system be used in commercial dealings. Upon his return home, he followed up on this resolve by writing to individual states, urging them to encourage teaching the metric system in the schools. In Connecticut the state legislature responded in 1864 by passing without opposition a resolution that both state and local school authorities "be urgently recommended to provide that the Metrical System of Weights and Measures, together with its relations to the legalized systems, be taught in all the schools which are under their charge."[12] By August of that year Hubert A. Newton, professor of mathematics at Yale College, had prepared an eight-page pamphlet explaining the metric system, designed as a supplement to existing arithmetics. In the preface Newton expressed the hope that discussion of the metric system would soon be incorporated into textbooks, rendering the booklet superfluous.[13]

Additional incentive for teaching metric units came from both outside and within the United States. In 1864 Parliament voted to legalize metric units in Britain.[14] Two years later a committee of the newly established U.S. National Academy of Sciences recommended that the country adopt a decimal system of weights and measures.[15] John A. Kasson's legislation legalizing metric weights and measures also passed both houses of Congress without discussion and was signed into law by President Johnson. Soon the Office of Weights and Measures was preparing metric standards for distribution to the states. Metric units were now legal, though not required, in the United States.[16]

Promoting the New System

Not long after it approved legislation making the metric system legal, the Committee on Coinage, Weights and Measures of the U.S. House of Representatives urged others, especially textbook author Charles Davies, to promote instruction on the subject in the schools. Davies promptly wrote a pamphlet on the topic, and he and other arithmetic textbook authors expanded the discussion of metric units in their books.[17] Prussia, which legalized the metric system in 1868, decided to make it mandatory by 1872 and introduced school apparatus to help students make the transition. Birdsey Northrop, a clergyman by training and secretary of the Connecticut Board of Education from 1867 to 1883, visited Prussia during the transition and was greatly impressed by the devices he saw.

During the same period several international scientific conferences were held in Europe to establish common definitions of metric units. In the United States the American Metrological Society was established in 1873 to promote the use of metric units. Frederick A. P. Barnard of Columbia University served as president. Responding to these changes, a few American scientific instrument makers introduced apparatus with metric scales. The 1871 catalog of W. & L. E. Gurley of Troy, New York, included a "French Standard Metre, divided on three edges to millimetres and on one edge to 5ths of millimetres." The rule was made of steel and sold for $10.[18] The 1874 catalog of James W. Queen of Philadelphia listed a diagram showing a scale 6 inches long, divided to eighths of an inch, next to a scale 15 centimeters long, divided to millimeters. This was included to compare measurements in inches with the "Metric Measure recently adopted by Congress."[19]

All of this activity persuaded Birdsey Northrop to plan an exhibit of metric teaching apparatus as part of Connecticut's contribution to the United States Centennial International Exposition, the World's Fair held in Philadelphia in 1876. Northrop was most interested in popularizing length measures, believing that they would serve as the "entering wedge" for the new system.[20] He arranged for the New Britain, Connecticut, firm of A. & T. W. Stanley (later Stanley Rule & Level Company) to produce foot-long wooden rules that were divided metrically along one edge and in inches along the other. Northrop sold these instruments to schools for five cents apiece, half the cost of conventional rules.[21] Stanley also offered a "Meter-Diagram" on linen paper; it showed a yard next to a meter, with the meter divided to millimeters. The diagram also included tables explaining the metric system generally. On the less expensive version of the chart, which sold for $2 per dozen and was 2¼ inches wide, the tables were on the opposite side of the paper from the meter and yard. More elaborate versions of the diagram, in color and illustrated with views of the Philadelphia fair, sold for as much as $3.50 per dozen. In an advertisement for this chart Stanley included recommendations from Northrop, H. A. Newton, Yale professor of physics and astronomy Chester S. Lyman, Frederick A. P. Barnard, and others.[22] Northrop arranged to have a copy of the more elaborate form of Stanley's diagram bound with his 1877 annual report. The state of Connecticut also sold the chart, printed on paper, to Connecticut schools and teachers "at half price, or fifteen cents per copy."[23] In addition to length measures the Connecticut Educational Exhibit in 1876 included metric scales made by the Fairbanks Scale Company as well as metric measures of mass, surface area, and volume, which Northrop apparently made no attempt to market. He was proud to report that the Connecticut exhibit was well attended, with thousands of visitors weighing themselves in kilograms. The display even received an award at the fair, the only exhibition of metric demonstration apparatus so honored.[24]

Northrop was well aware that some people might disapprove of his advocacy of metric units. For him "the certainty of appreciation by the rising generation, by all the children in the schools who are thus led to its study, will amply compensate for any such criticism or censure."[25] Nonetheless, the duties of state school officials generally did not involve arranging the design and distribution of teaching apparatus. The center of this activity was already moving to a private organization, the American Metric Bureau.

One visitor to the Centennial Exposition who paid careful attention to the exhibits of metric apparatus and collected available samples was the librarian and lifelong reformer Melvil Dewey. Dewey had been attracted to the system as a high school student in Adams Center, New York, when he realized how much use of the new units would simplify his arithmetic problems. As an undergraduate in the Amherst College class of 1874, he not only remained interested in metrification but embraced three other causes—simplified spelling, shorthand, and the decimal classification of books. Reforms in these four areas would occupy him for much of the rest of his life.

Dewey joined the American Metrological Society as a college senior. He spent his first years after graduation working in the Amherst College library but left Amherst in April 1876 to establish a metric department at Ginn & Company in Boston. Ginn was to publish a metric school chart designed by Boston architect John Pickering Putnam. Using Ginn's name, Dewey also arranged for the Fairbanks Company of St. Johnsbury, Vermont; the Keuffel & Esser Company of New York; and the G. M. Eddy Company of Brooklyn to provide metric scales and measuring devices at cost.[26] Dewey shared his plans at the May 1876 New York meeting of the American Metrological Society and won that organization's endorsement for his products.

With this backing Dewey, Putnam, and Cambridge High School classics master and mathematics textbook author William F. Bradbury decided that a distinct corporation selling metric supplies and publications was needed, and in July 1876 they incorporated the American Metric Bureau. Early officers included F. A. P. Barnard as president, Charles Francis Adams, William Watson, and W. F. Bradbury as vice presidents, Dewey as secretary, and Putnam as treasurer.[27] In mid-August the American Metric Bureau produced the first issue of a new periodical, the *Metric Bulletin*. By early 1877 the bureau had 140 members and offered 323 items for sale. Special sets of apparatus could be acquired for school use.[28]

When Northrop distributed rules and charts, he stressed the value of comparing metric units with those in common use. The materials distributed by the American Metric Bureau represented a different approach. They did not emphasize the relationship of old and new units but, instead, highlighted the interconnections among metric units. A four-sided wooden me-

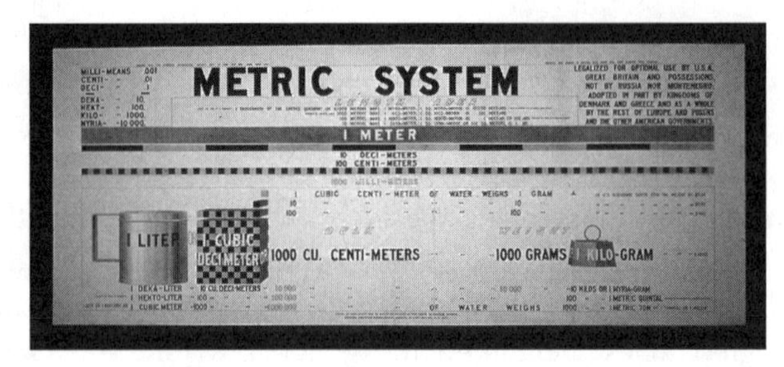

Figure 11.1. Chart of the metric system, 1890. This chart was distributed by the American Metrological Society and shows several pieces of teaching apparatus. National Museum of American History collections. Smithsonian Negative no. 92-14161.

Figure 11.2. Volumetric measures made for the American Metric Bureau, ca. 1880. National Museum of American History collections. Smithsonian Negative no. 92-14244.

ter stick made by Stanley and distributed by the Metric Bureau was divided, for example, into millimeters on one side, centimeters on another, and decimeters (10 centimeter units) on a third; the fourth side was blank. Three sides of the rule also appeared on a chart (Figure 11.1). From 1877 onward, Shaker residents of West Gloucester (later Sabbathday), Maine, made a set of seven wooden grain measures for the Metric Bureau (Figure 11.2). The measures came in sizes of 1 deciliter, 2 deciliters, 5 deciliters, 1 liter, 2 liters, 5 liters, and 1 dekaliter (10 liters). Metric Bureau officer and textbook author William Bradbury had even patented a liter demonstration apparatus to

Figure 11.3. Liter/kilogram demonstration apparatus like that of the American Metric Bureau, ca. 1880. National Museum of American History collections. Smithsonian Negative no. 92-14242.

show the interconnections between units of length, area, volume, and mass. It consisted of a wooden cube, 10 centimeters on a side, with a 10 × 10 centimeter grid of lines on each face. The cube fit snugly into a tin box marked "liter." When the cube was removed and the tin filled with water, the water weighed one kilogram. An example of this apparatus survives in the Smithsonian collections, but no maker's name is indicated (Figure 11.3). The Metric Bureau eventually would sell length measures marked with both metric and English units but did so reluctantly.[29]

The American Metric Bureau offered its products directly and through established instrument vendors. An 1881 catalog of J. W. Queen & Company listed several items it supplied, including the meter stick, the liter block with metal case, and a set of seven wooden volume measures that sounds like the set made by the Shakers. The Queen catalog also listed a set of eight copper volume measures, a set of thirteen brass weights ranging from one "gramme" to one "kilogramme," and a set of school scales that was sensitive to 2 grams.[30]

In order to make an impression, teaching apparatus must actually be seen by students. Surviving minutes of the Boston School Committee suggest part of the process of introducing the new apparatus into the classroom. By 1877 Boston had nine high schools, forty-eight grammar schools, and eighty-four buildings used as primary schools.[31] In February of that year the school committee adopted a uniform course of study for the high schools. During their first year all students were to spend some time studying "arithmetic including practical instruction in the metric system."[32]

To aid in this instruction, the Boston School Committee planned to appropriate funds to purchase metric apparatus for each high school. At the request of Harvard physiologist and committee member Henry Pickering Bowditch, the Committee on Text-Books of the Boston School Committee was asked to "consider the expediency of providing practical instruction in the use of the Metric System of weights and measures in the Grammar and Primary Schools."[33] After due deliberation a majority of the five members of the textbook committee concluded that advanced students in the grammar schools should receive some instruction in the metric system, given that they would encounter it in high school. Necessary apparatus should also be provided. They saw no need for such instruction in the primary schools.

Two members of the committee, physician John G. Blake and lawyer Godfrey Morse, filed a respectful but impassioned dissent. They argued that the intrinsic merits of the metric system, its use by all European nations except England, its sanction by several departments of the U.S. government, and its role in scientific measurement all made it an important topic for school instruction. Adoption of the system was coming soon, they predicted, and "its use will be general by the time the younger classes of our schools are prepared to engage in business."[34] Blake and Morse proposed that instruction in the metric system should be given to all classes of the grammar schools and, in the form of object teaching, in primary schools.

The Boston School Committee received these recommendations in June 1877. It voted to print both the majority and minority reports and tabled the issue until fall. At its September meeting a motion to accept the minority report failed by a vote of fourteen to six. The majority report passed, and Boston thus came to have metric apparatus in its grammar and high schools but not in primary schools.

Second Thoughts

The enthusiasm of Barnard, Northrop, Dewey, and their associates proved insufficient to bring about the adoption of the metric system in the United States. Dewey's handling of the funds of the American Metric Bureau and two other reform efforts he spearheaded, the American Library Association and the Spelling Reform Association, disappointed the leadership of all of these organizations. In 1883, when Dewey left Boston to become the librarian of Columbia University, the efforts of the American Metric Bureau ceased. Distribution of some metric apparatus was taken over by the American Metrological Society and continued at least through the end of the century (see Figure 11.1).

Many Americans saw little reason to abandon long-accepted units of measure for the new system. In areas such as electromagnetism, in which there were no established units, Americans happily adopted such metric units as the ampere, the volt, and the henry.[35] The United States already had relatively uniform measures of distance, area, volume, and mass, however, which were used throughout the nation. Mechanical engineers had agreed on standard sizes for machine parts and gauges.[36] Many saw no reason to give up standards developed through long practice for those designed by scientific experts. Others thought it unwise to abandon the English system of measures, when Britain ruled such a large part of the world. A few adopted the arguments of Scottish astronomer Charles Piazzi Smyth, who found ties between British units of measure, those used in building the pyramids of Egypt, and the measures of the Hebrew people. They were loath to abandon such a rich metrological tradition.[37] Smyth arranged to have the British instrument maker L. P. Casella make ceramic rules showing both British inches and the "pyramid inches" that he thought had been used in constructing the ancient monuments. At least one of these rules made it to the United States (Figure 11.4). Even those who did not accept Smyth's pyramid metrology might believe that existing ways of thought should not be put aside lightly. As one Philadelphia editor wrote in 1882: "Easy as it might be to furnish us with rules and sticks graduated off in metres and their parts, and with measuring glasses and weights accurately made . . . it would be by no means so easy to furnish us with the mental standards of reference which our English feet, inches and yards have become, after years of intimate knowledge of them."[38]

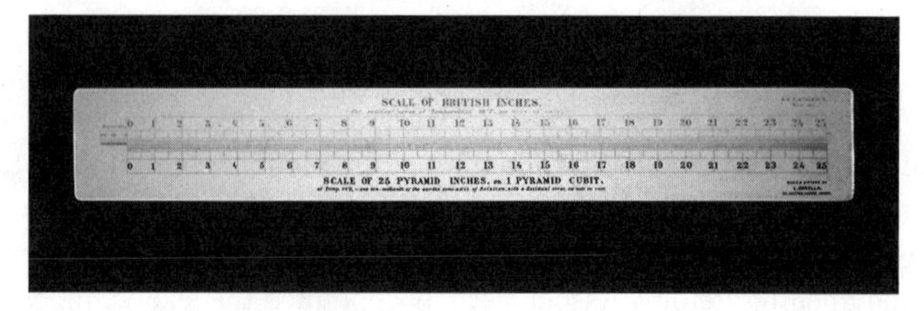

Figure 11.4. Scale of British and pyramid inches by L. Casella of London, 1867. National Museum of American History collections, gift of Augustana College, Rock Island, Ill. Smithsonian Negative no. 92-14162.

By the mid-1880s doubts about the metric system had reached the Boston schools. Mastering the measures seemed an unnecessary burden for children, not a key to their future understanding of the world. Near the close of the 1886–87 school year the school committee voted to modify arithmetic teaching in several ways. Some topics, such as finding the volume of common solids, compound interest, taking cube roots, and the metric system were eliminated altogether. Homework in arithmetic was to be given only in exceptional cases. Practical problems were to be stated in terms intelligible to the student (and, although the policy did not say this explicitly, the parent). In 1887 General Francis A. Walker, president of the Massachusetts Institute of Technology and a member of the Boston School Committee from 1885 to 1888, offered an explanation of the new policy in an article published in the *Academy* and reprinted as a pamphlet. Here Walker argued that an excessive amount of time had been devoted to the study of arithmetic. Methods used to teach it were directed to general mental training, not to the acquisition of facility and accuracy in numerical computation that students needed, and other subjects recently introduced into the curriculum were equally, if not better, adapted to mental training. Moreover, the undue difficulty of exercises offered in arithmetic classes often was positively injurious.[39]

Walker's comments did not specifically address the subject of the metric system. They were an early sign of a more general movement to condense and simplify arithmetic teaching. Publishers, not authors, increasingly determined what went into textbooks, and they found few incentives to include the metric system. Metric weights and measures increasingly were taught as a part of courses in science, not mathematics. Some of these courses used metric teaching devices quite similar to Bradbury's liter block. The 1912 catalog of L. E. Knott Apparatus Company of Boston offered a "dissectable" liter block for three dollars.[40] Versions of the liter block were offered by W. M. Welch Scientific Company in Chicago in 1931,[41] and they were available from Cambosco of Boston as late as 1956–57.[42]

At a more elementary level charts and objects were used to introduce schoolchildren to common units of measure. Like Ira Mayhew, the Michigan superintendent of schools who argued that students needed a palpable way of understanding weights and measures, Nathan A. Calkins proposed object lessons to introduce students to conventional measures of length, area, volume, and mass. He published these ideas in 1862, the year he was named the superintendent of the New York City primary schools.[43] By 1873 J. W. Schermerhorn & Company of New York offered weights and measures to be used in conjunction with his book, but this apparatus apparently did not become widely adopted.[44] Later charts were more popular. In 1889, for example, the Union School Furniture Company of Battle Creek, Michigan, advertised a series of arithmetical charts prepared by Francis W. Parker. Parker had been the superintendent of the public schools in Quincy, Massachusetts, and was then supervisor of public schools in Boston. His set of fifty charts, which sold for ten dollars, represented numbers by using both drawings and written numerals. The publisher claimed that it could replace ordinary textbooks in teaching primary school arithmetic, saving time and expense. In general Parker confined himself to basic arithmetic operations, as was appropriate for primary school students and consistent with the recommendations of the Boston School Committee outlined by Walker. He did include two tables of weights and measures and also devoted one side of the book's covers to metric weights and measures.[45]

In 1897 one R. O. Evans prepared a series of twenty charts, entitled, in part, *Evans' Arithmetical Study Applying the Object Method to the Entire Subject of Practical Arithmetic,* which were chromolithographed by Caxton Publishing Company of Chicago. As the title suggests, Evans's colorfully illustrated charts included not only an introduction to numbers and arithmetic operations but discussion of fractions, percentages, and the measurement of surfaces. Also considered were bookkeeping, business methods, and "commercial and legal" topics. One chart introduced the metric system, one lumber and timber measures, and a third common weights and measures. The third chart included troy, avoirdupois, and apothecaries' weights; dry and liquid measures; and units of length, area, and volume. The chart on the metric system was plain, decorated only by an image of the hand of a man holding a thermometer. In contrast, the sheet on common weights and measures was adorned with images of gold coins on scales, quarts brimming with milk and filled with fruit, and people going about common tasks of life.[46]

Perhaps the most enduring legacy of the metric school apparatus advocates, however, was not any chart but the common school ruler. Of course, foot-long rules had been available for decades and were discussed by authors such as Calkins. Melvil Dewey estimated in 1876 that millions of them were in use,[47] although most rules were sold to working people and to advanced students for drawing lines with ink. The wooden foot-long "student's rules" listed in an 1873 school supply catalog cost ten cents apiece, but the dealer promised a "liberal discount" on a gross (i.e., 144 rules). They were not prominently featured in the catalog, receiving roughly the same attention as blotting paper.[48] As already mentioned, by 1876 Birdsey Northrop was selling Connecticut schools rules 30 centimeters long for half the price. A few years later the American Metric Bureau listed such rules in boxwood for fifteen cents each, in maple for ten cents each, and on paper for four cents each, or two dollars per hundred. It had also obtained subsidies so that it could actually sell the rules for half of these prices, and teachers who could travel to one of the bureau's repositories could avoid the cost of shipping.[49]

Soon rules made from wood, paper, and then plastic were shown in numerous catalogs of both school supply houses and scientific instrument dealers. From at least the 1870s schools also could buy oversized rules that were sold for blackboard use.[50] Falling prices put such rules within the reach of every student and teacher. In its 1895 catalog the Boston firm of J. L. Hammett offered foot-long rules with both metric and inch scales. Prices ranged from a penny to three cents, with an additional penny for postage.[51] Milton Bradley Company's 1909 catalog listed foot-long rules divided to a quarter-inch for primary school students and to one-sixteenth of an inch for grammar school students as well as a 36-inch blackboard rule. It does not mention metric units. The rules for children cost 2 cents each or $1.75 for a gross. The blackboard rule cost 30 cents, plus 14 cents for postage.[52]

Although the efforts of the American Metric Bureau failed to bring about general metrological reform, they left a legacy of length measures in customary units to suit the needs and budgets of every classroom. Such rules still remain in the classroom. For the children of present-day immigrants, whose parents almost all come from countries that use metric weights and measures, these objects represent an introduction to a mode of measurement now used almost exclusively in the United States. Birdsey Northrop and Melvil Dewey would marvel at this unintended consequence of their endeavors.

CHAPTER TWELVE

Graph Paper
From the Railroad Survey to the Classroom

To secure for our young students a tolerable appreciation of the civilization of the twentieth century on the side of the theoretical and applied mathematical sciences, as teachers of mathematics we agree to shape our instruction in mathematics from the beginning from a point of view no older and no lower than that of the wonderful seventeenth century, and to this end, speaking theologically, we propose to
Canonize the Cross-Section Paper.

Graph paper became a common tool in American mathematics classrooms in the twentieth century. It had made its appearance in European scientific and engineering applications early in the nineteenth century, but it was not until the last decades of that century that influential British scholars advocated its use in education. This British example stimulated the emergence of American educational use of graph paper in the early 1900s.

The subsequent relatively rapid educational adoption of graph paper in the United States involved the confluence of several changes in mathematical pedagogy, which built through the last part of the nineteenth century and were linked by the increased promotion of visualization as a teaching tool. The impetus for some of these changes came from outside mathematics, notably from efforts to make mathematics education more supportive of science and engineering. Other contributory changes were internal to professional mathematics, especially the increased recognition of the value of the function concept for unifying large portions of the subject. By the early twentieth century the availability of low-cost special papers helped make it possible for these trends to affect the schools, allowing large numbers of students to create a variety of visualizations for themselves with modest effort. This chapter not only surveys these developments but touches on related topics such as the number line and nomography.

We will use the term *graph paper* here as a convenient way to designate any paper featuring a grid of intersecting lines, but in fact this terminology did not become popular until the 1930s. Before then, it was more common to see terms such as *profile paper, cross-section paper, cross-ruled paper, squared paper, plotting paper,* or *coordinate paper,* some of which have continued in use into more recent times.

Graph Paper in Nineteenth-Century Science and Engineering

William H. Brock and Michael H. Price in a 1980 article traced the manner in which graph paper came to be adopted in education, emphasizing the British case. They noted that the use of graphs by astronomers, chemists, and other European scientists can be documented in the late eighteenth and early nineteenth centuries but that such use was rare until the 1820s and 1830s.[1] They further observed that in Britain the cost of "squared paper" to facilitate such graphs was likely an inhibiting factor until the last quarter of the nineteenth century, after which the cost dropped rapidly.

Variants of graph paper were in use in the United States by the mid-nineteenth century, with a substantial impetus coming from surveying and civil engineering associated with the construction of roads, canals, and, above all, railroads. The country's first great railroad boom commenced in the late 1840s and lasted until the early 1870s. At the beginning of this period special paper with grids of intersecting lines was still novel enough that those promoting it felt compelled to describe its creation and use with some care. William Smyth, a professor of mathematics at Bowdoin College in Maine, published an applied trigonometry textbook in 1852 that included instruction in surveying for railroads. In particular Smyth described how to draw a "profile" of a railroad route, meaning a visual depiction of the vertical change of the route in relation to its horizontal progress, given a table of surveyed values at a series of "stations": "The differences of level are usually small in comparison with the distances between the stations. On this account it is usual to plot the former on a much larger scale than the latter. The proportion usually employed is 1 to 20 . . . To facilitate the plotting of levels, ruled paper is prepared, with parallel lines at a convenient distance from each other to represent a foot. These are intersected by perpendicular lines at a distance from each other equal to five times that of the former. On the scale employed for distances between the stations, each of these spaces will therefore be 100 feet."[2]

Smyth's wording suggests that the kind of paper he was recommending was not at that time universally available on demand but would sometimes have to be specially "prepared" by the individual user. An 1857 compendium "for the mechanic, architect, engineer, and surveyor" employed similar wording in discussing topographical drawings for railroad surveys: "profile paper is prepared, on which are printed horizontal and vertical lines." As in Smyth's book, the vertical lines were to be substantially farther apart than the horizontal. If, however, the prospective surveyor was engaged in measuring heights at right angles to the route—that is, "the plotting of sections across the line"—then "equal vertical and horizontal scales are adopted; these plots are mostly to determine the position or slope, or to assist in calculating the excavation. To facilitate these, cross section paper is prepared, ruled with vertical and horizontal lines, forming squares of 1⁄10 of an inch each."[3] Meanwhile, a handbook from 1854 noted more casually that designers of machinery "frequently employ cross-ruled paper, with horizontal and vertical lines equally spaced."[4] American astronomers were likely using such "squared paper" by this time as well. Truman Safford, presenting astronomical data in 1863, noted his use of squared paper with a minimum of elaboration.[5]

By the 1870s profile paper and cross-section paper were available as separate commercial products.[6] They were becoming a commonplace of engineering practice, as evidenced by the following extracts from treatises of 1870, 1872, and 1875, respectively:

> In practice, engraved profile-paper is generally used, which is ruled in squares or rectangles, to which any arbitrary values may be assigned.[7]
>
> But any preferred scale may be employed, or the cross-section paper in common use among engineers—which carries its own scale—and which will be found convenient in many respects.[8]
>
> The resemblance of Fig. 2 to ordinary cross-section paper is evident. In practice the latter is admirably adapted to this purpose and is always used for it [Figure 12.1].[9]

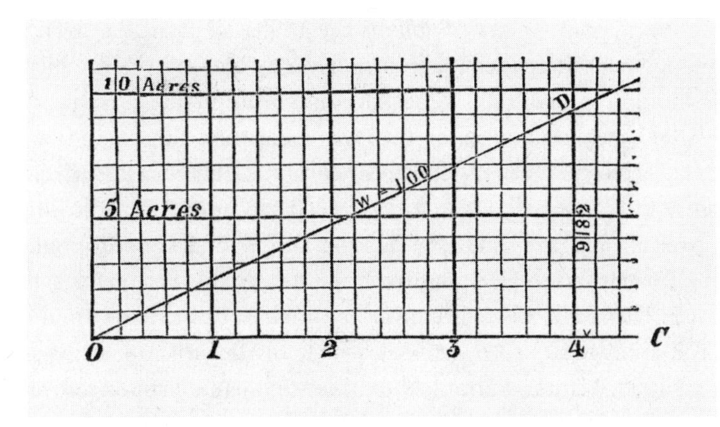

Figure 12.1. Diagram for a railroad survey, 1875. From Arthur Wellington, *Methods for the Computation from Diagrams of Preliminary and Final Estimates of Railway Earthwork* (New York: D. Appleton & Co., 1875), 3.

Indeed, the following passage from Mark Twain and Charles Dudley Warner's 1873 novel *The Gilded Age* indicates that the use of profile paper by engineers was diffusing into the consciousness of a wider public: "Harry would take off his coat, remove his cravat, roll up his shirt-sleeves, give his curly hair the right touch before the glass, get out his book on engineering, his boxes of instruments, his drawing-paper, his profile paper, open the book of logarithms, mix his India ink, sharpen his pencils, light a cigar, and sit down at the table to 'lay out a line,' with the most grave notion that he was mastering the details of engineering."[10]

Profile paper generally provided a grid with horizontal lines more closely spaced than the vertical lines, while cross-section paper often featured a grid of squares—hence, it was often equated with "squared paper." Some preferred *cross-section* as a generic designator for the paper, presumably because, apart from the original meaning in civil engineering, *cross-section* was seen as descriptive of any grid of crossing lines, however spaced or oriented. As early as 1867, James Kirkwood, the chief engineer of the Brooklyn waterworks, described his use of the paper: "a profile was then made, on cross-section sheets."[11] Similarly, the reference from 1875 to "ordinary cross-section paper" seems to be to squared paper, with the term *profile paper* reserved for a species of cross-section paper in which the grid was not composed of squares.

More exotic grids were available by the 1890s, in particular logarithmic and semilogarithmic papers: the former designated paper in which both vertical and horizontal lines were spaced according to a logarithmic scale; in the latter case only one set of lines was logarithmically spaced, with the other set spaced equally in the conventional way. George Sarton, in response to a query to his journal *Isis* in 1939, traced the initial use of logarithmic scales to Léon Lalanne in France in 1843 and W. S. Jevons in England in 1863.[12] The earliest American use documented by Sarton was that of William F. Durand, a Cornell University professor of marine engineering and naval architecture who wrote an article on "logarithmic cross-section paper" in 1893.[13] Durand arranged to have such paper printed for him by a firm in Buffalo, and in 1897 Keuffel & Esser (K&E), a New York City company already prominent in producing drawing instruments and other engineering tools, began to produce "Durand's logarithmic paper."[14]

As in England, the cost of the papers being discussed here appears to have declined in the late nineteenth century, although the effect was not especially dramatic until the paper began being marketed by firms specializing in school supplies in the early twentieth century. In 1850 the state of North Carolina paid $6 for twenty-five sheets of profile paper, and in 1866 the federal government paid $3 for a half-quire (twenty-four sheets) of cross-section paper.[15] This works out to 24 or 25 cents a sheet. By 1876 it was possible to buy "ruled" cross-section paper for as little as 5 cents a sheet, although a more accurate version "printed from steel or copper plates" still cost 20 cents per sheet.[16] Essentially identical prices continued to obtain in the 1880s.[17] In 1889 Davis Dewey, a college instructor of economic history describing his assignment of student graphing exercises, commented that "lithographed cross-section paper is too expensive for general purposes" and explained how he had created his own ruled paper to meet his needs.[18] The cost issue seems to have declined in intensity by 1897, when E. L. Nichols, professor of physics at Cornell, recommended without qualm "a supply of cross-section paper" for any teacher of high school physics.[19] In 1906 K&E was selling Durand's logarithmic cross-section paper for $1.75 a dozen, or 14.5 cents per sheet, and "co-ordinate paper" for between 4 and 7 cents a sheet, depending on the size. The company claimed high accuracy for its products while no longer making any distinction between lithography and ruling.[20] In 1909 the Atlas School Supply Company of Chicago, likely providing a less precise product for a less demanding clientele, was selling "Academic Cross Section Paper" for seven-tenths of a cent per sheet

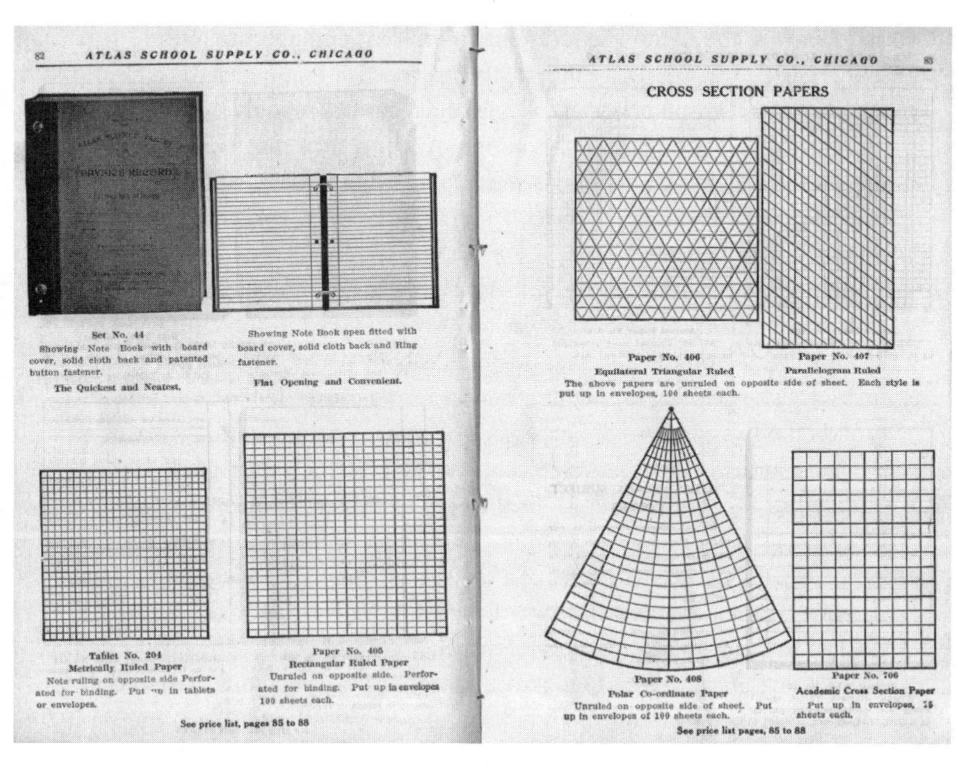

82 ATLAS SCHOOL SUPPLY CO., CHICAGO

Set No. 44

Showing Note Book with board cover, solid cloth back and patented button fastener.

The Quickest and Neatest.

Showing Note Book open fitted with board cover, solid cloth back and Ring fastener.

Flat Opening and Convenient.

Tablet No. 204

Metrically Ruled Paper

Note ruling on opposite side Perforated for binding. Put up in tablets or envelopes.

Paper No. 405

Rectangular Ruled Paper

Unruled on opposite side. Perforated for binding. Put up in envelopes 100 sheets each.

See price list, pages 85 to 88

ATLAS SCHOOL SUPPLY CO., CHICAGO 83

CROSS SECTION PAPERS

Paper No. 406

Equilateral Triangular Ruled

Paper No. 407

Parallelogram Ruled

The above papers are unruled on opposite side of sheet. Each style is put up in envelopes, 100 sheets each.

Paper No. 408

Polar Co-ordinate Paper

Unruled on opposite side of sheet. Put up in envelopes of 100 sheets each.

Paper No. 706

Academic Cross Section Paper

Put up in envelopes, 75 sheets each.

See price list pages, 85 to 88

Figure 12.2. Commercially available papers, 1909. From Atlas School Supply Co., *Catalog no. 22* (Chicago: Atlas, 1909), 82–83.

and "metrically ruled paper" for less than five-tenths of a cent (Figure 12.2). Atlas also sold equilateral triangle ruled paper, parallelogram ruled paper, and polar coordinate paper in the same price range.[21] This was the period when graph paper first became prominent in secondary schools.

Combating Compartmentalized Education with Visualization

Brock and Price emphasized that the utility of graph paper in science and engineering practice in Britain would not have sufficed to create widespread use in schools without concurrent changes in "educational philosophies."[22] This was also the case in the United States. The requisite changes came to the primary and secondary schools after precedents established both above and below in the nation's educational hierarchy: engineering training and kindergartens. As early as the 1850s, prospective engineers were clearly learning about graph paper in colleges and technical schools. No doubt others learned through on-the-job training. At almost the same time, some of the youngest American students were being introduced to simple versions of graph paper and related objects in kindergartens, with a great emphasis on tactile experience.

As noted in chapter 9, the kindergarten was the brainchild of German educator Friedrich Froebel and was actively promoted in the United States by Elizabeth Palmer Peabody of Boston. Peabody personally established the first English-speaking kindergarten in 1860 and proceeded to write and lecture on the subject. Private kindergartens soon appeared in many places in the East and the Middle West, and the first public kindergartens were founded in St. Louis in the 1870s.[23] In 1877 Peabody published a *Guide to the Kindergarten,* in which she wrote: "One essential furniture of the Kindergarten is paper ruled in squares of a sixteenth of an inch, which can be done wherever paper is ruled. Every child should have a piece of this paper, pricked in the crossings of the squares, and be taught to use the needle and colored thread, so as first to make parallel lines, then diagonals, then right angles, then squares; and then other more complex but still symmetrical figures."[24] A related object was the kindergarten table: "The tables for the children to sit at should be low; and it is a good plan to have them painted in squares of an inch; chequered, or ruled by lines, so that they may be able always to set their blocks down with perfect accuracy."[25]

It would take another quarter-century before corresponding notions would gain much foothold in American secondary schools. Part of the resistance was the standard nineteenth-century practice in these schools of making strict demarcations among mathematical subjects. Mathematician E. H. Moore decried the tradition in 1902, referring to engineers, significantly, in doing so: "Engineers tell us that in the schools algebra is taught in one water-tight compartment, geometry in another, and physics in another, and that the student learns to appreciate (if ever) only very late the absolutely close connection between these different subjects."[26] One consequence was that pictures and other visual aids—considered the accoutrements of geometry, not algebra—were rarely found in algebra textbooks. This was true of graph paper as it was of the cube root block.

An even more basic instance of this phenomenon was the number line, in some ways a prerequisite tool for appreciating graph paper. Until the late nineteenth century it was uncommon in an algebra textbook to encounter the now conventional depiction of positive and negative numbers arrayed on a horizontal line, with zero in the middle, the positive integers marked at regular intervals to the right, and the negative integers marked at regular intervals to the left.[27] One 1881 pioneer in using the number line for algebra was G. A. Wentworth.[28] In the same year Simon Newcomb was a self-conscious proponent, confirming the previous rarity of the number line's use. After introducing the number line, he advised in a footnote, "The student should copy this scale of numbers, and have it before him in studying the present chapter."[29] Newcomb went on to explain the elementary operations of addition, subtraction, multiplication, and division in terms of the number line and continued to make occasional appeals to it throughout the text.[30] Illustrating the previous prejudice against such visualization tools, Newcomb felt it necessary to rebut those who claimed it was mathematically naive to use geometry to assist algebraic understanding, explaining that precisely such geometric aids had been crucial to the great advances made by mathematicians in mastering "the complicated relations of imaginary quantities."[31] He was referring here to the use of the square root of negative one and other quantities that could not be identified with any point on the "real" number line. Carl Friedrich Gauss and others in the early nineteenth century had fruitfully introduced the notion of placing these so-called imaginary quantities on another number line, perpendicular to the first, with each point in the resulting plane representing an appropriate sum of a real and an imaginary number.[32]

Newcomb, a noted astronomer and mathematician, would later head the mathematics subcommittee of the important "Committee of Ten" of 1893.[33] He and his subcommittee placed less emphasis than their predecessors on the alleged "mental discipline" conferred by studying mathematics. They were also advocates of "objective" teaching methods derived from Froebel and his teacher Johann Pestalozzi. Such pedagogic views may have helped improve the prospects for graph paper in the schools, but the report of Newcomb's committee made no mention of this tool, although at least two of the members, Newcomb himself and Truman Safford, would surely, as practicing astronomers, have been acquainted with its utility.

It was not until a younger generation of mathematical educators rose to prominence that the conditions truly became ripe for widespread use of graph paper in the schools. Educators associated with the University of Chicago took the lead. One of the first was George W. Myers, who had done graduate work in mathematical astronomy there in the mid-1890s and had then gone to Germany for a Ph.D. degree from Munich in 1898.[34] On his return he became a professor of mathematics at the Chicago Institute, a teacher training institution, and stayed on when the institute was merged with the university to create the School of Education in 1901.[35] Writing on "Algebra in the High School" in a 1900 Chicago Institute publication, Myers emphasized the use of visualization tools, including both the number line and "cross-ruled paper."[36]

Even more significant than Myers was the chairman of the University of Chicago mathematics department, the aforementioned E. H. Moore.[37] Moore pressed explicitly for linking mathematics education with science and engineering, laying out his program for what he called a "laboratory method" of instruction in a December 1902 address as retiring president of the American Mathematical Society (AMS).[38] In this address he cited English engineering educator John Perry as a source of inspiration and in particular noted Perry's championing of "the many uses of squared paper."[39]

Perry had been stirring up controversy in England for some years, but it was only in early 1902 that one of his polemical statements on mathematics education was published on the other side of the Atlantic.[40] His varied career in industry and technical education led him to advocate a new engineering education that would synthesize the best elements of academic laboratories and industrial workshops. He found the role of mathematics in the education of engineers to be especially problematic. In particular, by the late 1890s Perry had injected himself forcefully into an already vigorous debate in Britain regarding the dominance of Euclid in mathematical training.[41]

Perry felt he had achieved considerable success teaching mathematics to engineers, and he concluded that his methods should be made widely available to all students of mathematics.[42] He especially insisted that mathematics should be taught as an inductive science; for instance, propositions from Euclid should be tested by careful measurements using "squared paper," a device Perry proceeded to promote vigorously.[43]

Perry achieved his biggest splash with a symposium he organized on "Teaching of Mathematics" in September 1901 at the meeting of the British Association in Glasgow. This symposium, held before a joint session of the Mathematics and Physics Section and the Education Section, began with his own forceful exposition of his views and was then followed by a round of comments and discussion, all later incorporated into a book. The result in

Britain was the long-sought victory of the anti-Euclid forces in the secondary schools, with some corresponding modifications at the university level as well.[44]

In addition to the individuals actually present in Glasgow, Perry sought responses to his views from other educators interested in mathematics, not confining himself to Britain. He sent a copy of his Glasgow address to the American David Eugene Smith, then recently appointed to a position at Teachers College of Columbia University.[45] Smith, who had been quickly building a reputation in the world of mathematics education through his textbooks and his publications on the history of mathematics, sent Perry his response to be included with the published account of the Glasgow meeting.[46] It is likely that it was via Smith that Perry's Glasgow address came to be printed in February 1902 in the *Educational Review,* edited by the president of Columbia, Nicholas Murray Butler. Later in the year another address by Perry was published in *Science,*[47] and then the expanded version of the Glasgow paper appeared as a book, *Discussion on the Teaching of Mathematics.* Moore cited both recent Perry writings, as well as others, in his December 1902 address.[48] Other American educators besides Moore also became intrigued by Perry at this time and began to speak of a "Perry movement."[49]

Promoting Visual Aids in Mathematics Education

In his 1902 address Moore endorsed Perry's specific proposals to emphasize "squared paper" and "graphical methods" as well as Perry's general philosophy that by building upon practice with "experiment, illustration, measurement," the student could be made "*familiar*" with abstract ideas and "thoroughly *interested*" in their use.[50] Moore admitted that he was calling for a "diminution of emphasis on the systematic and formal sides of the instruction in mathematics" and that "many mathematicians" would fear "irreparable injury to the interests of mathematics." But Moore stood firm, maintaining that the Perry approach would ultimately result in more students developing the all too rare "feeling that mathematics is indeed itself a fundamental reality of the domain of thought, and not merely a matter of symbols and arbitrary rules and conventions."[51]

For the primary schools Moore recommended that mathematics should always "be directly connected with matters of a thoroughly concrete character." He encouraged making models, drawing graphs of magnitudes associated with natural phenomena, and deducing conclusions about the phenomena from the behavior of the graphs. He alluded to the function concept and connected it to cross-section paper and to the kindergarten tables mentioned earlier: "The Geometry must be closely connected with the numerical and literal arithmetic. The cross-grooved tables of the kindergarten furnish an especially important type of connection, viz., a conventional graphical depiction of any phenomenon in which one magnitude depends upon another. These tables and the similar cross-section blackboards and paper must enter largely into all the mathematics of the grades."[52] By the time of Moore's 1902 address these tables were being specially built and marketed for the kindergarten market.[53]

For the secondary schools Moore detailed his "laboratory method," which involved combining mathematics and physics instruction. Phenomena were to be described "both graphically and in terms of number and measure." He again recommended "cross-section paper." He boldly claimed that the method he was proposing was the best one "for students in general, and for students expecting to specialize in pure mathematics, in pure physics, in mathematical physics or astronomy, or in any branch of engineering."[54]

After 1902 Moore promoted his pedagogic ideas vigorously for several years, both in Chicago and nationally, and gained a measure of support. In 1903 a committee of the AMS, including Moore's University of Chicago colleague J. W. A. Young, published a report on college entrance requirements in mathematics. Among its recommendations the committee endorsed the use of graphs in algebra instruction, in line with Moore's program.[55] In 1906 Moore endeavored to give his ideas greater specificity in a talk and subsequent paper titled "The Cross-Section Paper as a Mathematical Instrument."[56] In a footnote on the first page of this article Moore expressed his awareness of the various styles of graph paper commercially available, explicitly citing a specific company: "In the few diagrams to be given here only the square-ruled paper appears. I advise the use of the various styles of ruling—into squares, into rectangles, into parallelograms, into triangles, and with concentric circles and diverging radii—obtainable, for instance, from the Atlas School Supply Co., Chicago."[57]

Moore declared that the great virtue of the cross-section paper was that it brought together "the three phases or dialects of pure mathematics—*number, form, formula.*" In addition, it led directly to "the concept of *functionality*—a concept which since the seventeenth century has dominated advanced mathematics and the sciences; a concept which in the twentieth century, according to the auspices, will play a fundamental role in the reor-

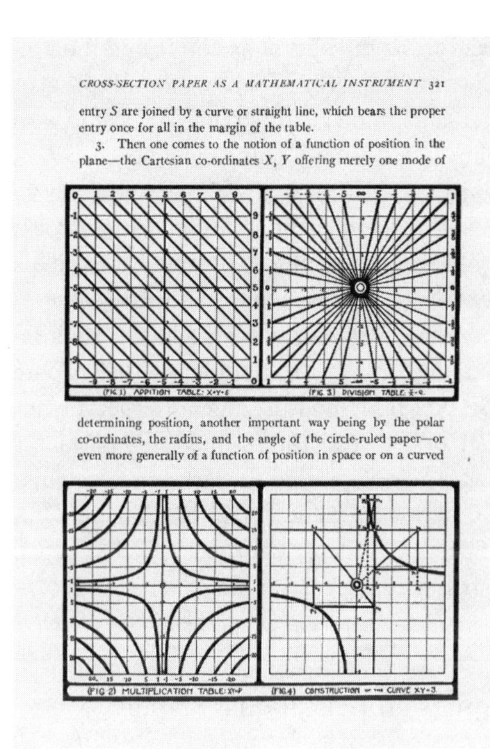

CROSS-SECTION PAPER AS A MATHEMATICAL INSTRUMENT 321

entry S are joined by a curve or straight line, which bears the proper entry once for all in the margin of the table.

3. Then one comes to the notion of a function of position in the plane—the Cartesian co-ordinates X, Y offering merely one mode of

determining position, another important way being by the polar co-ordinates, the radius, and the angle of the circle-ruled paper—or even more generally of a function of position in space or on a curved

Figure 12.3. Cross-section paper, 1906. From Eliakim Hastings Moore, "The Cross-Section Paper as a Mathematical Instrument," *School Review* 14 (May 1906): 321.

ganization of elementary mathematical education."[58] Moore used a striking chemical metaphor to emphasize the effectiveness of the cross-section paper in correlating algebra and geometry: "By maximizing the function of the cross-section paper we secure, to speak only of pure mathematics, intense reaction between geometry and algebra . . . releasing, as it were, abundant stores of sub-atomic energy."[59] There followed the heart of the paper, which consisted of detailed instructions for constructing graphs of certain algebraic equations and for using these graphs to facilitate computations (Figure 12.3).[60]

In his 1906 paper Moore went beyond Perry by introducing methods derived from French mathematician Philbert Maurice d'Ocagne, who in 1891 had systematized the field of graphical computation and dubbed it "nomographie" ("nomography" in English). The essential scheme was to read off answers to selected calculations by strategically laying one scale (often a straightedge) on other scales (possibly curved) appropriately constructed and oriented in a plane.[61] Although Moore's application of nomography was relatively elementary, the subject can be thought of as employing the ultimate generalization of the idea behind "squared paper," with multiple axes of arbitrary orientation, no longer constrained to be straight lines, and with scales derived, if desired, from any strictly monotonic function.[62]

Moore and others at the University of Chicago worked vigorously to spread the message. Directly following Moore's piece in the *School Review,* there appeared a commentary by Moore's student N. J. Lennes, based on his experiences as a high school teacher in Chicago, fully endorsing Moore's emphasis on graphs.[63] A few months later George Myers added further favorable remarks in the same periodical.[64] For the summer of 1907 Moore announced a course entitled "Graphical Methods in Algebra Especially for Teachers."[65] J. W. A. Young's book of that year, *The Teaching of Mathematics in the Elementary and the Secondary School,* praised the virtues of "squared paper," citing Moore.[66] A 1908 description of the curriculum for the University of Chicago Elementary School repeatedly cited cross-section paper, noting in particular that in the sixth grade "cross-section paper is used constantly."[67]

Graph Paper Becomes an Educational Staple

Despite Moore's energy, organizational skill, and high place in the American academic world, his larger program for totally remaking the American mathematics curriculum was not realized.[68] Nevertheless, his narrower goal of popularizing graphs and graph paper was successful; they soon became staples of secondary education and beyond. The years 1910–30 were a transition period. In 1911, for example, we find a university professor complaining, "In my own experience as an instructor of freshmen I have found but little evidence indicating previous instruction in graphics."[69] A short account in 1912 found no graphical treatment in American algebra texts before 1902 and attributed the topic's recent increase in popularity to the influence of John Perry and the 1903 AMS committee on college entrance requirements.[70] In the same year the college text *New Analytic Geometry* by Percey Smith and Arthur Sullivan Gale carefully noted that "the work of plotting points in a rectangular system is much simplified by the use of *coördinate* or *plotting paper,* constructed by ruling off the plane into equal squares, the sides being parallel to the axes."[71] An article of 1915 explicitly employed the phrase *graph paper,* citing the usefulness of this tool for instructing college freshmen in the application of mathematics to physics:

"This work is concluded by using ordinary graph paper to discover physical laws of the linear type, and logarithmic and semi-logarithmic paper to discover some other laws."[72] A 1917 survey of "Recent Advances in the Teaching of Mathematics" especially noted "the introduction into elementary algebra of the subject of graphs, including the plotting of equations, the graphic solution of equations, and graphic analysis."[73] Book reviews, survey articles, and textbook advertisements of the period frequently highlighted the role of graphs and graph paper. An advertisement in 1919 proclaimed that "diagrams appear consistently on cross-section paper; graphical methods are stressed, as well as graphical checks on computation."[74] By 1928 a university chemistry professor had more positive observations to make than the 1911 professor regarding the general level of student familiarity with graph paper, although not without reservations: "Most students can use graph paper satisfactorily but there are always a few who insist on starting the scales at zero and wasting most of the paper. If the volume of a gas is to be plotted against temperatures between 273 and 293, there are some who will start with zero rather than 270 or 273."[75] A 1937 historical survey of algebra textbooks found that the use of graphs in these books had been miniscule in the 1890s but had expanded steadily in the first three decades of the twentieth century.[76]

The 1930s was the decade when the term *graph paper* finally began to eclipse other terminology. In the K&E catalog for 1928 one finds "standard profile and cross-section papers and cloths" as well as "loose leaf co-ordinate paper."[77] In the corresponding catalog for 1937 one still finds these items, but the designation for the co-ordinate paper has become "K&E graph sheets."[78] A competing firm, Eugene Dietzgen Company, went through a similar evolution between 1931 and 1938.[79] In 1941 K&E published a pamphlet on its "graph sheets" in which it treated cross-section paper as a subcategory and never used the term *profile paper* at all.[80]

Both K&E and Dietzgen were by this time competing for the school market as well as selling to engineers, but how much commercial paper was actually used in schools is difficult to know. At least a few teachers took the opportunity to explore the wide universe of commercially available paper. A 1942 account by Illinois high school teacher Norma Sleight described her attempt to expand students' understanding of graphs: "So this year a sample of nearly every kind of graph paper produced by a well known company was purchased, and I was amazed to learn of the varieties and uses made of this simple discovery by Rene Descartes. These papers were thumb-tacked

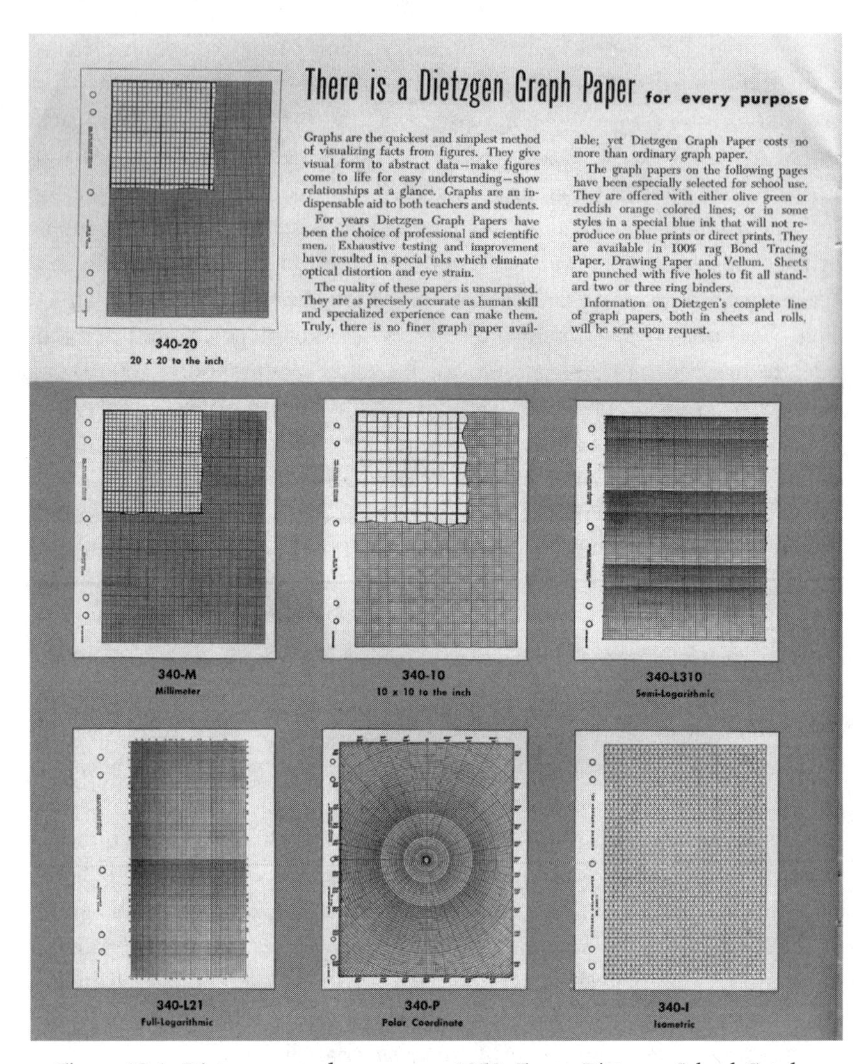

Figure 12.4. Dietzgen graph paper, ca. 1953. From *Dietzgen School Catalog* (Chicago: Dietzgen Co., ca. 1953), 32.

on the bulletin boards in the classroom; the wall was literally papered, because there were 59 varieties."[81] In addition to various rectilinear papers Sleight reported sixteen types of semilogarithmic papers, five logarithmic papers, polar coordinate paper, and triangular coordinate paper. (Figure 12.4 illustrates some of these varieties.)

Certainly, the picture-free algebra books of the nineteenth century had by the mid-twentieth century been swept away by a new generation of textbooks that effectively promoted graph paper through sections on graphing equations and other mathematical objects, with pictures of such graphs on grids of perpendicular lines and often with explicit mention of graph paper in one of its various designations. A sample of books from 1945 to 1965 provides evidence.[82] Judging from these books, it appears that most graph paper for school use was the simplest kind, a grid made up of identical squares. It is, of course, likely that the most popular graph paper of all was the crude form improvised by students from blank or composition paper by hand drawing and labeling perpendicular axes, perhaps without even the benefit of a straightedge. The philosophy expressed in a textbook from 1910 was doubtless widely prevalent at the time and since: "Though not an absolute necessity, cross-ruled paper is a great convenience in all graphical work. Excellent results, however, can be obtained with ordinary paper and a rule marked in inches and fractions of an inch for measuring distances. Hence the graphical work which follows should not be omitted even though it is found inconvenient to obtain cross-ruled paper."[83]

For the most part through the twentieth century graphs and graph paper appear to have played a noteworthy, though unspectacular, part in mathematics education. From the evidence of textbooks high school graphing exercises have usually focused on such basic objects as straight lines, parabolas, and circles, with occasional forays into higher-order polynomials and logarithmic, exponential, and trigonometric functions. The computational possibilities of graph paper have been largely confined to solving two simultaneous equations of at most second degree.

In particular E. H. Moore's hope of 1906 to make nomographic techniques a significant feature of the school curriculum went largely unfulfilled, unless we include the slide rule, which can be thought of as a nomographic device. Nomograms (also called alignment charts) were occasionally suggested as a school topic in the 1920s. MIT mathematics professor Joseph Lipka wrote on "Alignment Charts" in the *Mathematics Teacher* in 1921, noting the French origins of the field and asserting that the topic had remained largely a domain of French engineers until "the world war brought our ordnance engineers in contact with the French engineers, and the former have learned how the latter apply the principles underlying the alignment chart to the graphical solution of some of their problems in ballistics and allied subjects."[84] The purpose of Lipka's article was to introduce elementary aspects of the subject by connecting them with "ideas which the average secondary school student easily grasps." A similar purpose was advanced in a 1924 article by J. A. Roberts, a Connecticut secondary school teacher who noted the availability of textbooks treating the topic.[85] The cause was taken up again in 1945, when the National Council of Teachers of Mathematics (NCTM) included a chapter on nomography in its 1945 yearbook on *Multisensory Aids in the Teaching of Mathematics.* (Figure 12.5, from this yearbook chapter, shows an alignment chart for multiplying two numbers.) But the author, MIT engineer Douglas P. Adams, concluded with a retreat from egalitarianism: "It should be emphasized that the material in this article is not intended to be a basis for teaching the subject of alignment charts to the average high school student. It should, however, prove an adequate guide to those able, ambitious, and aggressive young people who enjoy wrestling with new problems once they know that a fair fight promises a liberal reward."[86]

Nomography in the schools remained an enrichment topic for a few rather than a standard topic for many. Its fate is confirmed by a 1999 article by Thomas Hankins, in which he describes the use of nomograms by Harvard physiologist L. J. Henderson in the 1920s, thus establishing that nomography did in fact spread beyond engineering. But Hankins, a distinguished historian of science who received a doctorate in 1964 and who has written often on mathematical topics, confessed frankly that until 1996 he had never heard of nomograms.[87]

Several of the curriculum reform programs of the 1950s and 1960s, known collectively as the "New Math," placed emphasis on graphing.[88] Mary Dolciani, a member of the largest New Math program, the School Mathematics Study Group (SMSG), wrote a highly popular high school text that placed the number line on the first page and featured transparent overlays bound into the book to encourage the graphical imagination of the students.[89] Perhaps the most ambitious effort in this direction was that of University of Minnesota mathematician Paul C. Rosenbloom, who around 1960 proposed making graph paper and graphical computation central to mathematics and science education, beginning at an early age. As he summarized it, "By presenting very early the concrete model of a scale on a line, we would make it possible for children to learn the operations on numbers and the properties of numbers in terms of observations on the real world."[90] Rosenbloom's resulting Minnesota School Mathematics and Science Teaching project, known as MINNEMAST, was taught to students in Minnesota

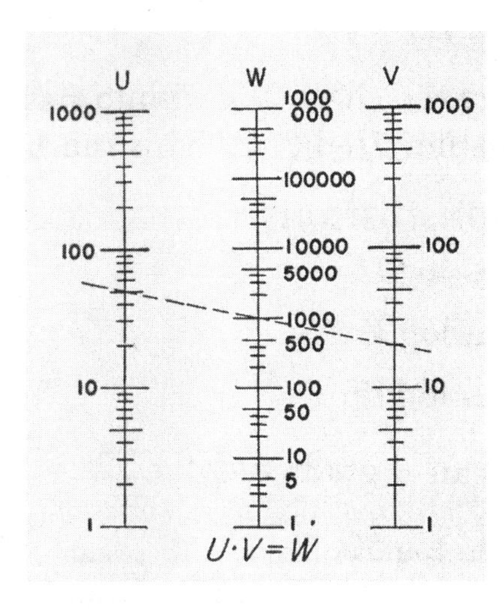

Figure 12.5. Nomograph for multiplying two numbers, 1945. From Douglas P. Adams, "The Preparation and Use of Nomographic Charts in High School Mathematics," in *Multisensory Aids in the Teaching of Mathematics,* ed. W. D. Reeve, National Council of Teachers of Mathematics 18th Yearbook (New York: Teachers College Press, 1945), 175. Smithsonian Negative no. 2006-21927.

and other locations across the country until funding ran out in the early 1970s.

More recent reform initiatives in mathematics education, as exemplified by the "standards" documents promulgated by the NCTM in 1989 and 2000, continue to highlight graphing, with a few new twists. Graph paper, now sometimes referred to as "grid paper," especially for the early grades, continues to be valued as a tool.[91] More encouragement than in the past is given to applications of graphing in data analysis and statistics.[92] But graphing is introduced as merely one species of "representation," and the role of graph paper is now usurped in some cases by the more dynamic capabilities of graphing calculators.[93]

CHAPTER THIRTEEN

Geometric Models
Ocular Demonstrations

> I take this opportunity to recommend the geometrical set as worthy of the first attention either for families, schools or Lyceums. These are probably the first articles to be procured, whatever might be the second.

> The graphical depiction of the phenomena serves powerfully to illustrate the relations of number and measure. This is the fundamental scientific point of view. Here under the term graphical depiction I include representation by models.

From at least the eighteenth century three-dimensional models have been used to convey scientific and technical information. These objects range from patent models of inventions to architectural renderings to reconstructions of sites of archaeological interest to the dioramas of natural history museums.[1] Some of the most striking among them represent mathematical surfaces. Such models have been made and sold for teaching students from the most elementary to the most advanced levels. Models of simple forms were available in the United States from at least the 1830s and are still widely used. Representations of more complex surfaces were first imported from Europe in the mid-nineteenth century and were sold most widely between 1890 and 1930.

Among mathematicians and mathematics teachers, the design and study of models has been of greater interest than the preparation of any other teaching tool, with the exception of textbooks and, in relatively recent years, computer software. The challenge of computing patterns and constructing models, their visual appeal, and their rich and diverse mathematical content have all contributed to this endeavor. Creating and interpreting mathematical models places a premium on mathematical knowledge not found, say, in designing protractors or making demonstration measuring apparatus.

Both the mathematical content and the intriguing appearance of models have led several scholars to explore aspects of their meaning and history. Mathematicians have offered accounts of the significance of forms created in the nineteenth and early twentieth centuries for contemporary mathematics and have also traced the intellectual lineage of types of models such as polyhedra and dissections. Advocates of recreational mathematics have prepared detailed instructions for making models. Historians of mathematics have used models to study the diffusion of mathematical ideas from France and later from Germany over the course of the nineteenth century. They also have explored the interconnections between mathematical model making and such other areas as modern art.[2]

A few mathematical models—the cube root block, simple geometrical forms, and a cone sliced to reveal conic sections—have been widely present in many mathematics classrooms for decades. Literally hundreds of other forms of models were made but had a much more limited distribution. The work of two twentieth-century American model makers, Richard P. Baker, a faculty member at the University of Iowa, and Albert Harry Wheeler, a mathematics teacher in Worcester, Massachusetts, suggests the richness of this tradition, its dependence on a few scholars, and the ambiguous status of these model makers in the American mathematical community.

Early Geometric Models in the United States

One of the first to advocate the widespread use of geometric models in elementary teaching in the United States was Josiah Holbrook, that maker of numeral frames we encountered in chapter 6. Holbrook designed and marketed a set of simple geometrical forms that sold widely from the 1830s (Figure 13.1). According to him, for any school or family investing in educational apparatus, geometric models were "probably the first articles to be procured, whatever may be the second."[3] Models helped a child to learn the names of different shapes. At the same time, blocks marked with a grid of lines on the surface made it easier to learn to calculate surface areas and volumes. For those studying art and engineering drawing, models of prisms, cylinders, cones, and other shapes were useful. Models of this sort, made of wood, paper, and later plastic, would sell through the nineteenth and twentieth centuries.

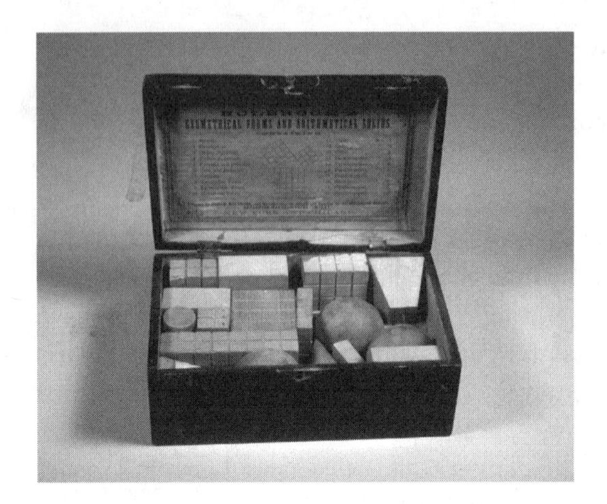

Figure 13.1. Holbrook's Geometrical Forms and Arithmetical Solids, 1859. National Museum of American History collections. Smithsonian Negative no. 2001-7651.

In the mid-nineteenth century a few schools such as Union College in Schenectady, New York, Harvard University, the U. S. Military Academy at West Point, and the Columbia School of Mines in New York City purchased string models in brass frames for teaching descriptive geometry. These models could be adjusted to demonstrate the surfaces generated by the motion of lines. Associated with the ideas of Gaspard Monge, they were designed by the mathematician Théodore Olivier and manufactured by the Parisian scientific instrument making firm of Pixii & Fils and its successor, Fabre de Lagrange. Made from brass and silk with wooden cases, these elegant and expensive objects were available to only a few.[4]

Descriptive geometry also attracted the attention of German scholars and model makers. Ferdinand Engel, a graduate of the Royal School of Arts in his native Magdeburg, taught drawing privately and prepared drawings for textbooks on descriptive geometry and optics. He also made a model of the wave surface in biaxial crystals that won him a prize at the Crystal Palace Exhibition, the World's Fair held in London in 1851. When the Revolution of 1848 failed, Engel, who had liberal sentiments, fled to England and then the United States. In 1855 he published *Axonometrical Projections of the Most Important Geometrical Surfaces: Drawings in Descriptive Geometry.* This volume includes a catalog of thirty-eight models in wood and plaster that are shown in the drawings. Engel's notion of important geometrical surfaces was similar to that of Olivier, but he made solid models, not string ones. He included second-order surfaces such as the ellipsoid, hyperboloid, paraboloid, and cone as well as helicoids, screws, and compounds of cubes. As one would expect, he also made models showing optical surfaces, such as the one that had won him a prize in 1851.[5] Although Engel's book contained testimonials from such well-known scholars as Wolcott Gibbs of New York and G. Lejeune Dirichlet of Berlin, he apparently did not find commercial success.

From the mid-1870s Alexander Brill, his eminent associate Felix Klein, and other German mathematicians designed series of models for university teaching. The structures ranged from common surfaces used in analytic geometry classes to string models of ruled surfaces mathematically similar to those made by Olivier. Brill and his compatriots also modeled other surfaces that had been introduced or reinterpreted relatively recently by mathematicians such as Carl F. Gauss, Charles Dupin, and Ernst Kummer. Alexander Brill arranged to have these models distributed by his brother, the publisher Ludwig Brill.

In 1893 the German government sent a large array of these models to the educational exhibit it sponsored at the Columbian Exposition, the first World's Fair held in Chicago (Figure 13.2). Felix Klein attended the fair,

Figure 13.2. Three Brill models exhibited at the Columbian Exposition, 1893. National Museum of American History collections, gift of Wesleyan University. Smithsonian Negative no. 2006-2047.

demonstrated the models, and then gave a series of lectures at nearby Northwestern University.[6] Several American colleges and universities, including Northwestern and Clark University in Worcester, Massachusetts, acquired collections of models distributed by Ludwig Brill and his successor, Martin Schilling. Baker and Wheeler would have access to these materials and reworked them for schools with more modest budgets.

In the late nineteenth century fundamental changes had also occurred in more elementary geometry teaching. As we have seen, early American geometry textbooks followed British translations of Euclid's *Elements*. From the 1820s English translations of French geometry books, particularly the text of Legendre, had been widely used. In the 1860s, as logically consistent non-Euclidean geometries came to be generally known and as other areas of knowledge expanded, many questioned whether Euclidean geometry should play such a large role in general education. Even those most devoted to the subject suggested that new textbooks were needed.[7]

This rethinking of geometry teaching tended not to change the order in which topics in solid geometry were discussed. Following books 11–13 of the *Elements,* most authors of solid geometry books wrote about lines and planes in space, polyhedra, the cylinder, the cone, and the sphere. Some of Euclid's theorems, however, were either omitted entirely or were included only as supplementary to the main argument. Textbook authors also felt free to present many of Euclid's results algebraically and to include exercises in both numerical computation and proof.[8]

Authors differed about how best to use visual images in their books. The mathematician, Unitarian minister, and short-time Harvard University president Thomas Hill thought solid geometry should give students opportunities to use geometric imagination. In the section of his text *A Second Book in Geometry* (1863) that was devoted to the topic, he purposely removed the figures from the text proper and encouraged students either not to use illustrations or to draw them themselves. For those without these abilities the diagrams were included on a separate plate at the back of the book.[9] More commonly, drawings were intermixed with text. Authors such as George Wentworth showed figures in outline and also drawn carefully in perspective with appropriate shadows.

A few American educators went further and proposed to replace proofs entirely with models, claiming that their visual impact would leave a more lasting impression than logical demonstrations did. In 1873, for example, Isaac Harrington of Huntington, Connecticut, patented an "improvement in apparatus for teaching mensuration." Harrington's apparatus consisted of a set of wooden forms, hinged or doweled so that they could be transposed into other shapes whose areas were known to students. He proposed "ocular demonstrations" for finding the areas of a rhombus and a regular polygon as well as the volumes of prisms and other solids. Mathematicians Hubert Anson Newton of Yale University and Charles Davies of Columbia commended Harrington's invention, and his "geometrical blocks" were offered for sale in the National School Furniture catalog for 1872.[10]

Harrington's patent did not guarantee him a full share of the market. From the 1880s a set of blocks designed by Albert H. Kennedy, an Oberlin College graduate who became the school superintendent of Rockport, Indiana, was used in some schoolrooms. Kennedy was interested in dissections of curved figures as well as rectilinear ones. He designed a dissected circle that could be (roughly) rearranged to form a parallelogram of the same area, a cylinder that could be considered as made up of triangular prisms, a cone made up of triangular pyramids, and a sphere cut up into pyramids (Figure 13.3). These wooden figures were held together with cloth and leather strips nailed to the outside. Versions of this apparatus were produced and sold into the twentieth century.[11]

At about the same time, William W. Ross, superintendent of schools in Fremont, Ohio, began to sell a set of geometric models for use in teaching

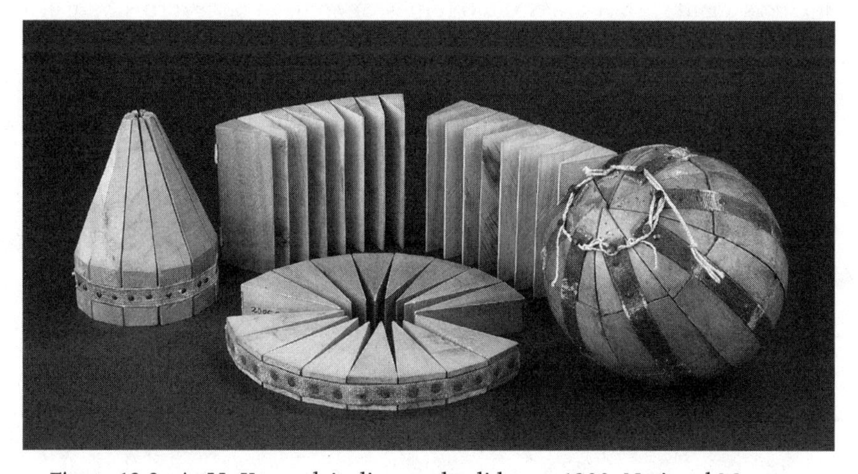

Figure 13.3. A. H. Kennedy's dissected solids, ca. 1900. National Museum of American History collections, gift of Jeremiah P. Farrell. Smithsonian Negative no. 2006-0201.

Figure 13.4. W. W. Ross's model of the Pythagorean theorem, ca. 1890. National Museum of American History collections, gift of Wesleyan University. Smithsonian Negative no. 2001-7654.

students about surface areas and volumes. Several of Ross's forms were dissected to show how a surface or solid might be considered the sum of more familiar parts. A rhombus, for example, could be divided and rearranged to form a rectangle of equal base and height. Similarly, a triangular prism could be divided into three triangular pyramids. Squares constructed on the shorter sides of a right triangle were dissected so that they could be rearranged to form a square equal in area to a square constructed on the hypotenuse (Figure 13.4). Ross argued that mensuration that was taught using objects, or "objectively," would greatly benefit students. The formulas for finding areas and volumes would become "the permanent property of the reason rather than the uncertain possession of the memory."[12] In later years Ross's business was taken over by W. D. Ross, also of Fremont, Ohio. An advertising leaflet from after 1905 indicates that by then W. D. Ross offered apparatus for comparing the relative volume of a cylinder, an inscribed cone, and an inscribed sphere. The device resembles a model patented earlier by A. H. Kennedy. Ross had garnered endorsements from colleges and normal schools throughout the country as well as recommendations from textbook authors William J. Milne, Adelia Roberts Hornbrook, and David Eugene Smith.[13]

Harrington, Kennedy, and the Rosses believed that it was more useful to show pupils the properties of surfaces and solids using models than to offer formal proofs of these properties. While this view was not generally adopted, many of those who preferred more rigorous mathematical thinking also believed that objects had a place in geometry teaching. E. H. Moore addressed the topic in an address he gave in 1902 as outgoing president of the American Mathematical Society (AMS) (see chap. 12). Here Moore advocated the use of "laboratory methods" in mathematics teaching. The well-fitted mathematics laboratory would include in particular geometric models.[14] One of Moore's students, Richard P. Baker, took up the suggestion.

R. P. Baker: "A One Man Shop at Best"

In an obituary tribute Richard P. Baker's colleague at the University of Iowa, Henry L. Rietz, described him as "a versatile man whose knowledge in all fields was far more than superficial."[15] While this may overstate the depth of Baker's learning, it well suggests the breadth of his interests. An Englishman by birth, Baker studied at Clifton College in Bristol and Balliol College in Oxford before taking a law degree at London University in 1887. The following year he moved to the United States, where he was admitted to the bar and practiced law in Texas. He also worked for a time as a chemist. By the spring of 1895 Baker had moved to Illinois, where he enrolled as a graduate student in mathematics at a relatively new institution, the University of Chicago. He took courses from a distinguished array of mathematicians but apparently stalled in his progress toward a degree (he would eventually receive his Ph.D. degree from Chicago in 1911).[16] By 1902 he was teaching a wide variety of high school courses at the Union Academy in Alma, Illinois. Some calculations relating to both his dissertation and his early geometric models were carried out on the back of student essays on topics ranging from mathematics to English literature to the classics.[17]

Baker was soon ready to draw attention to his models. In early 1903 Virgil Snyder, then an assistant professor of mathematics at Cornell University, published an article in the *Bulletin of the American Mathematical Society* entitled "On the Quintic Scroll of Three Double Conics," in which he examined the properties of a ruled surface of fifth degree. His discussion is entirely algebraic.[18] The paper caught Baker's eye, and he made a model of the surface that Snyder had described. After some correspondence, Snyder not only purchased a copy of the model but also published a note in the *Bulletin* indicating that Baker would sell further examples.[19] Marin Liljeblod, secretary to the eminent Swedish mathematician Gösta Mittag-Leffler, wrote in August 1904 to order a copy of the model, and Baker promptly complied.[20]

Encouraged by this and other responses, Baker decided to prepare a full line of models for sale. In the fall of 1904 he moved to Chicago to work on his first catalog, modestly entitled "A List of Mathematical Models" and featuring one hundred models. The first forty were for solid geometry, showing lines, planes, solid angles, polyhedra, cones, cylinders, spheres, and spherical segments. These models were listed in the order in which they were discussed in contemporary American textbooks such as George A. Wentworth's *Solid Geometry*.[21] Remaining models illustrated topics from such diverse areas as arithmetic (a cube root block), projective geometry, spherical geometry (including a slated globe), and analytic geometry (surfaces of second degree). Of course, Baker also listed Snyder's quintic scroll as well as a variety of other ruled surfaces. Also included were linkages, optical surfaces, and models for crystallography.[22]

Snyder had been immediately taken with the "neat and dignified appearance" of the model Baker sent him and, on careful inspection, concluded that Baker worked as proficiently as those who made the thread models published by the world-famous German firm of Schilling, successor to Brill.[23] Baker's models were constructed more simply than their European counterparts. In place of metal frames for string models, he used wood painted black. The bars of linkages were made from oak, not brass. Baker also avoided making models from plaster when possible and encased the plaster models he did make in wood so they would be sturdier. When he used metal, it was painted sheet metal, not brass.

Baker's small catalog was well publicized in mathematical journals of the day. Research mathematicians could read about it in the *Bulletin* of the AMS or in the *Jahresbericht* of the German Mathematical Union.[24] College teachers perusing the *American Mathematical Monthly* for April 1905 learned that Baker had received sufficient orders to devote all his time to model making. High school teachers saw an advertisement for the models for solid geometry in a relatively new American journal, *School Science and Mathematics*.[25]

E. H. Moore recommended Baker to at least two correspondents who had inquired about where they might obtain geometric models.[26] Of greater direct importance to Baker's model sales was Moore's colleague on the Chicago faculty, Jacob W. A. Young, associate professor of the pedagogy of mathematics. In 1907 Young published his influential book *The Teaching of Mathematics in the Elementary and the Secondary School,* new editions and printings of which appeared regularly through at least 1927. In his chapter on geometry Young emphasized the importance of models in teaching solid geometry. He thought students should be encouraged to make models themselves and noted sets of rods and joints that could be purchased for their use. Young also commented: "Of ready-made models for the propositions of solid geometry, I know of none manufactured in this country except those of Baker, who makes a very good set of about forty models for elementary solid geometry as well as a large number for more advanced work."[27] A helpful footnote gave Baker's address. For over twenty years teachers who read this comment would turn to Baker.[28]

Most nineteenth- and early-twentieth-century mathematicians who designed models for sale relied on others to manufacture and distribute their wares. Olivier had Pixii et Fils and then Fabre de Lagrange, Alexander Brill had Ludwig Brill, and Friedrich Schilling had Martin Schilling. At the University of Illinois in the 1920s Arnold Emch of the Department of Mathematics designed a series of fifty-one models and lantern slides. He assured readers that arrangements for duplicating them could be made with private firms.[29] Baker, however, not only designed models but made and distributed them himself.

Teachers who were accustomed to ordering ready-made goods, and who sometimes urgently needed models for immediate classroom use, were surprised and disappointed to discover that Baker had no stock of models at hand.[30] In 1905 alone, he received at least eleven orders for a total of roughly 250 models.[31] Filling these requests took months, and sometimes orders were canceled before they were filled.[32]

After 1905 Baker apparently did not try to advertise his models. With a steady paycheck from the University of Iowa, a backlog of orders, and a thesis to write, he may have found it was unnecessary. Moreover, any business so dependent on the work of one person was easily disrupted. On February 9, 1930, one of Baker's more patient customers, O. W. Albert of the University of Redlands in California, ordered two models, hoping to receive them by May 1. On July 16 Baker wrote to say that the models finally had been shipped. He commented ruefully, "I am sorry that I cannot be more prompt with shipments, but my shop is a one man shop at best."[33]

Perhaps more fundamentally, Baker was more interested in designing models than in selling them. To be sure, he seems to have not only computed designs for the 100 models listed in his 1905 catalog but to have sold one or more examples of at least 93 of them and made patterns for at least 4 others. Baker's first catalog showed models that were actually made. In later years, however, planning moved ahead of production. Of some 410 models Baker computed after the 1905 catalog, only about half actually sold.[34]

Several aspects of these later models merit attention. First, Baker assiduously copied some series from his predecessors, particularly Brill and Schilling. He designed, for example, paper models of surfaces of second degree using circular sections, just as Brill had done in one of the first series of models he issued. Baker also made models of cyclides, emulating several early Brill models. He developed models of the curvature of surfaces and of linkages, as had his predecessors.

Second, Baker made models that reflected the research interests and teaching needs of his contemporaries. We have already mentioned his interchange with Snyder. Later in his career he designed a model of a ruled surface of order eight that had been described in 1900 by one of his University of Chicago professors, Heinrich Maschke.[35] Working with Karl Eugene Guthe, a German-born physicist who came to Iowa the same year he did, Baker designed several thermodynamic surfaces. Such models had been envisioned by the American theoretical physicist and mathematician J. Willard Gibbs and built by James Clerk Maxwell but were not part of the usual offerings of model makers. Baker made copies of some of these models for both Columbia University and the University of Michigan.[36] He also designed and built models of Riemann surfaces, perhaps for courses he taught in complex analysis (Figure 13.5). He made several models for use with the dynamics textbook of the German mathematician Eduard Study.[37] After the statistician and actuary H. L. Rietz came to the University of Iowa, Baker designed a few models of statistical distributions. He even designed a model relating to opalescence that he thought might interest Albert Einstein. He sent an illustration of the model to Einstein, who found the picture was so good that no additional model was needed.[38]

Although Baker found that recent topics in physics and mathematics lent themselves to modeling, mathematicians generally were taking more abstract approaches in both research and teaching. Demand for mathematical models was on the wane, and the new catalog Baker issued in 1931 was

Figure 13.5. Richard P. Baker's model of a Riemann surface, ca. 1930. National Museum of American History collections, gift of Frances E. Baker. Smithsonian Negative no. 2001-8960.

greeted with widespread indifference. The financial crisis of the time and Baker's own poor health may partly account for this lack of interest. At least as important, however, were the changes in mathematics that had occurred over the previous thirty years.

Correspondence in the Baker Collection at the University of Iowa indicates that he sold models to a wide range of colleges and high schools. The largest collection sold in his lifetime apparently went to the University of Delaware, which acquired several hundred dollars worth of his models in the late 1920s and early 1930s. After Baker died, in 1937, his daughter Frances E. Baker offered to sell his remaining stock of models at reduced cost. The University of Arizona acquired at least thirty Baker models in this way.[39] Frances and Gladys Baker also lent over a hundred of their father's models for exhibition at MIT. These models were shown there from the late 1930s until 1956, when the Baker sisters gave them to the Smithsonian Institution. Thus, Baker's work continued to play a part in mathematics education well after his death.

A. H. Wheeler: The Stratification of American Mathematical Activity

Despite a few forays into higher academic realms, Albert Harry Wheeler spent the bulk of his career as a high school mathematics teacher. There is abundant evidence that Wheeler, unlike Baker, brought model building directly into the mathematics classroom. Many models in his collection (there are hundreds preserved at the Smithsonian's National Museum of American History) were signed by the students who made them. These models were made of materials befitting the modest resources of a teacher: paper, cardboard, balsa wood, ordinary thread, and inexpensive plastics. As far as we know, none of them was made for sale.

Wheeler's great love was creating models of polyhedra and instructing others in this art (Figure 13.6).[40] He presented his work in front of a very disparate set of audiences. The different reactions he encountered provide an enlightening portrait of mathematics in the United States in the first half of the twentieth century. In particular, his life gives insight into the strained relations growing within the hierarchy of mathematical educators, from schoolteachers to researchers; it exposes public perceptions of mathematicians and their work; and it displays something of the evolution of fashions in mathematical research, especially the relative decline of classical geometry as an active area of inquiry.

Wheeler spent nearly all his life in the vicinity of Worcester, Massachusetts, graduating from high school there and then proceeding to a bachelor of science degree from Worcester Polytechnic Institute (WPI) in 1894. It was not, however, the burgeoning engineering field that Wheeler entered im-

Figure 13.6. Albert Harry Wheeler with student and models, 1937. National Museum of American History collections, gift of Helen M. Wheeler. Smithsonian Negative no. 94-13511.

mediately after graduation but another rapidly rising field: high school teaching.

Only two years later, in 1896, he interrupted his career as a teacher of high school mathematics to enroll in graduate study at Clark University, another Worcester institution. Clark had opened in 1889, and for a brief time after its founding it was a leading center of pure mathematical graduate training in the United States. The Clark mathematics department was headed by William Story, who had been lured from Johns Hopkins. When the Clark faculty, beset by growing discontent with president G. Stanley Hall, was in turn raided in 1892 by the University of Chicago, Story was one of the few full professors who remained loyal to Hall and Clark. Story continued to head the department, in increasingly reduced circumstances, for the next thirty years. Thus, Wheeler became a student at Clark four years after its brief golden age. Nevertheless, Clark remained a bastion of pure mathematics, with an emphasis on research, and for Wheeler to join this department clearly marks him as one who was allied to these ideals.[41]

Why Wheeler left Clark in 1899 without a degree to return to high school teaching is not clear. Over the next fifteen years he wrote two algebra textbooks, which seem to have been moderately successful,[42] but it was geometry that most interested him, and within geometry he was most captivated by one narrow subfield, the study of polyhedra. He was fascinated by the goal of finding ways for easily constructing models of known polyhedra, especially out of paper, and he began to instruct his students in constructing their own models. Wheeler saw himself as following in the research tradition of Kepler in the seventeenth century and Poinsot in the nineteenth century, manipulating and generalizing the five Platonic solids to produce new polyhedra.[43] One can, for example, "stellate" a convex polyhedron by extending the faces until they meet again, or one can intersect convex polyhedra.

Wheeler's papers contain a notebook entitled "List of Models," which has handwritten entries from at least as late as 1938. Here he wrote down 796 numbers and described models for 678 of them (the remaining entries are blank). Wheeler and his students seem to have made most of the numbered models described, along with a few not described in the notebook but to which numbers were assigned. Wheeler took great pride in his ability to fold many polyhedra without needing glue to hold parts together.[44]

In 1920, at the age of forty-seven, Wheeler's research aspirations prompted his return to graduate study at Clark, where he earned a master's degree in

1921. When Clark decided to end its graduate program in mathematics, Wheeler petitioned to be allowed to continue to work toward a doctorate. He explained that he had been working for ten years on mathematical work, "which has proved more fruitful of results than I had at first hoped."[45] Because there were now no mathematicians at Clark who could serve as his thesis advisor, Wheeler suggested that perhaps Clark could arrange to have a professor from another university serve in this capacity, someone such as Julian Coolidge at Harvard. The Clark authorities did not dismiss this idea out of hand but did request an outline of Wheeler's proposed dissertation.[46] We have found no evidence that Wheeler ever prepared such an outline nor any evidence that he applied to other graduate programs.

Wheeler did begin, however, to look for venues in which he could display his geometric facility outside his high school base. His first major effort in this direction was in 1924 at the International Mathematical Congress held in Toronto, where Wheeler presented a paper and displayed some of his geometric models.[47] Also in 1924 he found part-time employment (supplementing his high school position) as an instructor of geometry at the college level, first at Brown University and then at Wellesley College. The Wellesley course catalog described his course as being on the "theory of polygons and polyhedra, with constant practice in the construction of models."[48]

With Depression era belt-tightening and a falling off of student interest, Wheeler's part-time college teaching did not persist into the 1930s. He did attract a number of speaking engagements during this decade, ranging from the Brown University Mathematics Club to the Maine Mineralogical Society.[49] He also made at least three presentations at meetings of the American Mathematical Society.[50] His models continued to attract attention up to the time of his death. In 1950 he was invited to exhibit part of his collection at the International Mathematical Congress in Cambridge, Massachusetts. He fell seriously ill before the conference, however, and was unable to take up this invitation.[51] He died a few months later.

A number of influences may have contributed to Wheeler's geometric interests, although the precise stimuli are sometimes obscure. At WPI he took synthetic geometry, analytic geometry, descriptive geometry, mineralogy, and free and mechanical drawing.[52] It is further intriguing to note that WPI had a long-standing tradition of producing wooden geometric models in its Washburn Shops, for marketing to art classes.[53] Quite possibly, Wheeler had been required to make drawings from some of these models in his descriptive geometry or drawing classes. At Clark University Wheeler would certainly have had some exposure to William Story's mathematical interests, which were indeed largely geometric, especially the theory of curves and surfaces and non-Euclidean geometry.[54] Another potential source of inspiration at Clark would have been the collection of geometric models owned by the mathematics department. At Story's request, Clark had in 1892 purchased a collection of the Brill models from Germany,[55] which included a set of polyhedral models made of paper, the medium that would become Wheeler's favorite for his own models. But polyhedra made up a very small part of the Brill collection, the bulk of which were plaster models depicting objects of interest in differential geometry and complex function theory, areas in which Wheeler exhibited little later facility. Even the paper polyhedra in the Brill collection, being in fact projections of four-dimensional figures, were more sophisticated than what Wheeler usually attempted.[56]

It is evident from Wheeler's 1924 Toronto paper that at some point after his mathematical education of the 1890s he came under the influence of one specific book, *Vielecke und Vielflache* by Max Brückner, published in 1900.[57] This book was entirely devoted to two- and three-dimensional figures. Using straightforward tools from geometry, trigonometry, and school algebra (tools that Wheeler would continue to exercise in his high school teaching duties), Brückner had described these figures in great detail, precisely computing angles and lengths, drawing pictures, and providing photographs of actual models constructed of paper.[58] Interspersed within Brückner's text were a plethora of historical remarks, in which he took pains to cite priorities not only for authorship of concepts but for construction of particular polyhedra. This historical narrative provided by Brückner clearly was a primary source for the importance that Wheeler later attached to constructional priority as a measure of mathematical achievement.

For most research mathematicians, Wheeler's so-called research was not worthy of notice. The gaps and misapprehensions in his mathematical knowledge would undoubtedly have been enough to limit his influence sharply. Furthermore, even had his exposition of his ideas been entirely rigorous, his goals for accomplishment in geometry were trivial and narrow compared to the vast abstract projects that had transformed geometric research beginning in the middle of the nineteenth century: the scrutiny of the axiomatic foundations of the subject; the search for higher dimensional analogs of theorems in two and three dimensions; and the attempt to capture geometric symmetry and transformation properties with algebraic constructs. Wheeler was far more comfortable with an older geometric tradition that gave a central place to individual results with a strong visual appeal. There remained champions of this more concrete point of view, mathematicians such as the aforementioned Julian Coolidge at Harvard and Frank Morley at Johns Hopkins, but their influence had become very attenuated by the 1920s and 1930s, when Wheeler was most active in reaching out to the research community.[59]

Wheeler did catch the attention, however, of one young enthusiast for the older geometric tradition, H. S. M. Coxeter. Coxeter, born in 1907, had become intrigued with Wheeler's models after being introduced to him in the early 1930s. The men entered into correspondence, some of which survives. Before the two had met, Coxeter had received his doctorate from Cambridge University in England. After doing postdoctoral work at Princeton, he settled in 1936 into a professorship at the University of Toronto, where he remained until his death in 2003. Coxeter became an internationally renowned research mathematician and the author of several significant books and well over one hundred research papers. All his life he championed classical geometry, very much against the tide of mathematical fashion for many years.[60] But Coxeter also fully mastered the new abstract tools that were becoming a standard part of twentieth-century mathematics. Wheeler, for his part, not only failed to master these tools but remained fixated on physically constructing models. A different mathematical goal is found in a letter from Coxeter to Wheeler in 1934: "Sometime soon I must write up a description of the method by which I derived the enumeration [of icosahedra], showing how one can obtain all the necessary information *without reference to actual models*."[61] Coxeter was here recasting the matter at hand to focus on what had become a standard research aspiration: a classification theorem. He was as charmed as anyone by individual models, more charmed indeed than most research mathematicians, but for him models were only the beginning; for Coxeter there could even come a point in an investigation beyond which models served more to obscure than to clarify.

Despite Wheeler's failure to grasp the aims of elite research mathematicians, he developed an easy familiarity with his beloved polyhedra, a quite literally constructive grasp of their properties and the interrelationships among them, as seen in the following account of a Wheeler presentation at a meeting of the Mathematical Association of America:

> Mr. Wheeler took his audience into an interesting and little explored field. He showed by means of paper models some very unusual transformations of solids. Archimedean solids were derived by dissecting actual models of the regular solids and removing parts . . . A cube was cut into eight congruent cubes whose parts were hinged together in such a way that when rearranged in a different order there were produced two congruent rhombic dodecahedra of the second species. One of these was then dissected, and from it were obtained two congruent rhombic dodecahedra of the first species, and when one of these was dissected and turned "inside-out" there were recovered two of the eight cubes into which the original cube was subdivided.[62]

Wheeler and his models were admired by many people with a general appreciation of mathematics, including future MIT president J. R. Killian and Harvard historian of science George Sarton.[63] Even members of the larger general public, possessing minimal mathematical knowledge beyond everyday uses, were on occasion awestruck by him as a performer, accepting without question the importance of his work and proclaiming him a brilliant educator and a major mathematician. Wheeler showed great skill in cultivating these spectators, by connecting his work with themes more appealing to them than pure mathematics. One newspaper report from 1937 proclaimed him as "one of the outstanding geometricians in the country" and touted his model building courses as providing exercise in manual training.[64] In reality there is no evidence of any significant connection between Wheeler's work and contemporary interest in industrial education.[65] It was, however, highly politic at the time to claim "social utility" as a benefit of any educational initiative.[66] Another journalistic account, from the *Wellesley College News* of 1926, emphasized Wheeler's performing skills and uncritically accepted an exalted view of his accomplishments:

> Mr. Wheeler carries with him a diminutive notebook which contains the essence of his art and enough to astound and amaze us. On one page he has what appear to be scraps of paper. He lifts the topmost gently, gives it a turn, and suddenly there is a full four sided parallelogram standing from the page, a real as life tri-dimensional figure. He takes from his pocket a piece of folded paper gives it a twist and holds in his hand a perfect sphere made of intricately interlocked circles.

> Mr. Wheeler has gone on from these more simple models to the very complex such as the six intersecting polygons a figure of which Mr. Wheeler has made and owns the only existing model. Two years ago he exhibited his models at the greatest of all mathematical congresses at Toronto and showed there models that had never been made before.[67]

Wheeler's story illustrates the interest and enthusiasm generated by geometric models but at the same time exemplifies the growing stratification of American mathematical activity in the early decades of the twentieth century. In the diversity of reactions to his work we can observe barriers of incomprehension between those with differing relationships to the use of mathematics, in particular the widening gap between higher mathematics and mathematics as experienced by the general public. Geometry might appear to offer a means for bridging the gap, but the research elite had so transformed this field that someone such as Wheeler could no longer command much respect. All evidence suggests that Wheeler was an effective and engaging teacher, enlisting his students in making a host of models. How much this work contributed to their mathematical learning is hard to assess. Very likely, he assisted many students to have a more positive view of mathematics, and he may well have induced a few to look further into the subject, but he could not possibly have functioned as an emissary from the research frontier.

Both Richard P. Baker and A. Harry Wheeler had ties to contemporary research mathematicians. Yet, despite the rhetoric of figures such as E. H. Moore and J. W. A. Young, models did not become a standard part of the mathematical experience of most students. Even at schools where they were kept in the classroom or hallway, teachers often preferred to rely on blackboard and textbook drawings. This reflects both a decline in the teaching of solid geometry and a preference of mathematicians and mathematics teachers for other approaches to their discipline.

CHAPTER FOURTEEN

Linkages
A Peculiar Fascination

Can you draw a straight line? Or do you copy one from a straight edge?

Briefly put, a linkage is a system of rods or bars connected by hinges or pivots allowing easy deformation. Many machines can be analyzed in terms of such devices. They also can be used to construct geometrical figures, either statically or dynamically. The topic is one that has periodically created enthusiasm among a minority of mathematical educators, beginning in the 1870s, but it has never attained an enduring niche in mathematics instruction at any level. Most educational applications have been confined to cases in which the rods are constrained to lie in the same plane, but this is not a necessary condition. Indeed, some mathematicians have generalized the concept into higher dimensional space and in other ways strayed far beyond the bounds of visualization and hands-on manipulation, which have been central to the appeal of linkages in their more elementary forms.

From James Watt to J. J. Sylvester and Franz Reuleaux

One could probably claim that many mechanical devices of ancient lineage have some connection with linkages, but most commentators have invoked the work of the celebrated Scottish engineer James Watt as formative for later developments in both engineering and mathematics.[1] In 1784, seeking to improve his steam engine, he devised a system of rods and pins that was able to convert circular motion into an approximately straight line motion.[2] This innovation allowed the piston of Watt's steam engine to perform work on both the up and down strokes and began a long line of development in mechanical engineering.[3] (See Figure 14.1 for an 1841 refinement of Watt's approximate straight line mechanism.) Eventually, however, the gen-

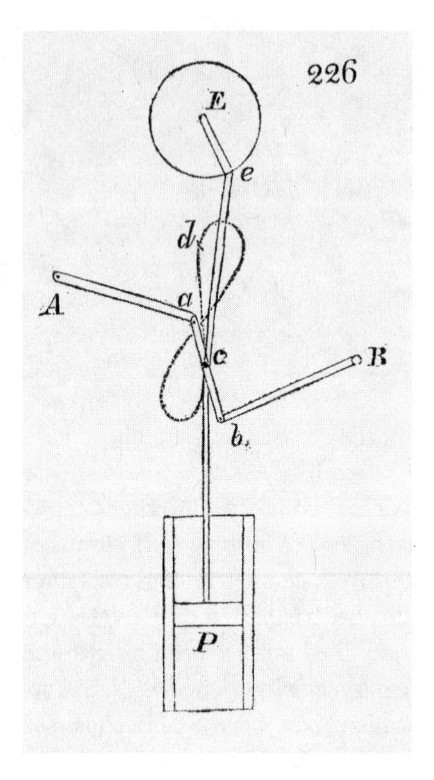

Figure 14.1. Straight line mechanism, 1841. Robert Willis, *Principles of Mechanism, Designed for the Use of Students in the Universities, and for Engineering Students Generally* (London: John W. Parker, 1841), 400. Smithsonian Negative no. 2006-2250.

eral problem of converting between rotating and reciprocating motion was more simply solved by the so-called slider-crank mechanism, central, for example, to the modern internal combustion engine. The slider-crank made attaining exact straight line motion a less pressing concern, but applying it successfully required advances in planing smooth metal surfaces and in lubrication that had not yet been achieved in Watt's time.[4]

In the nineteenth century, even as engineers were perfecting the slider-crank, Watt's linkage sparked interest among some mathematicians and theoretically minded engineers in a question that had seemingly escaped serious attention since antiquity: is it possible to construct a true straight line mechanically? As one commentator from 1922 remarked, this puzzle "exercised a peculiar fascination on the minds of geometers."[5] The historian Eugene Ferguson declared in 1962 that "the quest for a straight-line mechanism more accurate than that of Watt far outlasted the pressing practical need for such a device."[6]

The difficulty of the problem can be illustrated by comparing it with the corresponding problem of constructing a circle, the process of which is easily accomplished by the simplest of linkages: a single rod with no joints. By placing a marking device at one end of the rod and rotating it about the other end of the rod, one achieves the same effect as a compass. A similar mechanism for producing a straight line is not readily apparent. The problem entered mathematics primarily through the work of the prominent Russian mathematician P. L. Chebyshev, who had become fascinated by Watt's mechanism during a trip to England in 1852. Chebyshev analyzed the deviation from straightness achieved by the Watt linkage and then proceeded to construct a series of more complicated linkages in which this deviation was made smaller each time. In the course of this work he seems to have concluded that exact straight line motion was impossible, although he was unable to prove it.[7] Then in 1871 one of Chebyshev's students, Lipmann I. Lipkin, unexpectedly solved the problem with a seven-bar linkage (Figure 14.2). By holding one joint of the linkage fixed on the diameter of a circle and moving a second joint along the same circle, a third joint is forced to follow a straight line path. The action of this linkage is explainable in terms of inversive geometry, in which an inversion with respect to a circle in the plane transforms certain circles into straight lines and vice versa.[8] It was subsequently discovered that a French engineer, Charles Nicolas Peaucel-

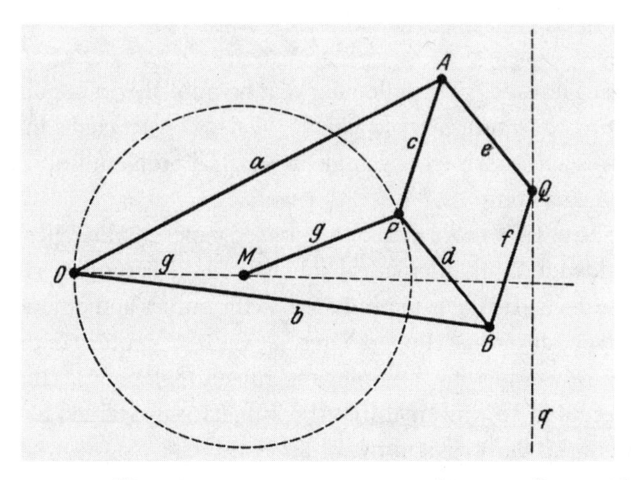

Figure 14.2. Peaucellier's Inversor, 1932. Reprinted in D. Hilbert and S. Cohn-Vossen, *Geometry and the Imagination* (New York: Chelsea Publishing Co., 1952), 273. Smithsonian Negative no. 2006-21926.

lier, had anticipated the Lipkin linkage in 1864, and henceforth it has often been referred to as "Peaucellier's cell," or "Peaucellier's inversor."[9]

Interest in linkages, also called "linkwork," caught fire in the 1870s.[10] In 1873 Chebyshev visited England, where he informed English mathematician J. J. Sylvester of Lipkin's breakthrough.[11] Sylvester proceeded to lecture on linkages, and indeed it was he who coined this term.[12] Referring to the fact that the topic had connections with both abstract mathematics and practical machinery, Sylvester declared, "Its head towers above the clouds, while its feet plunge into the bowels of the earth."[13] He often brought along a model of Peaucellier's cell to amuse his audience. Sylvester's description of the reaction of one such audience member illustrates the level of excitement that linkages could inspire: "Presently, after the speaker [Sylvester] exhibited the same model in the hall of the Atheneum Club to his brilliant friend Sir William Thomson of Glasgow, who nursed it as if it had been his own child, and when a motion was made to relieve him of it, replied, 'No! I have not had nearly enough of it . . . it is the most beautiful thing I have ever seen in my life.'"[14] Sylvester apparently excited many others in England, including Arthur Cayley, Harry Hart, and Alfred Bray Kempe. In 1874 Hart showed, again via inversive geometry, how a straight line could be constructed by a linkage consisting of only five bars (Figure 14.3).[15] Kempe in 1876 sketched a proof that it was possible to construct any *n*th degree curve

$$f(x, y) = 0$$

by means of a linkage.[16] The following year he published a popular book on linkages, *How to Draw a Straight Line.*[17] He also constructed a linkage with which he was able to trisect any angle, a classical problem impossible with straightedge and compass.[18]

From 1874 to 1879 more than one hundred papers on linkages were published worldwide in mathematical journals, with some investigators attempting to devise linkages to produce specific curves and others exploring more abstract questions.[19] Robert Yates, a twentieth-century enthusiast, explained the outcome with the following mixed metaphor: "The epidemic was so fierce and so universal that the subject was drained almost completely dry in the short span of five or six years."[20]

Meanwhile, several generations of scientists and engineers—including Monge, Ampère, Hachette, Willis, and Rankine—had been working on the problems associated with designing machines.[21] Linkwork in general and

Figure 14.3. Inversor of Hart, ca. 1900. National Museum of American History collections, gift of Department of Mathematics, University of Michigan. Smithsonian Negative no. 2006-3.

Watt's mechanism in particular (often described as producing "parallel motion") were standard topics in the nineteenth-century textbooks coming out of this tradition.[22] An important culmination of this line of development was reached just when linkage enthusiasm among mathematicians was reaching its height, with the publication in 1875 of *Lehrbuch der Kinematik* by the German engineer Franz Reuleaux.[23] This book was translated into English almost immediately, with the author's cooperation, by Alexander Kennedy of University College, London.[24] Reuleaux sought, with great success, to categorize all mechanisms into a few basic types, revealing in the process previously unrecognized commonalties, such as the slider-crank referred to earlier.[25] Kennedy used *link* for Reuleaux's *Glied* and *linkage* for *Gliederung*.[26] In the German original Reuleaux had cited Chebyshev's earlier contributions,[27] but he made no mention of the latest mathematical developments of the 1870s. Kennedy, who evidently had imbibed some of the excitement in London, added a note in his translation referring to "linkworks and cells (about which so much that is interesting has recently been written and done by Peaucellier, Sylvester, Hart, Kempe, and others)."[28] Reuleaux himself later took enough notice of these developments to incorporate a model of Peaucellier's cell in the collection of more than eight hundred models, which he had been assembling for many years. The German firm of Gustav Voight Mechanische Werkstatt, in Berlin, manufactured copies of these models, and in 1882 Cornell University acquired a collection of them, including the Peaucellier cell. Cornell has retained them to the present day and has made them conveniently available for dynamic inspection online.[29]

Beginning in the 1870s, however, engineering interest in linkages fundamentally diverged from that of mathematicians. Some textbooks of mechanics continued to include Peaucellier's cell in the early twentieth century,[30] but as a mid-twentieth-century mechanical engineer phrased it, for practicing engineers "it has been little used because of its complexity and because it came after the acute need for such a device had passed."[31] A recent engineering treatment of linkages mentions Watt and Chebyshev but not Peaucellier.[32] The notion of using Peaucellier's or Hart's cell instead of a straightedge to draw a line for some practical purpose has seemed far-fetched to nonmathematicians. One twentieth-century historian of technology scoffed at Kempe: "He did not weaken his argument by suggesting the obvious possibility of using a piece of string."[33]

Linkages in American Mathematics Education

Before the Civil War some American educators with enthusiasm for objects promoted a simple classroom device, the "gonigraph," which in retrospect can be classified as a linkage: "*The Gonigraph* is a small instrument composed of a number of flat rods connected by pivots, which can be put into all possible geometrical figures that consist of straight lines and angles, as triangles, squares, pentagons, hexagons, octagons, &c."[34] This device (Figure 14.4) was aimed at young students learning the most basic properties of geometric shapes. It appears to have been invented in the 1830s or earlier by Samuel Wilderspin, whom we met in chapter 6 in connection with numeral frames and infant schools.[35] Similar instruments would emerge again in the twentieth century under other names, and in some cases at-

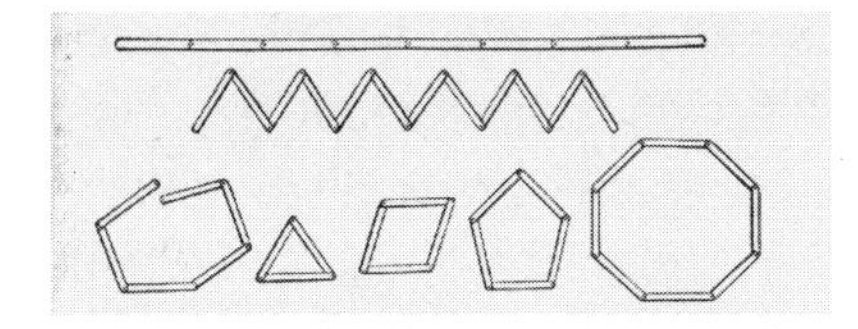

Figure 14.4. Gonigraph, 1851. From Henry Barnard, *Practical Illustrations of the Principles of School Architecture* (Hartford, Conn.: Press of Case, Tiffany & Co., 1851), 157.

tempts would be made to associate them with the more dynamic and mathematically sophisticated linkages described in the previous section. Most educational energy was expended on mechanisms of the latter type.

The 1870s linkage craze reached the United States almost immediately. Peaucellier's cell quickly entered general reference works.[36] W. W. Johnson, a professor of mathematics at St. John's College in Annapolis, wrote on linkages in 1875 and 1876, citing his reading of an exposition by Sylvester from 1873.[37] Sylvester himself seems to have lost interest in the subject by the time he arrived in person from England to take up his position as professor of mathematics at the newly founded Johns Hopkins University. Linkages formed no part of his teaching during his Hopkins tenure from 1876 to 1883.[38]

Nevertheless, there was a Hopkins connection to several of the American writers on linkages over the next fifty years, beginning with Sylvester's Hopkins student G. B. Halsted, who included a discussion of both the Peaucellier cell and the Hart cell in his college geometry textbook of 1885.[39] In an 1895 obituary of Chebyshev, Halsted recounted much of the history given here and declared his faith in the educational future of linkages: "Henceforth Peaucellier's Cell and Hart's Contraparallelogram will take their place in our text-books of geometry, and straight lines can be drawn without begging the question by assuming first a straight edge or ruler as does Euclid."[40] This prediction has not been fulfilled, although for a time in the early twentieth century there were efforts to incorporate linkages into college courses in geometry.[41]

The University of Chicago also generated interest in linkages. In 1906 mathematics department chairman E. H. Moore, aiming mainly at an audience of high school teachers, employed linkages in his paper on "The Cross-Section Paper as a Mathematical Instrument."[42] But Moore's "linkage diagrams" were less realizable as physical mechanisms than most of the linkages discussed heretofore because Moore required that some of the links be indefinitely stretchable. This style of abstract linkage does not seem to have gained any foothold in the schools, then or later. In 1907 Moore's colleague J. W. A. Young recommended the more standard physical linkages as a good source for presentations at college mathematics clubs.[43] Finally, Moore's student R. P. Baker evidenced interest in such linkages by including them in the collection of geometric models that he built for sale, although whether the two discussed linkages is unknown. Baker's linkages used wooden bars connected with metal pivots.[44] (For more on Baker and geometric models, see chap. 13). Model builder Martin Schilling in Germany was also incorporating various linkages in his product line about this time, including Peaucellier and Hart inversors constructed of metal parts.[45]

Another linkage enthusiast emerged just after World War I in the person of Frank Vigor Morley, son of Johns Hopkins University mathematics professor Frank Morley. The elder Morley had grown up in England, and the son returned there to get a D.Phil. degree at Oxford in 1923 under G. H. Hardy.[46] The younger Morley published papers on the history and mathematics of linkages,[47] and in 1933 he and his father published a book titled *Inversive Geometry,* which includes a concise treatment of the Peaucellier and Hart inversors.[48]

Linkages also found a brief place in each of three semipopular books emanating from the great tradition of Göttingen mathematics, all first appearing in English in the United States between 1939 and 1952: *Elementary Mathematics from an Advanced Standpoint: Geometry,* by Felix Klein; *Geometry and the Imagination,* by David Hilbert and Stefan Cohn-Vossen; and *What Is Mathematics?* by Richard Courant and Herbert Robbins.[49] The Morleys and the German works probably helped to keep linkages alive as a frequent topic of college mathematics club talks from the 1920s to 1950s.[50] Joseph Hilsenrath, whose article of 1937 has been noted, was a student at Montclair State Teachers College in New Jersey at that time and had become interested in the topic through hearing it discussed at a mathematics club presentation.[51]

Probably the single most energetic figure in promoting linkages as a teaching tool in the twentieth century was Robert C. Yates, who obtained a Ph.D. degree from Johns Hopkins in 1930. His dissertation suggests only a vague connection with linkages,[52] but in the mid-1930s, as a professor at Louisiana State University, he began to publish a series of elementary articles promoting linkages and related topics as appropriate for American high schools.[53] He eventually compiled his ideas into a short book titled *Tools: A Mathematical Sketch and Model Book,* apparently self-published in 1941 and republished in 1949 as *Geometrical Tools*.[54] In 1942 he joined the mathematics faculty at West Point, where he continued to champion linkages both at that institution and nationally.[55]

Yates contributed to a flowering of interest in linkages within the National Council of Teachers of Mathematics (NCTM) during the 1940s and early 1950s. In 1945 this organization, largely oriented toward secondary school instruction, published a yearbook on *Multisensory Aids in the Teaching of Mathematics,* which included three chapters concerning linkages, two by Yates.[56] The following year the NCTM's journal the *Mathematics Teacher* printed "Linkages as Visual Aids," by Bruce Meserve.[57] Clearly hoping to attract adherents, Meserve began by stating: "Although many fundamental concepts of mathematics may be demonstrated by the use of linkages, very few teachers have used them regularly in their classes."[58] In 1950 the *Mathematics Teacher* was advertising a "New Mathematics Kit" from Professor M. H. Ahrendt of Anderson College in Indiana: "Can your students draw an original straight line? Can you draw such a line? Euclid postulated the straight line, but he did not have any way to draw one. In fact it was less than one hundred years ago when the first theoretically perfect means of tracing a continuous straight line in a plane was discovered."[59] In 1951 the W. M. Welch Company was advertising in the same journal a set of linkages, in some ways an elaboration of the nineteenth-century gonigraph, calling them "The Schacht Instruments for Dynamic Geometry."[60] In 1952 a reprint of Kempe's 1877 booklet *How to Draw a Straight Line* was also advertised in the *Mathematics Teacher.*[61]

James Gates, longtime executive director of the NCTM, was active in teaching during the 1950s and also attended summer school at Teachers College, Columbia University. When interviewed in 2004 he recalled the activities of both Yates and Ahrendt during that earlier time: "Robert Yates . . . taught at West Point during the academic year, and would come to Columbia in the summers until he left West Point and established the math department at the University of South Florida. He gave very interesting lectures often using linkages and the proofs for the mathematics they were demonstrating . . . My predecessor as executive director was Myrl Ahrendt, and Myrl constructed physical models for the linkages that people like Robert Yates and others did the mathematics for . . . We had a box of them at the NCTM headquarters office for many years."[62]

As the image of this lonely box at NCTM headquarters suggests, linkages did not become commonplace in mathematics education. The reform movement of the 1950s and 1960s known as the "New Math" seems not to have been favorable to linkages. We are not aware of any standard textbook from that era that treats the topic. Very likely linkages did not fit well with the New Math's emphasis on abstraction,[63] nor have they been revived in more recent years. We find no allusion to linkages in the NCTM standards.[64] We are likewise unaware of any call for linkages by opponents of the standards.

The highest place the topic has ever achieved in the mathematics curriculum has been as an intriguing curiosity, outside the mainstream of development. Its standing is exemplified in the well-known book by Courant and Robbins, first published in 1941, in which the section on linkages is marked with an asterisk to indicate its status as among "parts that may be omitted at a first reading without seriously impairing the understanding of subsequent parts."[65] The 1996 revision of this classic text illustrates a subsequent decline from even this minor role, with the reviser of the index mistakenly conflating *linkages* with *links,* part of the topological theory of knots.[66] In fact, linkages have recently emerged again as a fruitful field for abstract mathematical research but at the price of removing them from contact with all but the most advanced student.[67]

PART FOUR

Electronic Technology and Mathematical Learning

CHAPTER FIFTEEN

Calculators
From Calculating Machines to the Little Professor

Should we, at this stage in our civilization, cease to teach addition and other processes except by machine?

The calculating machine, like the slide rule, took considerable time to move from the world of practical affairs into the mathematics classroom. Commercial calculating machines date from the mid-nineteenth century. It was only a century later that the efforts of a handful of American manufacturers led a few schools in the United States to try using calculating and adding machines as a part of arithmetic teaching. The programs garnered considerable publicity for the companies involved but had little lasting influence on mathematics education. The mid-1960s saw the advent of desktop electronic calculators. During the same period federal funds became available for purchasing apparatus to help teach students basic skills. Further attempts were made to try calculators in the classroom. Yet it was only in the early 1970s, with the introduction and decreasing cost of handheld electronic devices, that calculators found a place in general education. They first were widely used in advanced technical courses, displacing the slide rule. Mathematics teachers, parents, and school officials disagreed about how much, if at all, the new tools should be allowed in more elementary courses. Meanwhile, modified calculators were widely sold for arithmetic drill in both the home and the school.

Mechanical Arithmetic for Education

Machines that could add and subtract automatically through the interaction of gears and levers were introduced as mechanical marvels as early as the seventeenth century, but commercial calculating machines only began to sell in the mid-nineteenth century. In 1867 Frederick A. P. Barnard, the president of Columbia University, served as a judge at the Exposition Universelle, a world's fair held in Paris. There he saw for the first time a machine that could add, subtract, and (with suitable repeated addition) multiply. Barnard was enchanted. He wrote in his account of the fair: "To most persons the process of calculation involves a species of mental labor which is painful and irksome in the highest degree; and to such, no part of their educational experience recalls recollections of severer trials, or of burdens more difficult to bear. That this toil of pure intelligence—for such it certainly seems to be—can possibly be performed by an unconscious machine is a proposition which is received with incredulity; and even when visibly demonstrated to be true, is a phenomenon which is witnessed with unmingled astonishment."[1] Barnard arranged to purchase a calculating machine for his personal use, and several other scientists followed suit.

The manufacture and sale of calculating machines soon spread to the United States. By the 1890s American machines for solving routine arithmetic problems were available for a few hundred dollars.[2] By the 1920s they had become commonplace in commerce, government, and science. A few business machine manufacturers offered their own training classes, and even more sold machines for use in bookkeeping classes at high schools and commercial colleges.[3]

Commercial adding and calculating machines were rarely used in mathematics education. They were heavy, expensive, and, particularly for multiplication and division, not terribly efficient. To be sure, as early as 1903, George W. Myers proposed equipping a high school mathematics laboratory with a single calculating machine.[4] It was apparently only after World War II, however, that the devices were seriously proposed as aids to teaching elementary arithmetic. In late 1949 and early 1950 teachers at the Hunter College Elementary School in New York City, with the aid of New Jersey–based Monroe Calculating Machine Company, experimented with supplementing usual arithmetic teaching with problem solving on the newly named Monroe Educator calculating machine (Figure 15.1).[5] These efforts were sufficiently successful for the company to encourage further studies. Monroe lent twenty-eight desktop machines for ten weeks to a class of third-graders in the Glenwood Public School in nearby Short Hill, New Jersey. Mildred D. Schaughency, a teacher at Glenwood, reported in the February 1955 issue of *Arithmetic Teacher* that students enjoyed using the machines to check their work.[6] Schaughency's article prompted Ben A. Sueltz, the editor of

Figure 15.1. Monroe Educator calculating machines at the Memorial School in Cedar Grove, N.J., 1956. Courtesy of Monroe Systems for Business.

Arithmetic Teacher, to raise several more general questions concerning the role of machines in arithmetic teaching. He called for more detailed studies of precisely what aspects of arithmetic teaching might be aided by machine—and what might not be. At the same time, Sueltz asked whether schools should "cease to teach addition and other processes except by machine."[7] Although this query seems to have been largely rhetorical, it would later come to loom large in discussions of less expensive, more powerful handheld electronic calculators.

The following fall and winter Howard Fehr, the head of the mathematics department at Teachers College of Columbia University, spearheaded a more systematic study of the effect of using Monroe calculating machines in teaching. Fehr and his associates compared a group of fifth-graders at Memorial School in Cedar Grove, New Jersey, who used the machines for four and a half months, with a control group taught in the usual manner. They found that working with calculators increased both the computational skills and the reasoning ability of students. Reports of this work appeared not only in the *Arithmetic Teacher* but in various newspapers.[8] Indeed, on January 31, 1957, Fehr and some of the students who used the Educator appeared on television on Dave Garroway's *Today* show.[9] Impressed by this work, two aerospace companies agreed to underwrite the use of Monroe Educators in elementary schools in Santa Monica and Culver City, California.[10] At the same time, Sueltz, in an editor's note appended to Fehr's 1956 article, suggested that scholars should consider more broadly how the interest generated by machines might be used not only to assist in learning arithmetic operations but to further more general mathematical understanding. He encouraged readers "to seek the optimum relationship of paper and pencil arithmetic to mental arithmetic to machine arithmetic."[11]

As we have seen, the National Defense Education Act (NDEA) of 1958 made U.S. government matching funds available to schools to purchase scientific and mathematical apparatus. Educators recommended spending some of the money on various forms of the classroom abacus, geometric models, and measuring devices.[12] The Riverside City Schools in California, however, used NDEA funds to purchase Monroe Educator calculators for use in fourth-, fifth-, and sixth-grade classrooms. Lois L. Beck of that school system presented a preliminary report on the project in the February 1960 issue of *Arithmetic Teacher.* According to Beck, students readily learned to operate the machines and appeared to enjoy the work. Investigators hoped the pupils would come to appreciate connections between basic arithmetic processes such as addition and multiplication and would better understand place value. They also thought using machines might foster good work habits. The scheme was not sufficiently successful to prompt numerous purchases of Monroe Educators, but the manufacturer remained very interested in the educational market.[13]

The Minnesota firm of Ken, Inc., thought that its much simpler, portable, four-wheeled, stylus-operated adding machine, which cost only $6.95, would be well suited to school use. An account of the Ken+Add Pocket Adding Machine appeared in *Mathematics Teacher* in December 1952 (Figure 15.2). The reviewers noted that while the device was a practical adding machine, it was also "fascinating as a toy and very useful in many arithmetic situations at school." They went on to suggest specific applications and recommended that "a few of these calculators should be in the teaching collection of each school to enrich the teaching of computation."[14] The instrument was popular enough to be advertised in *Arithmetic Teacher* as late as

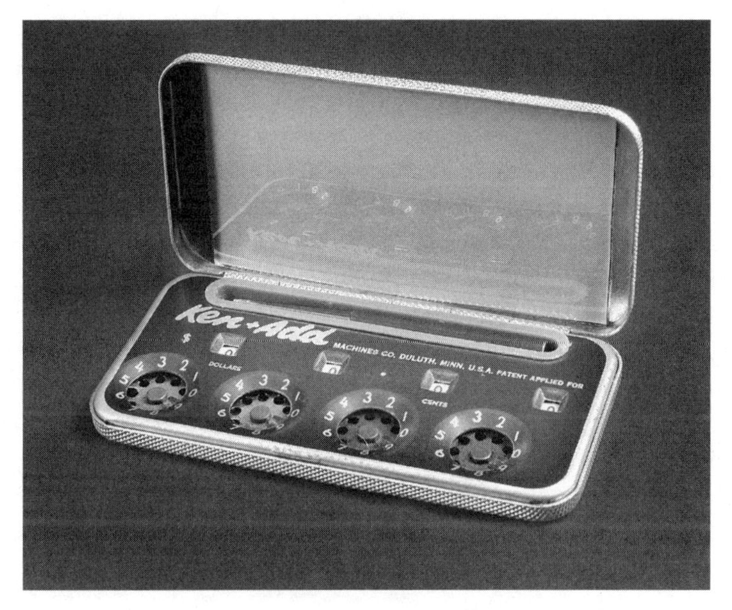

Figure 15.2. Ken+Add adding machine, ca. 1956. National Museum of American History collections, gift of Robert K. Otnes. Smithsonian Negative no. 2006-2061.

1956, when the maker claimed that "experience in manipulating numbers with the KEN+ADD adding machine stimulates [a] pupil's interest in arithmetic." The device was particularly useful for visually demonstrating carrying and for teaching counting by twos or threes. According to the advertisement, the machine had been tested and approved by both the Association for Childhood Education and the National Council of Teachers of Mathematics (NCTM).[15] Apparently, it was never widely adopted in teaching.

The Burroughs Corporation of Detroit, the largest manufacturer of adding machines in the world, also made a brief foray into the educational market. Andrew F. Schott, an educator who had a Ph.D. degree from the University of Wisconsin and taught at the Wisconsin State College in Milwaukee, led the project. His preliminary studies began in 1952. By January 1954 teacher training was under way. Burroughs provided funds and key-driven adding machines to seven schools. Children in grades 1–3 used a special form of abacus that Schott had designed and dubbed the "Numberaid." Students in the fourth grade through high school used a Burroughs adding machine called the Instructor. Initial reports of gains in test scores achieved by students during the 1954–55 school year encouraged Burroughs to continue the program. By the fall of 1955 experiments had expanded to include schools in Michigan, California, and Virginia,[16] although the project seems not to have lasted much longer. This was a period of upheaval for Burroughs. The firm had changed its name from the Burroughs Adding Machine Company to Burroughs Corporation in 1953 and increasingly sought to be known as a computer manufacturer. This fact, the cost of Burroughs machines, and Schott's lack of prestige within the mathematics community may have persuaded Burroughs to leave classroom adding machines to others. Then, too, the appeal of adding machines as technical marvels was beginning to fade. A new device, the desktop electronic calculator, would soon be seen as a better way to excite children's interest.

Desktop Electronic Calculators for Underachievers

The development of transistors and then integrated circuits in the 1960s made it possible to build relatively compact, reliable desktop electronic calculators. These machines were initially designed for use in business and in scientific laboratories. Yet new federal legislation led manufacturers to consider possible educational uses of their products. More specifically, Title I of the Elementary and Secondary Education Act of 1965 provided special funds for school systems in areas having a large concentration of low-income families. Some of this money was spent on programs for students in these areas who were having difficulty in mathematics.[17] By 1968 the Massachusetts firm of Wang Laboratories had developed a supplement to ordinary mathematics curricula intended to "multiply student achievement" at all levels. Not surprisingly, the course involved heavy use of Wang's programmable calculators (Figure 15.3). The company advertised the plan in *Mathematics Teacher* for several months in 1968 and 1969.[18] Attracted by the promise of this program, the Columbia High School in Maplewood, New Jersey, acquired four Wang calculators in the spring of 1970. In December of the following year Peter A. Tenewitz, the chairman of the school's mathematics department, reported that the machines had been an instant success. They were used in algebra and statistics classes as well as for science homework and teaching elementary programming.[19]

Wang was not alone in developing school programs. In late 1969 the

Figure 15.3. An early programmable calculator, the Wang LOCI-2, ca. 1965. National Museum of American History collections, gift of Computer Sciences Corporation. Smithsonian Negative no. 80-14510.

Friden division of the Singer Company suggested that students would greatly benefit from studying "Computer Math" and "Business Computer Education" with its Computyper. An advertisement in the November 1969 issue of *Mathematics Teacher* stressed the importance of giving secondary level and college students "the creative command of the computer they'll need in our widely computerized society." Singer Friden offered to provide teacher training, lesson plans, texts, and a Computyper at a cost of $15,250 per course. The Computyper was the size of an entire office desk, although it could be used by only one student at a time and had limited capabilities.[20] Few schools seem to have tried it.

In the early 1970s both the California firm of Hewlett-Packard and the American branch of Sharp Electronics Corporation suggested that their desktop calculators offered special promise to young people having problems with mathematics. Advertisements published in *Mathematics Teacher* featured handsome but cynical teenagers. Thus, the November 1970 issue of the journal included a photograph of a boy engrossed by an HP 9100 desktop calculator sitting at a table in front of a blackboard diagram of the Pythagorean theorem. The ad was entitled "Pythagoras, He's Not." Further copy explained that a student who had fallen behind in mathematics could now perform "dreary computations" on an electronic calculator. Operating the HP 9100 had restored his confidence and awakened his interest in theory: "Mathematics has come alive at his fingertips."[21] Similarly, in December 1971 Sharp ran an advertisement describing "something new to help you turn unrest into interest." The left side of the two-page spread showed the face of a glowering youth, cracked into jagged pieces. On the right the boy was shown whole, transformed into a smiling student. The change, readers learned, came from work with a Sharp desktop calculator and a set of activity cards describing the mathematics problems that could be worked out on the machine.[22]

In 1970 the HP 9100 cost thousands of dollars, the Sharp calculator hundreds.[23] At most schools could afford a handful of the machines, which had little immediate effect on the existing work of classrooms. Both Hewlett-Packard and the Monroe Calculator Company did suggest in advertisements that appeared later in the 1970s that their programmable desktop calculators were well suited to individual instruction in schools, particularly for teaching mathematics and programming (Figure 15.4).[24]

Handheld Calculators as Electronic Slide Rules

From the late nineteenth century American science and engineering students had relied on portable, individually owned slide rules. The engineer Bill Hewlett, cofounder of Hewlett-Packard, decided in 1970 that it should be possible to build and sell an electronic version of the familiar instrument. In the spring of 1972 Hewlett-Packard introduced the HP-35 pocket calculator (Figure 15.5). It was one-tenth the volume of the desktop HP 9100, ten times as fast, and, at $395, about one-tenth the price. It performed basic arithmetic to ten significant figures on numbers as small as 10^{-99} and as large as 10^{99}. Algorithms built into the calculator allowed it to compute logarithms, exponentials, and trigonometric functions. In other words, as Hewlett-Packard was quick to point out, the HP-35 did the work of a slide rule and much more, with much greater precision.[25] The manual for the machine informed customers that it would fulfill their wildest fantasies: "We thought you'd like to have something only fictional heroes like James Bond, Walter Mitty or Dick Tracy are supposed to own."[26] The futuristic instrument was sturdy enough to withstand a trip through a snowblower (the case broke, but the calculator still worked), bore up under a wide range of temperatures, and fit in a (large) shirt pocket. All that was required was cash,

Figure 15.4. Students using Monroe Classmate 88 desktop calculators, ca. 1976. Courtesy of Monroe Systems for Business.

nimble fingers, frequent recharging, and the ability to master a distinctive way of entering problems.[27] At the end of 1972 the United States was contending with the ongoing conflict in Vietnam and early questions about connections between the White House and a burglary at Washington's Watergate apartment complex. Hewlett-Packard was happy to point out in its advertisements that "some things are changing for the better."[28]

The HP-35 was both a commercial success and a treat for mathematics teachers. Advertisements for the HP-35 appeared in journals read by scientists and engineers, and the manufacturer soon reported a backlog of orders.[29] A May 1972 account of the instrument in *Mathematics Teacher* was entitled "The Ultimate Slide Rule?" It began, "Hewlett-Packard has recently announced the availability of an almost unbelievable electronic pocket calculator named the HP-35." After describing the instrument in some detail, the review indicated how readers could learn more about "this amazing de-

Figure 15.5. Hewlett-Packard's HP-35 electronic calculator (*left*) and Texas Instrument's TI SR-10 electronic calculator, 1970s. National Museum of American History collections, gifts of Nicholas Grossman and John B. Priser, respectively. Smithsonian Negative no. 2006-21178.

vice."[30] Other mathematics teachers were equally impressed. From 1973 until 1975 high school students who won a place on the United States team for the International Mathematical Olympiad received an HP-35 calculator as a prize.[31]

The HP-35 was not the first handheld electronic calculator. Pocket-sized calculators that performed simple arithmetic had been introduced in the United States in 1971 by firms such as Bowmar, Busicom, and Canon. The chips used in Bowmar and Canon calculators were made by Texas Instruments, and TI would begin selling four-function calculators in the fall of 1972.[32] The HP-35 and later scientific calculators were the first electronic instruments to displace a well-established mathematics teaching tool, namely the slide rule. Texas Instruments introduced the SR-10, its answer to the HP-35, in 1973 (see Figure 15.5). This calculator did not give values for trigonometric functions but cost only $150. The TI-50 (introduced in 1974 for $170) and the HP-21 (introduced in 1975 for $125) both did the work of a slide rule for a somewhat more reasonable price.

Calculators soon appeared in classrooms on an experimental basis. In

1974 the Berkeley public schools provided handheld calculators to two hundred underachieving junior high school students. That same year New York State distributed donated calculators to fifty-five sixth-grade students in schools not far from Albany. Teachers found that the instruments allowed students to do more realistic problems that involved larger numbers.[33] In March 1975 *Mathematics Teacher* reported on a National Science Foundation (NSF)–funded project that made HP-45 pocket calculators available to students in nine mathematics, science, and business classes at Menlo School and College in Menlo Park, California. The professors thought highly of the results of the experiment, although no controlled study was attempted.[34]

Calculators were more common at the college level. Cadets entering the U.S. Military Academy at West Point in 1974 were issued calculators, not the traditional slide rules.[35] Many more students bought calculators on their own. In May 1975 Lowell Leake Jr. reported that he had polled his freshman calculus class at the University of Cincinnati and discovered that fifty-one of the sixty-four students owned pocket calculators, but there was no uniformity in the models they had.[36] As the price of calculators declined still further, use of the slide rule decreased precipitously. Dietzgen of Chicago, a longtime slide rule vendor, was selling electronic calculators by January 1973.[37] Three years later Keuffel & Esser was advertising Texas Instruments calculators. It stopped manufacturing slide rules entirely, donating one of the special machines used to make them to the Smithsonian Institution.[38] Calculators also gradually displaced the printed tables of trigonometric, logarithmic, and exponential functions that had been a standard part of textbooks.

A Cacophony of Voices

In the mid-1970s mathematics teachers abandoned instruction in the slide rule and toyed with the electronic calculator. Commentators offered diverse views on how the new tool might best be used, if at all, to teach mathematics. In some cases entertainment clearly outweighed instruction. The California teacher Patrick J. Boyle, for example, published an article about calculator "charades." This game encouraged students to develop equations whose solutions spelled out words when a calculator display screen was inverted.[39] More generally, publications discussed how the new technology might alter both the content and the techniques of mathematics teaching.

To encourage discussion of the proper role of electronic calculators in schools, the editorial panel of *Mathematics Teacher* polled "a sample of teachers, mathematicians, and laymen." It reported in October 1974 that respondents favored continuing extensive instruction in computation. Over two-thirds of those who replied thought that arithmetic computation was the major goal of elementary and junior high school mathematics teaching, and 84 percent thought that speed and accuracy in this area was still essential for "a large segment of business and industrial workers and intelligent consumers."[40] Asked whether "every seventh-grade mathematics student should be provided with an electronic calculator for his personal use throughout secondary school," 28 percent of those polled agreed, and 72 percent disagreed.[41] Some worried about cost, others thought calculators should be provided only when students had demonstrated their proficiency in pencil and paper computations, and others thought providing calculators would promote laziness and inefficiency.

Despite such doubts, discussion of how best to use calculators continued, much of it growing out of a report by the National Advisory Committee on Mathematics Education (NACOME) of the Conference Board of the Mathematical Sciences. Work on the so-called NACOME report began in 1974, with funds from the NSF; its purpose was to evaluate two decades of reform in mathematics teaching. The 157-page document, completed in late 1975, was entitled *Overview and Analysis of School Mathematics, Grades K–12*.[42]

Perhaps the most controversial recommendation of the report concerned electronic calculators. The committee urged "that beginning no later than the end of the eighth grade, a calculator should be available for each mathematics student, during each mathematics class. Each student should be permitted to use the calculator during all of his or her mathematical work, including tests."[43] Parents, teachers, and pundits expressed doubts about the wisdom and practicality of making calculators available on a blanket basis. Some objected generally to classroom calculator use, considering it a form of cheating. Others thought that many secondary students might benefit from calculators but that some classes might well be devoted to reviewing basic computation with whole numbers and decimals, without recourse to calculators. Students would not always have working calculators at hand and should not become entirely reliant on them.[44]

Those who advocated calculator use sought to spell out more precisely just how the devices might contribute to mathematics instruction. The Instructional Affairs Committee of the NCTM prepared a list of nine kinds of problems for which the calculator might provide reinforcement and motivation. This list, with sample problems, was published in both *Arithmetic Teacher* and *Mathematics Teacher* in early 1976. Some problems required calculating very large numbers, others pointed up patterns in numerical relations, and others involved finding answers (such as cube roots) by trial and error. In these cases the calculator allowed students to go beyond their usual arithmetic skills and ask new kinds of questions. Other exercises involved applying arithmetic as a consumer, reviewing basic number facts, and checking numerical examples.[45] Generating enthusiasm for such problems would have required a skillful teacher indeed.

The NACOME report also emphasized the importance of providing more widespread instruction in statistics. Richard S. Pieters of the Groton School in Groton, Massachusetts, discussed this aspect of the report in the August 1976 issue of the *American Statistician*. Pieters traced efforts to encourage high school courses in statistics and probability to the mid-1950s. To be sure, concern for science education in the immediate post-*Sputnik* years had led to more emphasis on introducing calculus into high schools. More recently, however, there had been renewed emphasis on the possibility of teaching school students about data analysis and probability. According to Pieters, "The emerging availability of electronic handheld calculators will be exploited in order to make more realistic examples and projects possible."[46]

Pieters was not alone in asking how the use of electronic calculators might alter the broad outlines of the school curriculum. In an article published in *Mathematics Teacher* in April 1977, Henry O. Pollak of Bell Laboratories took up this theme. Pollak argued that anyone considering a new aid to instruction should ask what the most difficult outstanding problems of school mathematics were and how the device might help to solve them. Without asking this question, he argued, one might well end up with material that represented no substantial improvement over chalk and blackboard. Educational film, for example, had not been used as creatively as it might have been. In this spirit Pollak suggested that the concepts of function and inverse function were difficult to teach but easy to illustrate with a calculator. Like Pieters, Pollak also thought that calculators might make it possible to use a much wider array of experiments to illustrate ideas relating to probability as well as making statistical data analysis more accessible. Indeed, probability and statistics might now be taught much earlier in the curriculum than had previously been the case.[47]

The cost of calculators, the need to provide students with a basic introduction to numbers, and long habit may have combined to encourage NACOME to confine its recommendations concerning calculators to grades 8 and up. Yet some educators proposed a more radical transformation of mathematics teaching in the wake of the calculator. In a paper presented to the Association for Computing Machinery in October 1976, David Moursund of the Department of Computer Science at the University of Oregon explored the implications of the new technology. Moursund had received his doctorate in mathematics and was very active in the International Society for Technology in Education. He began his paper with the assertion by the distinguished applied mathematician Richard W. Hamming that "the purpose of computing is insight, not numbers." Assuming that this statement was true for education at all levels, Moursund listed three insights that elementary school children might be expected to glean from their study of mathematics. The first was a general arithmetic and geometric sense—that is to say, sufficient understanding to do mental arithmetic and to make mental approximations. The second was learning to present and deal with a wide range of arithmetic and geometric concepts and problems. These skills included the use of mathematical notation both to represent and to manipulate data. Finally, Moursund said that students should develop abilities and attitudes conducive to further learning and use of mathematics.

None of these goals was greatly advanced by extensive paper and pencil drill on routine arithmetic. Instead, Moursund argued, students should learn to solve problems in their heads and to formulate solutions that could then be carried out mentally, by hand, or with calculators if needed. The ability to formulate and solve problems in this way would not only introduce children to mathematics, but it would provide them with skills they could use for a lifetime. Mindful of the contemporary "back-to-basics" movement in mathematics education, Moursund claimed that problem solving should be seen as the basic skill central to elementary mathematics education. He acknowledged that others might not share his view. The existing school curriculum, the conservatism of educational establishments, and the need to retrain teachers also might work against his proposals. Nonetheless, he asserted that the ready availability of calculators should reshape the mathematics curriculum at even the elementary level.[48] Thus, by the end of 1977 the introduction of the electronic calculator had not only provided a new tool for advanced students but had raised challenges to the existing mathematics curriculum from the elementary level through high school.

Despite encouragement from professional educators, many teachers adopted calculators only gingerly, sometimes because of school policy and funding, sometimes because of teacher and parent preferences. Certainly, the novelty associated with the instrument faded, particularly with the advent of the personal computer. Calculators were first allowed on the Advanced Placement calculus examination in 1983, but their use was suspended two years later. It was only in the early 1990s that calculators came to be required on some College Board examinations and became customary in elementary and middle school classrooms.[49] Long before this time, however, Americans used integrated circuits to recast teaching machines in the form of electronic toys. We now turn to this vision of electronic technology.

Integrated Circuits in Educational Toys

Early advocates of teaching machines had agreed that frames of a programmed course should be carefully ordered by the person who wrote them. Later inventors suggested another approach. In 1971 Jerome C. Meyer and James A. Tillotson III of Sunnydale, California, received a patent for a "teaching device having means [of] producing a self-generated program." In the teaching machine they described, frames were not ordered by a programmer but were selected using a random signal generator. Meyer and Tillotson thought that such a machine might have many uses but were specifically interested in demonstrating an instrument that could generate simple arithmetic problems. They claimed that using problems produced randomly, rather than proceeding through a fixed program, would create a less complicated, less expensive, and more efficient teaching machine. Given a problem, a student entered the answer. The machine then checked it, with a correct answer eliciting a new problem.[50] William R. Hafel, also of Sunnydale, refined these ideas in the "mathematical problem and number generating systems" for which he was granted a patent in 1976.[51] Ideas used in both of these patents were reflected in an electronic teaching machine designed for drilling children in basic arithmetic called the Digitor, a device introduced by the California firm of Centurion Industries in 1974 (Figure 15.6).

The Digitor was an appealing plastic instrument with a body roughly the size of a grapefruit, mounted on four metal legs. As reviewer Thomas Rowan wrote in *Mathematics Teacher:*

Figure 15.6. Digitor teaching apparatus, 1975. National Museum of American History collections, gift of Centurion Industries. Smithsonian Negative no. 2005-26075.

> It looks much like a luxury calculator, but cannot be used to perform calculations. Its circuitry is designed to present the student with an exercise, indicate whether the student's response is right or wrong, and keep a record of the number of correct responses. The teacher or student can control the magnitude of the addends (or factors), the operation (+, −, ×, ÷), and the number of exercises (10, 25, 50, or 100). When the student's response to an exercise is correct, a green smiling face lights. If the student is incorrect, a red sad face lights and the correct answer is shown. The student must give a correct response before a new exercise is presented. If the student gets all assigned exercises correct, the smiling face flashes happily.[52]

The Digitor offered the reinforcement of a teaching machine and its insistence on correct answers, without the apparent sequencing of frames. In 1976 it sold for $249.50 and was intended for grades 3 through 8. Rowan reported that children liked to use the Digitor, and he thought it might be useful to those having difficulty with basic arithmetic. Yet he pointed out two drawbacks to the machine: it was expensive and was an attractive target for theft.

These weaknesses notwithstanding, the Digitor proved a successful product. By 1983 Centurion was manufacturing ten similar machines, which it called "educational computers." Four were for mathematics, five for language skills, and one for general drill in several subject areas. The mathematical instruments all were designed to provide drill on arithmetic operations with small integers. They differed in the number of problems available, the extent to which users could alter the problems displayed, and the time allowed for each problem. Prices were somewhat lower but still ranged from $139.50 to $199.50. Centurion also sold related classroom materials and provided an instruction booklet.[53] The devices are remembered in some circles for having incorporated an Intel 4004 microprocessor, the chip used in early microcomputers.[54]

A second electronic drill master, the Novus Quiz Kid (later called the National Semiconductor) Quiz Kid, was for the home market (Figure 15.7). An advertisement published in the *New York Times* just before Christmas in 1975 indicates that this small, four-function instrument sold for only $15. The calculator had no display, but the keyboard was decorated with an image of an owl with two large eyes, one green and one red. Children entered both a problem and their answer to it. If the answer was correct, the green eye flashed reinforcement. If not, the red eye lit up. The ad proclaimed that "the Novus 'Quiz Kid' just might make a Whiz Kid out of Jr [*sic*]!" At least it would "provide hours of fun and interest."[55] A report from late May 1976 indicates that by then roughly 600,000 of the toys had been shipped.[56]

Texas Instruments had responded to the popularity of four-function calculators by producing the Datamath 2500 and to the HP-35 with the SR-10. Its answer to the Quiz Kid and similar toys was the Little Professor (Figure 15.7).[57] Introduced in mid-1976, it was a calculator that had been altered to present simple arithmetic problems to a child. A correct answer led to another problem, a wrong answer to the message "EEE." The keyboard was decorated with an image of a bewhiskered and bespectacled professor holding a book. Questions and answers appeared on a red, light-emitting diode (LED) screen that, in combination with the top of the instrument, looked like a mortar board. In early examples of the toy, the on-off switch was on the right side near the professor's face and looked rather like a mortar board

Figure 15.7. Two educational games, the National Semiconductor Quiz Kid (ca. 1976), *left,* and the Little Professor (late 1970s). National Museum of American History collections, gifts of John B. Priser. Smithsonian Negative no. 2006-21181.

tassel. The machine sold for about $18 early in 1977, with the price dropping to $13 by the middle of the year. The Little Professor sold in the millions and continues to be produced, in modified form, to this day.[58] The Quiz Kid and the Little Professor were later joined by a range of similar mathematics teaching toys that included the Dataman (1977), Mathemagician (1977), and the Math Marvel (1980).[59]

Entertaining electronic arithmetic teaching machines found a ready market in the home. Debate about the appropriate use of calculators in the classroom continued far longer. In a 1999 interview Jeremy Kilpatrick of the University of Georgia, who had long been interested in the use of technologies in the classroom, observed that "calculators stir up emotions among parents and even some mathematicians that computers do not. And, yet, at least in our secondary program, we've had a fair amount of success in getting secondary teachers to use calculators in instruction in the secondary schools around here because they're much easier to manage."[60] We now turn to consider the less manageable, but perhaps less controversial, electronic computer as it entered the classroom.

CHAPTER SIXTEEN

Minicomputers
Drill, Programming, and Instructional Games

In approaching the problem of the interaction of educational technology with society, I shall first present a vision of technological possibility deliberately unclouded by economic or temporal realism. I shall then explore both some effects this vision, if realized, might have on the fabric of society and some of the factors likely to inhibit its realization.

Most mathematical teaching devices have a specific function. The cube root block was designed to explain an algorithm for finding cube roots. The protractor measured—and helped in drawing—angles. The slide rule assisted in calculation, especially multiplication and division. The textbook, the blackboard, the teaching machine, and the overhead projector all were used to teach a broader range of skills but basically were instruments for presenting information.

The computer is a different sort of object—a universal logic machine. The first computers, now often called mainframes, were room-sized instruments built during and after World War II to carry out calculations. They cost millions of dollars, were made to unique plans, and required a considerable staff to design, build, program, and maintain them. In the early 1950s companies such as Eckert-Mauchly Computer Company (later UNIVAC) and then IBM began manufacturing computers for both computation and data processing (Figure 16.1). These machines generally leased for thousands of dollars each month. Changes in electronics, most notably the discovery and standardization of the transistor, led to the introduction of somewhat smaller "minicomputers" in the early 1960s. By 1965 minicomputers such as the Digital Equipment Company (DEC) PDP-8 sold for about $18,000 and took up roughly the space of a closet (Figure 16.2). Some minicomputers could be used by several people simultaneously, with terminals

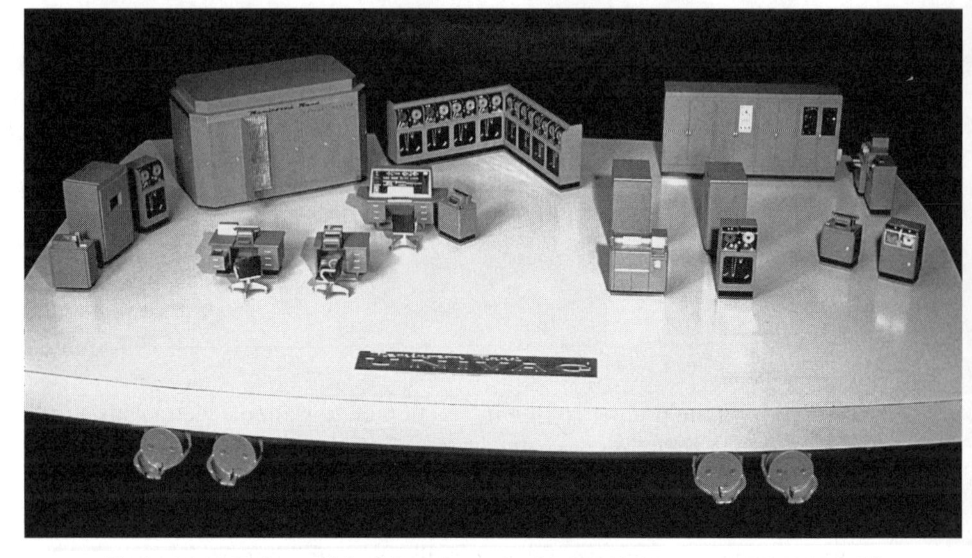

Figure 16.1. Scale model of the room-sized UNIVAC I computer, 1950s. Cf. the size of the computer components to that of the desks and chairs. National Museum of American History collections, gift of Univac Division of Sperry Rand Corporation. Smithsonian Negative no. 73-3061.

in distant locations. The terminals were sturdy enough to withstand even the sometimes rough treatment by students. Considering the expense, the unreliability, and the limited capabilities of even these "time-sharing" computers, it is perhaps remarkable that their educational potential was considered at all. Most mathematics educators confronted the electronic calculator long before they encountered such computer-based networks.[1]

Although computers would change radically, ideas that were developed with the time-sharing systems of the 1960s and early 1970s shaped the thinking about classroom use of the devices for decades. Four sorts of projects suggest the complexity and richness of these early encounters between computers and mathematical learning. The first kind built on military, industrial, and psychological interest in teaching machines. Computers were to serve as tutors and drill masters, offering patient, carefully ordered, individually tailored instruction to students. Users sat at terminals or teletype machines linked to a central processing unit. This style of learning, often termed computer-assisted instruction, was associated with the early work of the University of Illinois's Project PLATO (Programmed Logic for Auto-

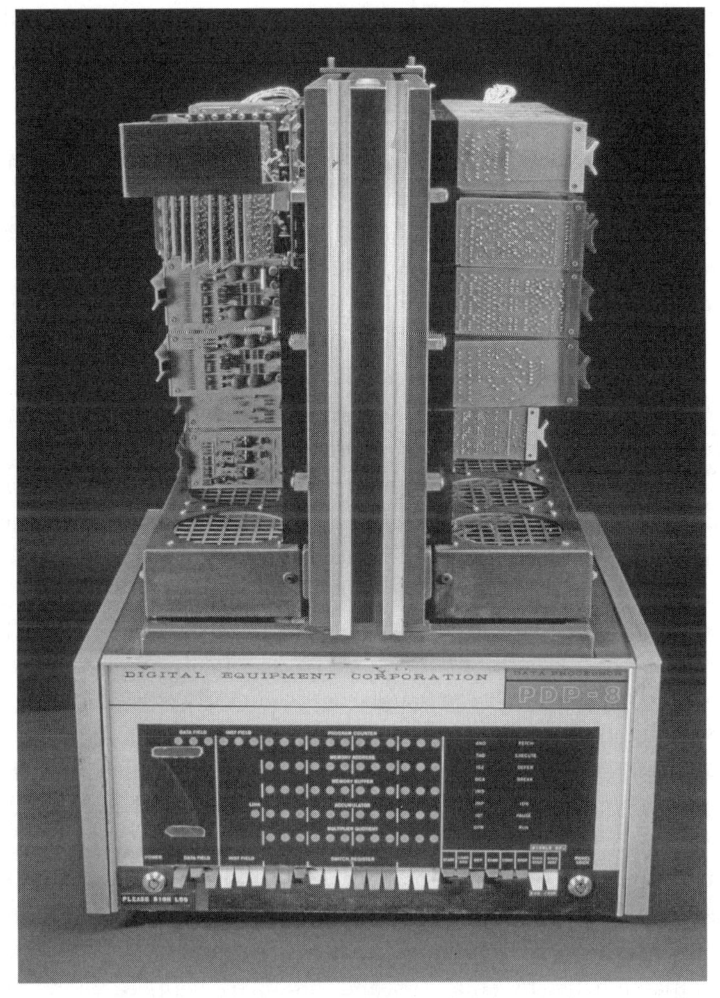

Figure 16.2. PDP-8 minicomputer, ca. 1965. National Museum of American History collections, gift of Digital Equipment Co. Smithsonian Negative no. 90-5950.

matic Teaching Operations), with research done at Stanford's Institute for Mathematical Studies in the Social Sciences, and with projects at several other universities. Project PLATO developed close links to Control Data Corporation (CDC), and even after the project ended in 1976, CDC would continue to market programmed instruction. The Stanford work developed in collaboration with IBM, which briefly manufactured the commercial System 1500 based on this trial.

A second group of mathematicians, educators, and computer scientists believed that students would benefit from learning to program computers themselves. The advent of computer time-sharing made the idea an expensive but not entirely unrealistic possibility. Work done at Dartmouth College, at the consulting firm of Bolt Beranek and Newman, and at the Massachusetts Institute of Technology took this approach. The programs resulted in a general view that computer terminals should be readily available to students and faculty at a college and perhaps a more elementary level. This assumption, as well as the graphical devices and relatively accessible programming languages that grew out of these studies, would prove important to early users of microcomputers.

A third group, including Anthony G. Oettinger at Harvard and Joel Moses and his associates at MIT's Project MAC, worked on developing systems to assist in solving more advanced mathematical problems. Using such programs initially required considerable expertise in computer science—far more than the average student might be expected to have. These programs were the starting point, however, of what would become widely accepted tools for both research and teaching in the 1980s.

Fourth, and finally, the first educational computer games date from this era. Attempts to teach computers to play checkers and chess were part of research on artificial intelligence from its early years. Ready access to time-sharing terminals led a wider swath of users to devise and share other games for their own entertainment. By the early 1970s a few games were designed to teach mathematical concepts, often using graphical displays. Some of this software was developed within larger curricular projects, by groups that included mathematicians, educators, and computer scientists. Because minicomputers were never widely used in education, these programs had only a limited influence in their original form. Some of them would prove highly influential in the early years of the personal computer.

General Considerations

Before considering specific uses of the computer in education, three general comments are in order. First, mathematics was at the center of much of this effort. The discipline had retained its long-standing role as a part of American education from the most elementary to the most advanced level.

Simply because of the ubiquity of mathematics teaching, it was a likely place to try innovations. Moreover, as the very word *computer* suggests, early computers were intended for mathematical use. To be sure, much American mathematical research of the 1960s emphasized highly abstract problems rather than computation. Nonetheless, many early computer users were trained as mathematicians, took an active interest in pedagogical matters, and were delighted to test the capabilities of the new machines. Nonmathematicians training future computer programmers and other technical specialists also needed to introduce mathematical concepts. Computerized lessons offered a way to do this. At the same time, United States government spending was particularly directed toward mathematics education. As we have seen in the case of calculators, after the launch of the USSR's *Sputnik* satellite in late 1957, special funds were directed to the purchase of equipment for teaching modern mathematics, sciences, and foreign languages. In the mid-1960s the focus of funding shifted to encouraging knowledge of basic skills, including elementary arithmetic. As in the case of calculators, manufacturers, university scholars, and school systems all responded to the opportunities provided by these grants.

Second, writing about the many programs designed to make use of computers in mathematics education offers numerous challenges. One difficulty is the diversity of programs, participants, and publications. University mathematicians, educators, psychologists, engineers, and computer scientists collaborated with colleagues, manufacturers, publishers, and widely scattered teachers. Systems were tried out not only in the United States but also in Canada, Central America, Britain, and Continental Europe. Programs were written for topics ranging from arithmetic to calculus and for students from elementary schools through universities. Another problem is in distinguishing between ideas advocated and systems actually in use. Thus, for example, the introduction of minicomputers did not immediately mean the end of educational programs for mainframe computers. Similarly, even when a manufacturer supported an educational system for only a relatively short time, as with the IBM System 1500, it might be retained by some institutions for years. At the same time, the mere availability of technologies did not mean that they played a role in the classroom. As with earlier devices such as the slide rule and the protractor, the existence of a technology was no guarantee that it would be applied in education. This point may seem less surprising for the expensive, relatively unreliable computers of the 1960s than for the slightly later electronic calculator.

Third, any discussion of such recent technologies necessarily lacks historical perspective. Now that the blackboard and the textbook have been available and widely used for well over 150 years, one can reasonably say that their introduction was of considerable importance to American mathematics teaching. It is far more difficult to judge which aspects of computer use will prove to be of enduring importance in learning mathematical ideas. The ready availability of inexpensive computing tools has led to questions about much of the traditional mathematics curriculum. How much basic arithmetic should students know by rote, when calculators are readily available? What should be the thrust of courses in algebra, when symbolic manipulation can easily be done by machine? How these questions will be answered is not yet clear. At the same time, networked computers made it easier for students to study at home or at work. Experiments with this approach took place in the late 1960s and early 1970s but only in a very small way. The long-term influence of "distance learning" and specific programs designed for the home or workplace on general educational practice is impossible to gauge at this point.[2]

Computer-Assisted Instruction

In March 1958, a few months after the launch of the Soviet *Sputnik* satellite, the Institute of Radio Engineers (forerunner of the Institute of Electrical and Electronics Engineers) began publishing a new journal called *Transactions on Education*. Many papers published here concerned specific topics relating to American engineering education. The new periodical also was a forum for discussing "automatic teaching devices," more generally known as teaching machines (see chap. 5). In its second issue *Transactions on Education* reprinted an article on these instruments by the physicist Simon Ramo, cofounder of the Ramo-Wooldridge Corporation in California. Here Ramo predicted that, just as new technologies had transformed transportation, production, and communication, they would alter education. Motion pictures and television would largely replace lectures, while special machines would be built for drill, memorization, and review.[3] The same issue of the journal contained an article by the physicist Paul K. Weimer of RCA Laboratories in Princeton, New Jersey. This paper, a reworking of a company memorandum that Weimer had written several years earlier, stressed potential economic and social benefits of automated teaching. As Weimer put it, "An exciting opportunity exists for a cooperative effort between educators and the electronics industry which could be of untold benefit to society and at the same time might open a vast new market for electronic equipment."[4]

Ramo and Weimer did not describe the machines they envisioned in detail. By December 1961, however, *IRE Transactions on Education* was able to offer more specifics. Of particular interest is a short article by Donald Bitzer, Peter G. Braunfeld, and Wayne W. Lichtenberger of the Coordinated Science Laboratory at the University of Illinois at Urbana. Bitzer and Lichtenberger were electrical engineers, while Braunfeld had a doctorate in mathematics from Illinois and would spend his career in the mathematics department there. The three authors described the PLATO system, which they had developed beginning in late 1960. It was a teaching machine that connected a student sitting at a "keyset" and television monitor with the Illinois Automatic Computer (ILLIAC), the university's mainframe computer. Questions appeared on the monitor screen, and students typed in their responses. As with B. F. Skinner's teaching machine, a correct answer led to a new question. A student who was totally stumped could press a help button that brought up slides offering further instruction. When he or she had mastered the material, pushing an aha button returned the student to the main program. The article describes an algebra program for introducing numbers expressed in arbitrary bases.[5]

By November 1962 Braunfeld and another colleague, the computer scientist Lloyd D. Fosdick, could report that PLATO had been extended to cope with two ILLIAC users simultaneously. They described a program for teaching computer programming to university students. Once again, the program included both text and type-in-the-blank problems. Only a small fraction of the students taking the course used PLATO, and their test scores differed little from those of other students. The prospect of teaching students by machine was sufficiently enticing, however, to encourage further experiments.[6] As a result of the efforts of Bitzer and his colleagues, the program was expanded in both size and scope. By 1966 PLATO III had twenty individual terminals and offered courses in algebra as well as languages, psychology, anatomy, and life sciences. Control Data Corporation provided one of its computers rent-free. A special programming language, TUTOR, made it possible for teachers to write their own courses.

PLATO originally had been financed using funds siphoned from grants to the Coordinated Science Laboratory for programs sponsored by various U.S. military agencies. Beginning in 1967, the project received funding from the National Science Foundation (NSF) and moved into its own Computer-

Figure 16.3. Students from Parkland College at PLATO III terminals, ca. 1970. Parkland is a community college in Champaign, Ill. From Records Series 39/2/20, Photographic Subject File, University of Illinois at Urbana-Champaign Archives. Photo courtesy of University of Illinois Archives.

Based Education Research Laboratory. By 1972 PLATO IV was in place, serving 146 locations and students from elementary school through college (Figure 16.3). The program developed much closer ties to CDC, which eventually acquired rights to develop and market PLATO software. By this time Bitzer and his associates had broken from the model of programmed instruction and devised methods of learning that students could more readily control. The computer was used to assist students in the development of logical, algebraic, and geometric proofs. It also could be—and was—widely used both to track student performance and for playing games. A few of these games were specially designed for teaching mathematics.[7]

At about the same time the PLATO project began at Illinois, scientists at IBM started using computers as experimental teaching machines. Early trials centered on forms of the IBM 650 computer, connected to typewriters. In addition to mathematical topics such as binary arithmetic and statistics, this early programmed learning included instruction in stenotyping and German reading. By 1964 IBM had developed the Coursewriter language for writing new curricular materials and was using computers in training its field engineers.[8]

Meanwhile, Patricia Suppes, the oldest child of Stanford philosopher Patrick Suppes, had entered elementary school. For Professor Suppes, as for Skinner, having a child in school led to new thinking about teaching. Suppes turned his attention particularly to mathematics, writing textbooks on geometry for primary school students and also preparing elementary texts on sets and numbers.[9] In the fall of 1962 Suppes and his colleague Richard Atkinson applied to the Carnegie Corporation for funds for a computer-based laboratory to study learning and teaching. The application was successful and was soon supplemented by money from the NSF. In 1964 Suppes and Atkinson received a large grant from the U.S. Office of Education to develop a program that offered supplementary drill in arithmetic and reading to elementary school students at Grant School in Palo Alto. By the 1965–66 school year students in grades 3 through 6 at the Grant School took a series of drills using teletype machines at their school that were linked to a PDP-1 minicomputer at Stanford.[10] The published log of the project suggests both the excitement students and parents felt at having access to a computer and the frustration of teachers and project staff with the unreliability of the system.[11] The following year Suppes and his associates opened the Stanford-Brentwood Computer-Assisted Instruction Laboratory at a school in East Palo Alto. By this time an IBM computer and peripherals were available for the project.[12] Federal legislation encouraged this emphasis on teaching basic skills. Both the Economic Opportunity Act of 1964 and the Elementary and Secondary Education Act of 1965 provided funds for improving the education of disadvantaged children.[13]

The educational system pioneered in East Palo Alto later sold as the IBM 1500 instructional system (Figure 16.4). It included a cathode-ray tube (CRT) display, light pen, and keyboard for each student; a central processing unit; and a central control unit. There were also components for projecting film and other images, providing sound, punching and reading cards, storing data, and printing. An installation with only one student station cost well over $600,000. IBM introduced the commercial form of the 1500 in 1967 and installed it in over thirty locations, most of them at colleges and universities, although the machine also was tried out in public schools in Kansas City, Missouri; Montgomery County, Maryland; Philadel-

Figure 16.4. A mathematics lesson on the IBM 1500 instructional system, 1966. National Museum of American History collections, gift of Science Service. Smithsonian Negative no. 2005-28924.

phia; and Pittsburgh. Although IBM had invested an estimated $30 million dollars in instructional systems, high costs, combined with cutbacks in federal education programs and a general decline in interest in programmed learning, led the company to begin phasing out the machines in 1970. Most users soon followed, although Pennsylvania State University ran mobile IBM 1500 stations as late as 1977, and the University of Alberta in Canada kept its model running until 1980.[14]

Suppes and his colleagues would go on to consider computer-assisted instruction for the deaf, focusing on mathematics and languages. He later worked on mathematics instruction in other media such as radio, carrying out a project in Nicaragua between 1974 and 1977. Suppes also experimented with computer-aided teaching at the university level. He argued that by using computers, universities could offer courses on topics of interest to only a small number of students and thus retain their intellectual richness in a time of diminishing resources.[15]

In addition to carrying out experimental studies, Suppes and his associates organized Computer Curriculum Corporation. The firm, formed in 1967, marketed software for teaching basic skills to students at work stations linked to a central computer. The company struggled at first and then saw several years of prosperity. In 1990 it was acquired by the publisher Simon & Schuster; eight years later the educational part of this firm was acquired by Pearson PLC and became part of its new Pearson Education group.[16]

CDC and IBM were by no means the only American companies who tried to build computers for educational use. A Chicago program to teach reading, language arts, and mathematics ran on a UNIVAC 418-III computer. In New York City an RCA Spectra 70/45 time-sharing computer was used for drill in elementary arithmetic. Hewlett-Packard also entered the business. With federal funds increasingly being diverted to spending on the Vietnam War, however, the enterprise faded.[17]

Learning Mathematics by Programming Computers: Dartmouth

To teach students as efficiently as possible, educational programs written as part of PLATO and for the Stanford computer-assisted instruction project were prepared by programmers and teachers, not students. In 1978 Stewart A. Denenberg of the Department of Computer Science at the State University of New York at Plattsburgh published a personal evaluation of the PLATO system. He found much to like about the teaching network, including not only the instructional software but also the library of games and the ability to post notes on electronic bulletin boards. Yet there was no provision that allowed students to program lessons themselves, as this took too much time on the central computer. Denenberg cited the specific example of a student named Bob Sawyer, who sought a lesson on plotting functions in polar coordinates. Finding no lesson in the PLATO library, Denenberg made special arrangements for Sawyer to learn to write such a program. The student gained a much more thorough understanding of the topic than he would have acquired from following a canned program. But PLATO could support only a few programmers. As Denenberg put it, "Here we have an instance of 'the student as captive' rather than 'the computer as pupil.'"[18]

As Denenberg well knew, a few computer users had long promoted computer programming as part of education. This approach was advocated for college students by John G. Kemeny, Thomas E. Kurtz, and their colleagues at Dartmouth College in New Hampshire. It was used with younger children by Seymour Papert of MIT and his collaborators at the Cambridge, Massachusetts, consulting firm of Bolt, Beranek, and Newman (BBN). Many of those involved in these projects were mathematicians, and their work had special bearing on mathematics education.

Kemeny, a Hungarian by birth, had come to the United States in 1940 to escape Nazi persecution. After wartime army service in Los Alamos, he completed an undergraduate degree at Princeton and stayed on there to obtain a doctorate in mathematics. He taught at Princeton for a short time and then joined the Dartmouth mathematics faculty in 1953. Kurtz, an Illinois native, received his B.A. degree from Knox College in 1950 and then also went to Princeton, earning his doctorate in statistics there in 1956. He joined the Dartmouth faculty the same year.[19]

Both Kemeny and Kurtz were interested in mathematics education. Kemeny and his Dartmouth colleagues wrote two textbooks designed to introduce students to a wide range of topics in modern mathematics. The first of them, coauthored by James Laurie Snell and Gerald L. Thompson and aimed principally at freshmen students in the social sciences and business, was entitled *Introduction to Finite Mathematics* (1957). It proved to be an enduring success, was widely used at Dartmouth and elsewhere, and spawned a host of imitators.[20] Kemeny also coauthored a text that introduced sophomores in engineering and physics to modern mathematics. Although the book never became as popular as the other, it was used at Dartmouth for some years.[21] Kemeny served on the Committee on the Undergraduate Program (from 1959 the Committee on the Undergraduate Program in Mathematics) of the Mathematical Association of America (MAA) from 1955 until 1963. One activity of this group was preparing experimental textbook materials at summer conferences. Dartmouth hosted such a conference in 1958, producing a volume on mathematical analysis for the social sciences that was partly written by Kemeny and distributed by the MAA.[22] In 1962 Kemeny and Snell formally published a text on this subject.[23] Kurtz was less active as an author, but in 1963 he published a statistics textbook that was a sequel to Kemeny, Snell, and Thompson's *Introduction to Finite Mathematics*.[24]

In addition to these activities, between at least 1955 and 1961 Dartmouth held a series of summer conferences on topics in mathematics education.[25] The school's mathematicians also kept abreast of contemporary developments in computing and artificial intelligence, bringing people to campus and traveling elsewhere. Kemeny had first heard of computers as a soldier at Los Alamos, when he attended a lecture given by John von Neumann. He saw computers in action as a summer advisor at the RAND Corporation in 1956. Kurtz's introduction to the new machines was at the 1951 Summer Session of the Institute for Numerical Analysis at UCLA. He later succeeded John McCarthy as Dartmouth College's representative to the New England Regional Computing Center at MIT. When Dartmouth acquired its first computer in 1959, Kurtz became head of the Dartmouth College Computer Center while retaining a position in the mathematics department.[26]

Faculty members who had left Dartmouth also influenced the course of computer development there. John McCarthy, another mathematician with a doctorate from Princeton, taught at Dartmouth from 1955 until 1958. Even after he moved on to teach computer science at MIT, he retained close ties with his former colleagues, keeping them abreast of his new ideas about time-sharing. This was a method for giving users of large computers the illusion that the machines were at their command. In the late 1950s most computer users submitted programs to be run on a computer in the form of a stack of punched cards. To make maximum use of limited and expensive computer time, cards were run in batches, with programmers returning the next day to find out how their efforts had fared. Frequently, a typographical error or other computer bug prevented the program from running at all. Kemeny, Kurtz, and McCarthy were frustrated by the slowness of batch processing, spurring McCarthy and his MIT colleagues to design a computer system that allowed several people to use a machine simultaneously. Kurtz thought such a machine would be helpful for both undergraduates and faculty, and with Kemeny he convinced NSF officials to fund an experimental program that provided much wider access to computers for both faculty and students. Dartmouth purchased two relatively small General Electric (GE) computers, one for communications and program scheduling, the second to run jobs. Students and faculty entered their programs into this time-sharing system via teletype machines and learned the results of their programs almost at once.

By the fall of 1964 Dartmouth had replaced one of its General Electric computers with a more advanced model, and the original two teletype terminals had grown to thirty-two ports, some at Dartmouth and a few at off-site locations. That winter GE programmers reviewed the code that students had prepared to run the system and increased its reliability significantly. GE would go on to sell a time-sharing system of its own. Kemeny, Kurtz, and their students also experimented with programming languages that would be easier for students to use than existing languages such as FORTRAN and ALGOL. After a few preliminary attempts, they developed BASIC (Beginner's All-purpose Symbolic Instruction Code), which first ran successfully in 1964. They also prepared BASIC manuals for use by faculty and students.[27]

Kemeny and Kurtz hoped to have large numbers of students use computers, not only for science and mathematics courses but also for business and social science applications. They believed that students should gain a general appreciation of the way computers functioned and were happy to have students contribute to the improvement of BASIC and to develop a variety of computer games. By 1972 Kemeny could boast that 90 percent of Dartmouth undergraduates knew how to use a computer and that most of them used the machine freely throughout their college careers. The heaviest use was by students of business administration and social sciences (Figure 16.5).[28] They encountered programming in courses in calculus and finite mathematics. A glance at a BASIC manual published by Dartmouth in 1965 gives examples of a wide array of computational problems that the young men were supposed to solve.[29] Dartmouth also encouraged students to use computers for nonacademic purposes. Mt. Holyoke College soon joined the Dartmouth time-sharing system so that—years before the general availability of electronic mail—men at Dartmouth could share files with women in South Hadley.[30] Dartmouth students and staff also wrote a host of computer games, some of which survived well into the era of microcomputers.[31]

Both the Dartmouth time-sharing system and BASIC were widely publicized at the time. As early as 1966, the National Council of Teachers of Mathematics (NCTM) endorsed BASIC as the language to use in education. Alterations in BASIC continued to occur, both at Dartmouth and at other institutions that adopted it. Particularly noteworthy for the educational use of computers were the versions of BASIC written for minicomputers produced by firms such as Digital Equipment Company, Data General, and Hewlett-Packard.[32] With the advent of microcomputers, those engaged in educational projects would adapt such programs for the new machines.

Figure 16.5. John Kemeny (standing, at *right*), Thomas Kurtz (*center*), and Dartmouth College students, with terminal, ca. 1968. Courtesy of Dartmouth College Library.

Learning Mathematics by Programming Computers: BBN and MIT

The programs developed at Illinois, Stanford, and Dartmouth all relied in part on ideas about computer time-sharing that had been pioneered in Cambridge, Massachusetts, by John McCarthy at MIT and his associates at the consulting firm of Bolt, Beranek, and Newman. BBN had originally developed out of work done at Harvard University on psychoacoustics and electroacoustics for the U.S. armed forces during World War II. After the war several Cambridge area scientists founded a firm that consulted on such topics as concert hall acoustics, noise control for air force jets, and noise reduction for U.S. Navy submarines. By 1957 Leo Beranek, a former MIT acoustics professor and the BBN president, wished to move the company into consulting in the general area of man-machine systems. To lead this new effort he hired J. C. R. Licklider, who had worked on the Harvard psychoacoustics project at Harvard during World War II and then led a group of psychologists at MIT's Project Lincoln. This group had developed the radar display for the SAGE air defense system.[33]

While at Project Lincoln, Licklider had also learned of a transistorized computer, the TX-2, which was being developed at MIT. This machine was unusually powerful for its day—users could program and interact with it in real time—and Licklider was enchanted. Once he was installed at BBN, he persuaded Beranek to try out computers at the company. The first was a Librascope LGP-30 and then, when this proved unsatisfactory, the prototype DEC PDP-1. The DEC computer worked sufficiently well that in early 1960 BBN decided to buy its own copy of the machine. Licklider was very interested in possible educational uses for computers and wrote programs to teach his own children spelling and vocabulary. Mindful of the importance of graphing in mathematics, in 1961 he prepared a program on exploratory learning with graphs; a user specified the coefficients of a polynomial, and the computer prepared a graph. In this very modest effort the student directed the computer, rather than simply supplying answers.[34]

The advent of time-sharing encouraged not only the developments at Dartmouth but a flurry of activity in Cambridge. By 1965 BBN had developed a small network of computer terminals, linked to a central computer by telephone lines. That year it established an Educational Technology Department under the direction of mathematician Wallace Feurzeig. One of the department's first projects was to develop an interactive computer language called Stringcomp, which was specifically designed to allow students to express and solve problems in algebraic language. Students in eight elementary and middle school classrooms tried out the language, preparing and running programs. The entire project was funded by a grant from the U.S. Office of Education. In the early 1960s only a handful of children had parents who, like Licklider, had access to a computer. For computer programming to play a role in mathematics education, using the machines had to be much cheaper. One way to reduce costs somewhat was for several people to use one computer simultaneously. To promote this possibility, and for other reasons, Licklider kept a careful eye on the work of John McCarthy. By 1961 McCarthy had prepared and demonstrated a new interactive programming language called LISP, which he implemented in a time-sharing environment on MIT's IBM 704.[35] BBN staff thought that they could use time-sharing on their much more modest PDP-1 computer, and by the summer of 1962 they had successfully done so. By the following summer MIT had established its Compatible Time Sharing System (CTSS), using a new IBM 7094 computer.[36]

Stringcomp was a modification of a programming language that had been developed at RAND Corporation for use by the professional programmers who worked on the JOHNNIAC computer. Experience with the language persuaded BBN staff—and their patrons—that students could learn to program. It also suggested that an entirely new programming language, specifically designed for children, was needed. In preparing this language, Feurzeig and his colleagues sought to keep programming conventions to a minimum but to allow for algebraic manipulation as well as numerical calculation. Toward this end they proposed a modified version of the LISP programming language developed by McCarthy at MIT. Feurzeig named the language LOGO, from the Greek *logos*, meaning the word or form that expresses a thought as well as the thought itself.[37]

In developing LOGO, BBN drew not only on its own staff but also on the advice of Seymour Papert. Like Kemeny, Kurtz, and McCarthy, Papert was a mathematician by training. A native of South Africa, he had obtained his B.A. and Ph.D. degrees from the University of Witwaterstrand, completing his dissertation in 1952. Papert then spent several years in England, France, and Switzerland, obtaining a second Ph.D. degree in mathematics from Cambridge University in 1959. He was particularly influenced by psychologist Jean Piaget, working with him from 1958 to 1963 at the Center for Genetic Epistemology in Geneva. Piaget, a great admirer of Maria Montessori, firmly believed that students learned most successfully through guided exploration of the world around them. This assumption would be fundamental to Papert's later work with children and computers.[38] It also influenced some early instructional computer games.

In 1963 Papert left Geneva to become a research associate in electrical engineering at MIT. Two years later he was a sufficient authority on artificial intelligence and neural networks to write the introduction to Warren S. McCulloch's pioneering book *Embodiments of Mind.* During these years he worked closely with his MIT colleague Marvin Minsky. In 1967 Papert became professor of applied mathematics at MIT and succeeded John McCarthy as codirector of the Artificial Intelligence Laboratory. That summer the Office of Naval Research sponsored a pilot study using LOGO to teach fifth- and sixth-grade students at the Hanscom Field School in Lincoln, Massachusetts. This effort led to a reworking of the language and a more extensive trial at public schools in Newton, Massachusetts. The NSF-funded project lasted from September 1968 through November 1969. In a 1970 report on their work, Feurzeig, Papert, and their collaborators explained in some detail why they believed that computer programming enhanced mathematical learning among children. They argued that the process of figuring out algorithms not only facilitated rigorous thinking and expression but also related formal procedures to intuitive understanding. Students were not simply told the meaning of algebraic procedures but actually applied them and saw the results. Programming also supplied a language with which to discuss problem solving. Moreover, mathematical learning could be tied to other subjects. Early topics for LOGO programs included the use of substitution ciphers in cryptography, logic games such as Nim, and translation of English into pig Latin.[39]

Papert recast many of these ideas in a series of mimeographed chapters he prepared in the course of 1970 for a Smithsonian Institution project on the technical augmentation of cognition. Here he argued—as he would in later published work—that LOGO gave students access to a new way of learning. Children in France learned French simply by listening and imitating the language they heard around them. Similarly, elementary school children who programmed in the LOGO "environment" lived in a "Mathland" and would learn the language of mathematics. Student difficulties in mathematics could be attributed to the way they had been taught previously, not to any genetic limitations in their capabilities.[40]

The first programs that children wrote in LOGO generated text and numbers but did not move any objects or icons. By March 1970 Papert suggested that students might learn particularly well if their programs controlled the motion of a robotic turtle.[41] MIT investigations of artificial intelligence had involved such turtles from the late 1940s. By 1972 the LOGO project had moved from BBN to the MIT Artificial Intelligence Laboratory, and laboratory staff had designed a robotic turtle that children could program to move in desired patterns (Figure 16.6). Figuring out how to program the motion of the turtle—and have it draw patterns—was a central part of the project.[42] The robotic turtle figure would soon give way to icons on a screen, although much of the playful spirit suggested by "turtle geometry" endured. So did the conviction that children in grade school and older could learn a great deal about mathematics and other subjects by preparing and running LOGO programs that allowed them to explore what Papert came to call "microworlds."[43] With the advent and widespread sale of the microcomputer in

Figure 16.6. Seymour Papert and students using a Turtle, 1973. Courtesy of MIT Museum.

the 1980s, LOGO would become available to a much larger group of children, both at school and at home.

Computer Tools for Mathematical Manipulation

While mathematicians and programmers at Dartmouth, BBN, and MIT's Artificial Intelligence Laboratory explored the possible role of computer program writing in mathematics education, others took a somewhat different tack. They wrote programs that students and faculty could use to solve specific mathematical problems. One early program of this sort was Project TACT (Technological Aids to Creative Thought) at Harvard University. Anthony Oettinger of the Division of Engineering and Applied Physics at Harvard and his colleagues developed an online computer system known as THE BRAIN (The Harvard Experimental Basic Reckoning And Instructional Network). Operating from 1964 through 1968 under a grant from the U.S. Department of Defense's Advanced Research Projects Agency, the project offered university students and faculty an opportunity to explore problems in mathematics and the mathematical sciences. Participants initially used MIT's Compatible Time Sharing System and then relied on a graphical interface with an IBM 360 Model 50 computer. A subsidiary project, begun in April 1967 under the direction of Harvard faculty member Ivan Sutherland, used a PDP-1 computer with attached display scopes as its main facility and concentrated on computer graphics and computer networking.[44]

Meanwhile, computer scientists at MIT had begun work on a new interactive computer known as MAC. The initials MAC were variously interpreted as an acronym for Multiple Access Computer or Machine Aided Cognition, suggesting the link that had developed between time-sharing and the use of computers as aids to knowledge. One part of Project MAC focused on developing programs for symbolic manipulation. Beginning in 1968, Joel Moses, William A. Martin, and Carl Engleman led a team that planned a series of programs written in LISP and related languages with wide mathematical capabilities. This software not only performed symbolic integration and differentiation but manipulated vectors and matrices, solved systems of equations, and computed a wide range of special functions. These programs were distributed as MACSYMA (Project MAC's SYmbolic MAnipulator). Even when Project MAC ended, MACSYMA lived on, taking different forms at MIT and other research institutions. From 1982 until 1999 it was distributed as a commercial product.[45]

The results of these projects were not primarily designed as teaching tools. The programs were intended to assist the working scientist, engineer, and mathematician, not to provide instruction. Nevertheless, they provided a foretaste of the computing capabilities that would become available in the classroom with the advent of the microcomputer and the graphing calculator.

Instructional Games

In the late 1970s and early 1980s, as microcomputers entered the home, educational games became a familiar product. These tools, too, had their roots in efforts of the 1970s. In planning the PLATO IV computer, Bitzer and his colleagues at Illinois designed a plasma display terminal with which students could interact using a keyboard.[46] One use for this new apparatus was playing games. The PLATO library came to include games for teaching students from elementary school through community college. Robert B. Davis, who had been active in New Math projects at Syracuse University, became a professor of education at Illinois in 1972 and soon became deeply involved in Project PLATO. Davis and his colleagues rejected frame-by-frame computer-assisted instruction, preferring to take advantage of the graphical capabilities of PLATO and to allow more self-directed learning. Like Papert, they were influenced not so much by Skinner's behavioral psychology as by Piaget's view that students constructed their own mathematical knowledge from personal experience. They also emphasized interaction among students, not strictly individual students encountering machines.[47]

Interactive games played a particularly prominent role in the development of curriculum materials for elementary school students in grades 4 through 6. Davis organized three teams of developers for the project, one to work on the teaching of whole numbers, one on graphing, and the third on fractions. The curricula prepared by all three of these groups included interactive games. To teach about whole numbers, there was the game How the West Was Won, a variation on the board game of Chutes and Ladders. Players were given three single-digit numbers that could be combined using any of the four arithmetic operations. The value of the result determined how far the student moved forward or back. As in the board game, there were various shortcuts along the way.

This PLATO game attracted the attention of John Seely Brown, Richard Burton, and their colleagues at BBN, who designed a similar game called West.[48] A version of the game would be available for Apple II computers in the late 1980s.[49] Donald Cohen and Gerald Glynn of the graphing curriculum group developed a set of Guess My Rule games for that project. Another program—Sharon Dugdale and David Kibbey's Darts—which would receive wide distribution in later years, was developed for the fractions curriculum. This game allowed students to explore the placement of rational numbers on a number line. The computer display showed balloons tied to points on the line, and students typed in the digit represented by the point. If the answer was correct, a dart moved across the screen, and the balloon popped. If not, the dart landed on the line (Figure 16.7).[50] Students found the game absorbing and developed strategies for estimating positions on the number line quite different from those that staff members had anticipated. They also observed and adopted techniques used by other students.[51]

PLATO instructional games were not confined to elementary schools.

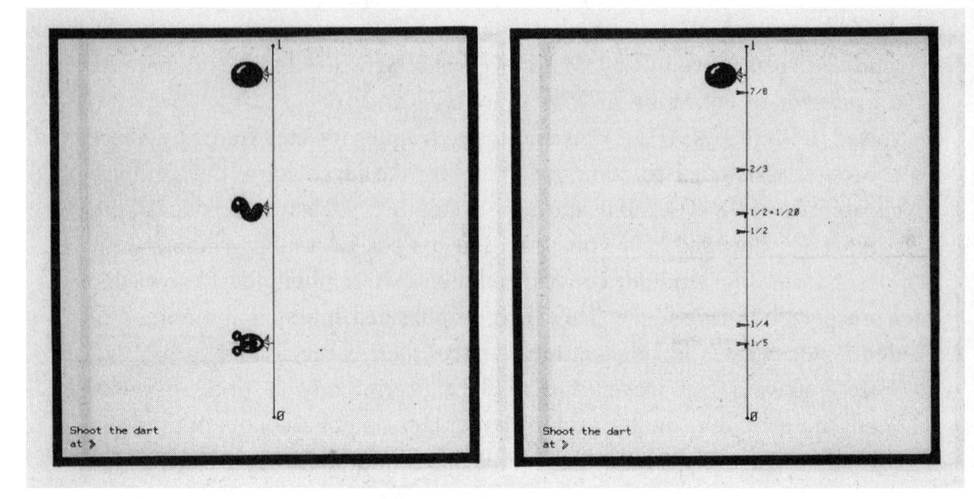

Figure 16.7. Darts, as described in Sharon Dugdale and David Kibbey, "The Fractions Curriculum, PLATO Elementary School Mathematics Project," Computer-based Education Research Laboratory (CERL), March 1975, Frances Day Papers, Record Series 15/14/32, University of Illinois at Urbana-Champaign Archives. Photo courtesy of University of Illinois Archives.

Patrick H. McCann of the U.S. Navy's Personnel Research and Development Center in San Diego tried using some of the PLATO IV mathematical games in the instruction given at the basic electricity and electronics school at his center and reported that gaming was more attractive to the young enlisted men than conventional drill.[52] Game development would continue at Illinois and elsewhere well after the advent of the microcomputer.

CHAPTER SEVENTEEN

Early Microcomputers
The Lure of Novelty

> Set in context, then, the computer will be *an* agent of change. It is and will continue to be one in a series of very important technical, intellectual, political, economic and social agents that affects who learns what, how, why and where. It is not *the* agent of change, let alone *the only* agent of change in casual or personal learning or in formal education.

By 1975 it was possible to place the central processing unit of a computer onto a small silicon chip. In January of that year a small electronics company in New Mexico introduced the Altair, a commercial "microcomputer" kit that hobbyists could order and assemble. Microcomputers soon were available that were sufficiently sturdy and accessible to sell as consumer products, and by 1985 millions of them were in homes and offices. Parents, teachers, computer manufacturers, educators, computer scientists, and software publishers suggested that the new machines could and should be put to use for educational purposes. Literally thousands of programs covering a wide range of subjects—from counting and arithmetic to calculus and computer algebra—were written for mathematics education. Costs ranged from very little to hundreds of dollars. Programs were distributed in many forms—as published code to be laboriously copied by users, on punched paper tapes, on both audio and "solid state" cassettes, and on floppy disks. Distributors were diverse, including book and journal publishers, educational consortia, microcomputer manufacturers, and commercial vendors of software and other educational materials. A few companies established in the late 1970s and early 1980s survived into the next millennium; most did not.

Many of the early programs that were sold for mathematics teaching with microcomputers followed patterns established with minicomputers. Existing programming languages, especially BASIC, were widely used. Some programs were specifically designed for drill and practice, often in the form of games. Others promised to offer tutoring as well as drill. Both BASIC and LOGO were also touted as ways to train students in programming and logical thinking. Finally, expert systems offered assistance in such topics as symbolic manipulation.

Gauging the impact of this software is exceedingly difficult. Few statistics are available. Bootleg copies of programs circulated freely. Programs were often sold with machines or as one of a group, and there is no guarantee that they were ever opened, let alone studied. Moreover, much of this software was designed for home use by young children. In 1984, for example, Future Computer, Inc., estimated that parents spent five and a half times more ($110 million dollars) on educational software than schools did.[1] This activity generated relatively few public records. Furthermore, while some microcomputer applications such as word processing and spreadsheets were adopted quite quickly and have received historical attention, the widespread application of microcomputers to mathematics education proved a more elusive goal.[2] There were few major success stories to report. Finally, for historians accustomed to working with objects, computer software is notoriously ephemeral. Programs changed rapidly and are not as easy to preserve or study as such tangible artifacts as textbooks, geometric models, or graph paper.

This chapter discusses examples of early mathematics teaching software for microcomputers, primarily as they happen to survive in the collections of the Smithsonian Institution. Most of these programs were collected with the microcomputers on which they ran. This software includes a version of BASIC for the Altair, a few educational programs written in that language, a form of LOGO for microcomputers, and an early computer algebra system written in LISP. While this selection is far from exhaustive, the stories of these programs demonstrate the enthusiasm of early microcomputer users, their intense interest in the new machines, and their hopes of transforming educational practice. These efforts would do much to make the microcomputer commonplace, whether or not they were classroom or entrepreneurial successes.

Both early microcomputers and software written for them were to a large extent products of people living in the United States. Design and manufacture of the computers, as well as writing and distribution of mathematics software, were to an unusual degree an American activity.[3] These new tools of American mathematics teaching were largely forged at home. American hobbyists, schools, and families also were major consumers of the new products.

BASIC and the Early Microcomputer

In 1978 the Special Interest Group on Programming Languages (SIGPLAN) of the Association for Computing Machinery sponsored its first conference on the History of Programming Languages. Thomas E. Kurtz of Dartmouth College spoke at the session on BASIC, describing the origins of the language and the six editions of it that had been introduced through 1971. Kurtz noted that BASIC might better be considered a class of languages, not a single language or even a language with several dialects. Both minicomputer manufacturers and hobbyists wrote their own versions of BASIC, incorporating their own innovations. Efforts to standardize at least some aspects of the language were under way. Meanwhile, Kurtz was delighted by its wide use. He could "only wildly guess that perhaps five million school children have learned BASIC on some minicomputer."[4]

Although Kurtz did not mention the matter, BASIC also was the language of choice for many users of the newly introduced microcomputer. Recall that the Altair was introduced in January 1975. By the summer of that year Harvard University students Bill Gates and Monte Davidoff, working with Honeywell employee Paul Allen, had written a version of BASIC for the Altair (Figure 17.1). This would be one of the first products sold by a firm they established that would be known as Microsoft. This company licensed its software not only to the makers of the Altair but to numerous other computer manufacturers.[5] Others soon followed suit.[6] By 1977 the author of an article on another programming language described BASIC as "the universal language of the microcomputer hobby."[7] BASIC, at least in modified form, could be accommodated in the small memories of early microcomputers. It was designed to be used by students and others not trained as computer programmers. For both of these reasons it was well suited to the new machines.

The first microcomputers were designed for hobbyists, not schoolchildren. They required considerable assembly, were programmed by flipping switches, and had no monitors. Despite such failings, a few schools tried to put them to educational use. One of the first to do so was John F. Kennedy High School (later John F. Kennedy Catholic High School) in Somers, New York. The school had relied on free access to a minicomputer through the

Figure 17.1. BASIC tape for the Altair, 1975. National Museum of American History collections, gift of Forest M. Mims III. Smithsonian Negative no. 1995-4900.

Board of Cooperative Educational Services of the state of New York. In late May 1975 it was informed that this service would not be available the following year. To teach courses in calculus and in programming scheduled for the fall, the school needed a computer. Alexander S. Lett, the father of student Christopher Lett, was an MIT graduate in electrical engineering and a computer engineer at IBM. He had read about the Altair and suggested to Sister Christopher O'Toole of the mathematics department that the high school should buy its own computer. She agreed and placed an order for an Altair kit, with two memory boards, an interface board, and BASIC on paper tape. After considerable delays, the kit arrived in October. Christopher Lett later recalled the sight: "The kit consisted of bags and bags of electronic components, more than four empty circuit boards, a metal case, and more wire than I had ever seen before."[8] The Letts took two weeks to assemble the machine and to attach it to a rented teletype that provided a paper tape punch and reader for input and output. In a June 1976 article in *Byte* magazine Christopher Lett described his experience ordering and assembling the computer. He also reported that the Altair was kept turned on from Monday morning to Friday afternoon, not only because it took twelve minutes to load BASIC but because users wished to avoid wear and tear on the paper tape. Alexander Lett had modified the Altair slightly so that students could not crash the system easily, but Christopher was called out of class "on more than one occasion" to reboot the system.[9] Students either loaded their own programs or selected from a library of routines in such subject areas as "mathematics, chemistry, and physics problem solving, lab simulations, text editing, and puzzles and games."[10]

The Altair remained in use at John F. Kennedy High School for six years, while Christopher Lett went off to begin a career in computer science. In 1977 a second generation of microcomputers was introduced onto the market. These machines boasted such features as monitors, keyboards, disk drives, and considerable libraries of subroutines. Prominent among them were Tandy Radio Shack's TRS-80, Apple Computer Company's Apple II, and the Commodore PET. John F. Kennedy High School, for example, purchased five Radio Shack TRS-80 IIIs in 1981. BASIC was soon available on these machines as part of the read-only memory, obviating concern about wearing out paper tapes.[11] Indeed, by the middle of 1979 CompuSoft Publishing could sell a handbook that claimed to explain in detail "the BASIC language as used in over 50 micros, minis and mainframes."[12] The success of these products in business as well as the home would lead IBM to enter the microcomputer market in 1981 with the IBM PC.

From June 1977 through June 1979 Apple II computers were sold with a version of BASIC known as Integer BASIC that was written by one of the firm's founders, Steve Wozniak. Apple also sold a few educational programs written in this language. For children twelve years and up, there was a set of four programs sold under the title Elementary, My Dear Apple. Three of the programs concerned mathematics and problem solving. The first, Lemonade, was a game for one or two players that simulated the operation of a lemonade stand. According to a description provided by Apple, a player "tries to make profits each day from his or her lemonade stand by estimating the number of glasses that will be sold, setting the price per glass, forecasting the cost of material and advertising, and anticipating the effect on sales of changes in the weather."[13] The game had its roots in a program developed earlier in the 1970s for minicomputers by the Minnesota Educational Computer Consortium. Initially sold on cassette, by 1982 Lemonade for the Apple was offered on diskette (Figure 17.2).[14] Versions of Lemonade were also sold for the Commodore 64 and an early Atari microcomputer. The Pleasantville, New York, firm of Sunburst Communications soon developed a similar program, sold as Hotdog Stand, for both the Apple II and the TRS-80. More elaborate versions of Lemonade have been used in teaching economics as well as mathematics.[15]

Figure 17.2. Mathematical software on cassette and floppy disk: Lemonade for the Apple II (1979), Graphing Package for the TRS-80 (1978), Edu-Ware Algebra (n.d.), muMath (ca. 1980), In Search of the Most Amazing Thing (1983), and Apple Logo II. Lemonade was bundled with the program Hopalong Cassidy; this photograph shows that side of the cassette. National Museum of American History collections. Smithsonian Negative no. 2005-26065.

Lemonade was designed to teach problem solving, not specific arithmetic skills. Supermath, another of the mathematics programs in Elementary, My Dear Apple, provided simple but colorful drills in arithmetic. It was written internally by Apple's Bruce Tognazzini. The third mathematical game in the Apple package was a version of Darts, the program developed by Sharon Dugdale and her associates in the PLATO project at the Computer-Based Education Research Laboratory of the University of Illinois (see chap. 16).[16]

These three programs represent only a small fraction of the BASIC programs developed for mathematics education. Computer magazines often published program listings, either in their text or as separate books. In 1979, for example, Jim Perry and Chris Brown published a book giving the listings for eighty programs in TRS-80 BASIC. Their volume was published, in part, for charter subscribers to the magazine *80 microcomputing,* who presumably might wish to have something to try out on their microcomputers. Fourteen of the eighty programs were educational. Five of them concerned aspects of arithmetic, and a sixth related to geometry.[17] Other BASIC programs were written and distributed by computer users. In 1985 Bertram Gader and Manuael V. Nodar published *Apple Software for Pennies.* According to the cover, the book listed over 2,000 programs for the Apple that were available in the public domain; they were distributed by some of the 270 Apple clubs in existence at the time the book went to press. Some 375 of the programs listed were classified as "educational," with 118 of them relating to mathematics. The largest distributor of this public domain mathematics software, with 76 of the 118 titles, was the Apple Corps of Dallas. Some of these programs were written by members of the corps, some were distributed on disks by the International Apple Core of Santa Clara, California (the umbrella organization of the Apple clubs, founded in 1979), and some were written by members of other clubs. The titles that can be dated seem to have been written between 1980 and 1983. Programs taught aspects of arithmetic, weights and measures (especially the metric system), algebra, geometry, trigonometry, graphing, calculus, linear algebra, logic, and number theory. Most of the programs were written in Applesoft BASIC, with some in the earlier Integer BASIC.[18]

Other programs were distributed by commercial companies. In 1979 the Maynard, Massachusetts, firm of EDU-WARE began selling a set of four programs under the name of MATH-PAK I. An advertisement in *80-U.S.: The TRS-80 Users Journal* described the program as "an absolute must for every TRS-80 user with elementary school age children." These programs introduced the basic operations of arithmetic. In addition to checking for errors and automatically positioning digits to simulate handwritten procedures, the package offered users a selection of difficulty levels, scoring, and games. Math-Pak I sold for only $14.95 and was aimed at young children.[19] Not long thereafter, another firm, Edu-Ware Services, Inc., of Woodland Hills, California, introduced programs for teaching not only arithmetic operations, fractions, and decimals but counting, algebra, and statistics.[20] Edu-Ware Services also published other titles relating to reading, spelling, computer programming, and perception. All of its programs sought to offer not only drill but also instruction—the company's motto was "unique software for the unique mind."[21] The prices for programs varied considerably. A set of six statistical programs sold in 1980 for $9.95 (on cassette) or $15.95 (on diskette).[22] Two-diskette packages on fractions and on decimals sold that same year for $39.95 each.[23] By 1982 the first of six diskettes offering instruction in algebra was available for $34.95.[24] The company's programs ran on the Apple II, the Commodore 64, and later the IBM PC. Edu-Ware left much of its marketing to long-established Britannica, Inc., but the alliance did not lead to overwhelming sales. The company soon was acquired by Management Sciences America, better known as the owners of Peachtree accounting software.[25] Not all computer owners felt compelled to purchase the software Edu-Ware sold. A diskette in the Smithsonian collections apparently contains a backup—or bootleg—copy of Edu-Ware's first algebra program (see Figure 17.2).

Educational programs written in BASIC could be relatively elaborate fantasies. In 1983, for example, the Cambridge, Massachusetts, firm of Spinnaker Software Corporation published In Search of the Most Amazing Thing (see Figure 17.2). Versions of this computer game would be available for the Apple II, Atari, Commodore 64, and IBM PC, for a price that ranged from $22.95 to $32.95, depending on the make of computer. The vendor promised that the software would "sharpen your ability to estimate distances and quantities, become aware of distance, direction and time, solve problems through trial and error, and develop a knack for economic and monetary principles."[26] These goals were relatively far from conventional drill and practice in arithmetic. They suggested, however, an emphasis on logic and reasoning skills that would loom large in the literature on late-twentieth-century mathematics education. Producing such games required more than the efforts of a lone teacher or hobbyist. According to literature received with the game, the program was created by Thomas F. Snyder, with programming support by Omar H. Khudari, editorial development by Karen Whittredge, and program graphics by Gabrielle M. Savage. A short book by yet another author outlined the story presented to players and was included in the package. Spinnaker also invested considerable sums in marketing its products to consumers generally.[27]

Meanwhile, there was considerable interest in applying electronic technology to graphing. As early as 1978, the magazine *Creative Computing* published a set of six microcomputer programs for use in showing graphs on the TRS-80. They were written in Level II BASIC and sold for the modest price of $7.95 (see Figure 17.2).[28] Further graphing software came from long-established projects. In 1980, with support from the National Science Foundation (NSF), Sharon Dugdale and her colleague David Kibbey at the University of Illinois wrote the program Green Globs. Students were shown a monitor with coordinate axes and thirteen green globs scattered randomly at points on the grid. They entered equations—if the equation intersected one or more of the points, the glob(s) burst. If the graph missed, feedback gave diagnostic information. The system of scoring used in this instructional game encouraged hitting more than one glob at once. Dugdale reported that students who played Green Globs took greater interest in a more routine program on graphing equations and, sometimes, in the textbook. Green Globs initially was programmed in a version of the PLATO project's TUTOR language known as microTUTOR and used on a short-lived Control Data Corporation microcomputer known as the Color MicroPLATO system. Kibbey soon prepared a version of Green Globs written in BASIC for the Apple II+. This was one of four programs designed as part of a package on studying graphs called "Graphing Equations." A second program plotted functions, a third offered drill and practice in graphing, and the fourth was another game. The software was distributed from the University of Iowa by the nonprofit organization CONDUIT and later sold by Sunburst Communications.[29] Much more elaborate graphics, generally written in languages other than BASIC, were soon available and came to be widely used in both business and education.

Meanwhile, John G. Kemeny and Thomas E. Kurtz viewed the commercial development of diverse forms of what they called "street BASIC" with some concern. In 1984 they encouraged the establishment of a new software company that offered as one of its major products what they called "True BASIC."[30] The language gained some popularity, particularly among academic users. Other languages, however, were rapidly becoming popular in mathematics education applications. Owners of the TI-99 home computer, for example, could buy instructional programs written in Graphic Programming Language, a language designed especially for Texas Instruments.[31] Versions of the languages LOGO and LISP also were implemented on microcomputers, as shown in the following sections.

LOGO and Turtles for Microcomputers

Microcomputer owners were not exclusively interested in BASIC. As manufacturers, teachers, and parents considered how children might use the new machines, a few MIT students explored the possibilities of the work of Seymour Papert and his colleagues (see chap. 16). In May 1978 the *Christian Science Monitor* reported that the newly established firm of Terrapin, Inc., planned to build and sell robotic, microcomputer-controlled "Turtles." By August *Byte* had received the first press release for the new Boston company. This document heralded the coming of "the world's least expensive commercially sold mobile general purpose robot."[32] Further details, along with a photograph, appeared in an advertisement the magazine ran in October. Turtles were "small home robots controllable by your computer." They sold as a kit for $300 or fully assembled for $500. An interface to a microcomputer cost another $40. The beasts reportedly could walk, talk, blink, draw, and feel. The advertisement suggested that buyers might use a turtle to "map rooms, solve mazes, dance, explore Artificial Intelligence, [and] teach geometry or programming."[33] As this list suggests, mathematics played only a small part in the role envisioned for the new robot. Indeed, at least one early account suggests that assembling a turtle was considered more of a project for a class in electronics than for mathematics students. James A. Guptan Jr. described the experience of students at Union County Career Center in Charlotte, North Carolina, who assembled a Terrapin Turtle. His 1979 article in *Byte* was replete with circuit diagrams but indicated that the class did not yet have a computer to order the turtle around.[34]

By 1981 Terrapin Turtles could be driven by Apple or Atari computers or indeed any microcomputer with a relatively standard architecture (an S-100 bus).[35] Efforts had also begun to run the programming language LOGO on home and school microcomputers. In 1980 Seymour Papert presented his views on the potential of the language in *Mindstorms: Children, Computers, and Powerful Ideas.* The same year several of his articles were republished in a volume edited by Robert Taylor entitled *The Computer in the School: Tutor, Tool, Tutee.*[36] Papert's publicity paid off—Texas Instruments (TI) introduced a version of LOGO for its TI-99/4 in 1981. This product, jointly developed by MIT's Artificial Intelligence Laboratory and TI, was sufficiently successful that Papert and colleagues founded Logo Computer Systems, Inc. (LCSI). This software company designed Apple Logo for enhanced versions of the Apple II, Smart Logo for a game-playing microcomputer called the Coleco Adam, and PC Logo for the IBM PC. MIT also licensed other software companies to prepare versions of LOGO. In December 1984 Richard Roth reported in *Creative Computing* that: "turtles and mice are the 'in' animals in the personal computer field this year. As in nature, microcomputer mice have both supporters and detractors, but almost everyone agrees that turtles are beneficial beasts—on the screen as well as in the stream."[37] Roth compared the several available versions of LOGO, noting variations in price, sound, parameters used, turtle shape, and documentation. In addition to commercial products, he mentioned Dave Smith's Ladybug Logo for the IBM PC, which was distributed for free with a request for donations from users (reportedly, few who ordered the program responded with cash). In all of these microcomputer versions of LOGO the robotic turtle gave way to an icon moving across a screen, as in the original version of the language created by the Cambridge, Massachusetts, firm Bolt, Beranek, and Newman.

Early advertisements by LCSI, as well as Papert's publications and lectures, stressed that LOGO would allow children not only to program but also to think about logical processes, setting it apart from the usual drill and practice routine. To quote the advertisement:

> In most contemporary educational situations where children come into contact with computers, the computer is used to put children through their paces, to provide exercises of an appropriate difficulty, to provide feedback, and to dispense information. The computer is programming the child. In the LOGO environment the relationship is reversed. The child, even at preschool ages, is in control: The child programs the computer. And in teaching the computer how to think, children embark on an exploration about how they themselves think. Thinking about thinking turns the child into an epistemologist, an experience not shared by most adults.[38]

A slightly later advertisement did not use the term *epistemologist* but claimed, "Working in the LOGO environment can inspire the programmer to 'think about thinking.'"[39] Several scholars wrote specifically about the use of LOGO in mathematics teaching.[40] By the time Apple Logo II appeared in 1984, however, the documentation for the program made no mention of epistemology or mathematics. Instead, it emphasized LOGO as a way of generating graphics (see Figure 17.2).

Research on LOGO continued, and the language is still available from LCSI and Terrapin, Inc. In some cases students were urged not only to program the motion of virtual turtles but to construct robots from LEGO blocks. Yet LOGO has not come to play the vast role in learning—at school or at home—that Papert and his associates envisioned. Many parents and school systems were left without support for LOGO when Texas Instruments decided to stop making the TI-99/A in 1983.[41] Microcomputer programs such as MacPaint produced graphics with considerably less effort. More generally, the programming of computers increasingly was left to professionals, and a knowledge of programming languages did not seem necessary as part of every child's education.

muMATH and Computer Algebra Systems

Microcomputer users did not confine their interest in programming languages to BASIC and LOGO. Versions of FORTRAN, COBOL, Pascal, and other established languages soon appeared for the small machines. One of the first complex languages to be implemented on a microcomputer was that favorite of artificial intelligence research, LISP. Two computer enthusiasts in Hawaii, Albert D. Rich and David R. Stoutemyer, firmly believed that many mathematicians and scientists who could benefit from using LISP and related computer algebra systems were ignorant of their existence. They set out to develop an inexpensive version of the language for these users. Before describing what they did, a bit of biographical information is in order.

Rich and Stoutemyer's collaboration suggests both the growth of the community of computer scientists in the United States and the interaction between microcomputer enthusiasts and established computer experts that shaped the development of early microcomputer software. Rich, a California native raised in Texas, had become fascinated with LISP as an undergraduate at the University of Texas. While there, he collaborated with computer science professor Laurant Siklossy on a paper describing an algorithm that Rich had invented for proving theorems in propositional calculus. Rich graduated in 1971 with a B.A. degree in mathematics and entered the U.S. Navy, spending much of his time in service as an officer on the USS *James Monroe*. This submarine had its home port in Honolulu, and Rich became very fond of Hawaii. On his discharge in 1976 he settled there. Being very interested in computers, he purchased and assembled an Imsai 8080 and set out to write a LISP interpreter for his machine. By the end of 1976 the interpreter was running.[42]

Rich's version of LISP soon came to the attention of David Stoutemyer, a professor of computer science at the University of Hawaii. Stoutemyer was a graduate of the California Institute of Technology who had received his master's degree in engineering from MIT and had been one of George Forsythe's last doctoral students in the computer science department at Stanford. He received his Ph.D. degree in 1971 and soon was deeply involved in research on computer algebra systems. By this time computer scientists at MIT and elsewhere were working on several powerful languages, most of them written in LISP, that made it possible to manipulate symbols. In addition to versions of MACSYMA, there was REDUCE (a program developed at the University of Utah especially for use in calculations for high energy physics) and SCRATCHPAD (a program developed at IBM). Stoutemyer had learned MACSYMA at MIT and had published articles relating to REDUCE. He was well aware that all of these programs required far larger memories than were then available on microcomputers.[43]

Rich and Stoutemyer hoped, as they wrote in a 1979 paper, "to infuse computer-algebra awareness throughout the math-science curriculum, starting at the undergraduate level and working downward."[44] For such purposes they needed programs that would run on much less expensive computers than those used at university centers—hence, the interest in microcomputers. In 1978, as part of an NSF-sponsored research project, Rich and Stoutemyer implemented a computer algebra system for microcomputers, which they dubbed muMATH (the Greek letter *mu* (μ) was and is used in the metric system as an abbreviation for *micro*). It ran on a relatively accessible version of LISP that they named muSIMP-77 (here SIMP stands for Symbolic IMPlementation). The following year they formed the partnership Soft Warehouse to market the copyrighted muLISP-77 (a fuller microcomputer version of LISP) and muSIMP-77 as well as the public domain software muMATH. They first brought their new product to the attention of computer scientists, speaking in June to a European conference on symbolic and algebraic computation. In July and August they published a description of muMATH in *Creative Computing*.[45] In August an article by Stoutemyer on LISP-based symbolic mathematics systems appeared in *Byte*. It described muMATH in detail and also discussed REDUCE, MACSYMA, and SKETCHPAD. Stoutemyer explained that he had "spent many fascinating hours using the four most actively supported and publicized LISP based systems and it seems likely that increasing numbers of students, scientists, engineers, and mathematicians will want an opportunity to try some of these systems."[46]

The partners also publicized their products in advertisements. A blurb in the November 1979 issue of *Byte* billed muMATH as the route to "X-Rated Revolutionary Computerized Math." It described the programs sold by Soft Warehouse and showed a computer screen with several problems solved by muMATH, including a solution to a quadratic equation, a simplification of a trigonometric expression, symbolic integration, matrix inversion, and an exact solution to a complicated arithmetic expression.[47] A second advertisement, which appeared in January 1980, featured muLISP. It was headlined, "Is Your Computer LISPless?" and showed a doctor in a lab coat with stethoscope, sensing the ailments of an anthropomorphized microcomputer. The ad generally recommended muLISP but also prescribed muMATH for those suffering from math anxiety.[48]

Software for microcomputers was still a novel product in 1979, and Soft Warehouse apparently had some difficulty deciding what to charge for its products. A note at the end of Stoutemyer's August 1979 *Byte* article indicated that object listings for muSIMP-77 and muLISP-77 cost $85, and machine readable versions of either program required an additional $95. The muMATH-79 source code came free with muSIMP-77; there is no mention of distributing this in machine-readable form. Documentation was available for all three systems "at a cost of approximately $.10 per page."[49]

To better promote its products, Soft Warehouse arranged to have them sold by Microsoft. This move allowed Microsoft to add LISP to its offerings, which already included the programming languages BASIC, COBOL, and FORTRAN. A March 1980 advertisement gave a price of $200 for the new muLISP-79 and $250 for a new muSIMP/muMATH package. Both programs were available on the CP/M and the TRSDOS operating systems (see Figure 17.2).[50] An April 1980 advertisement emphasized muMATH for the TRS-80 microcomputer, describing a muMATH/muSIMP package on disk with documentation that cost only $74.95. An edition of muMATH for the Apple II was promised for later in the year.[51] A November 1980 account of muMATH-79 by *Byte* editor Greg Williams described the functions of the latest version of software, indicating that it was available on disk for $290 in 8080 machine language and reported that two versions for the TRS-80 were being sold by Microsoft, one at $250 and one "slightly diminished" for $75. Documentation, in the form of 175 pages in a three-ring binder, came with the package.[52]

According to Williams, the intended audience for muSIMP/muMATH consisted of "high school and college students, educators, [and] programming language enthusiasts."[53] He reported that the package had been used in mathematics teaching from the high school level on up. Although the program could solve problems encountered in algebra, trigonometry, and even calculus, "educators need not fear: muMATH-79 does not provide a solution's derivation, only the final answer."[54]

Knowledge of computer algebra systems and of microcomputers spread rapidly during the early 1980s. In 1981 articles about computer algebra appeared in both *Scientific American* and *Science Newsletter.* The *Scientific American* article emphasized powerful software such as MACSYMA but noted that the computer algebra systems available for microcomputers, though slower and less comprehensive, could "still perform far more complex calculations more accurately than many mathematicians." The article reported that the "most sophisticated and widely available such system is called muMATH."[55]

Mathematicians also were informed about the new systems. An abstract of Stoutemyer's August 1979 article from *Byte* appeared in *Mathematics Magazine* in 1980. A more thorough introduction to the new tool appeared in January 1982. That month the *American Mathematical Monthly* published an article about the latest version of muMATH by Herbert S. Wilf, a distinguished mathematician at the University of Pennsylvania. He chose the provocative title "The Disk with the College Education." Here Wilf warned that the dilemma of whether or not to use electronic calculators, previously a concern of elementary school mathematics, was headed toward college teaching. He owned a personal computer that he used mainly for word processing, programming, and games. Having explored the capabilities of muMATH, however, he felt it only appropriate to alert college professors that the program knew calculus; in fact, he cautioned them, "some of your students may be doing your homework with it."[56] He went on to describe muMATH's ability to do mathematics at many levels—elementary school arithmetic, high school algebra, and college calculus, including both symbolic integration and differentiation. Wilf also showed how muSIMP could be used to write a program to print out successive derivatives of a function. Symbolic manipulation programs were not new but had never been widely available. Wilf predicted that programmable pocket calculators would probably be able to handle programs such as muMATH within a few years. How much would students be allowed to use them? Would college professors follow the advice that they had been giving to primary school teachers encountering calculators and "teach more of concepts and less of mechanics?"[57]

Two reviews of muMATH, with speculations about how such programs would influence mathematics teaching, appeared in the September 1983 issue of *Mathematics Teacher.* Beverly Mugrage of Wayne College at the University of Akron in Ohio hailed the product as "a package that should be used in every secondary school mathematics classroom in the country, at least in calculator mode."[58] Sharon Burrowes of the Wooster City Schools in Ohio reported that "the package is easy to use and breathtaking in its capabilities." After describing some of the tasks the software could perform, Burrowes, like Wilf, raised questions: "How are we, as mathematics teachers, going to react to computer programs that can do in seconds what it takes our students many minutes to do by hand? How should we react?"[59] Like Wilf, Burrowes suspected that the new software would raise questions for advanced mathematics teachers not unlike those the calculator had posed for those who taught arithmetic.

Reponses to such questions took several forms. A few entrepreneurs introduced other computer algebra systems. In a June 1983 article in the "Computers and Calculators" column of the *Two-Year College Mathematics Journal,* Stoutemyer described eight widely available computer algebra systems, three of which were available commercially. One was muMATH, which required 48 kilobytes of memory and was available not only for the TRS-80 and the Apple II but also the IBM PC. There also was PICOMATH, sold by the British firm of Acorn Atom for a microcomputer known as the BBC machine. As the name of the product suggests, it required only a paltry 4 kilobytes of memory. The third package was a form of MACSYMA sold by Symbolics, Inc., of Santa Monica, California, for a minicomputer known as the LISP.[60]

During the early years of the microcomputer diverse hobbyists, educators, and computer scientists explored the possibilities of the new machines. Educational programs did not prove as popular as applications such as games, spreadsheets, databases, or word processing. Yet they suggested new ways of learning, both at home and at school. Some emphasized the advantages of learning programming to promote logical thinking. Others sought to provide drill and more open-ended games and tutorials. Still others showed that electronic assistants such as computer algebra systems might offer a powerful tool that challenged long-accepted assumptions about the content of high school and university mathematics courses.

Of course, not all mathematics teachers embraced microcomputers. Some of them made do with programs for recording grades and attendance on minicomputers. Others—especially as clerical staff decreased—embraced word processing as a way of preparing examinations, course policy statements, and other classroom materials. Some of them developed assignments that used the software described here, and some learned programming for their own research or as an extension of their teaching duties.

CHAPTER EIGHTEEN

Graphing Calculators and Software Systems
The Media with a College Education

The single most important catalyst for today's mathematics education reform movement is the continuing exponential growth in personal access to powerful computing technology.

The more computer power we have, the less the students know what they're doing.

During the last fifteen years of the twentieth century many of the tools described in this book underwent major changes. Often, though not always, these transformations were associated with electronic technologies. General tools of pedagogy were affected. Textbooks not only became heavier and more visually elaborate but often taught about, and indeed included, computer software. A new round of school construction encouraged the replacement of chalkboards with dry erase marker boards. Sometimes electronic sensors were also associated with erasable surfaces, though such innovations were not generally adopted in schools. Standardized tests—as well as classroom quizzes and examinations—increasingly were graded, at least in part, electronically. Indeed, some tests were generated and administered using computers. Teachers adopted word processing for preparing classroom materials. In institutions with sufficient resources, overhead projectors were supplemented by "document cameras" or "visual presenters" that could project digital images of text and drawings from ordinary paper. Some professors experimented with computer programs such as Microsoft PowerPoint and Corel Presentations in lectures. New electronic capabilities also challenged the primacy of the classroom as a center for learning.

Within mathematics teaching, tools of calculation still included paper and pencil, manipulative devices such as blocks, and transparencies for overhead projectors. Representation of angles using protractors, of weights and measures using classroom demonstration apparatus, and of simple surfaces using three-dimensional models remained common. Professional engineers, architects, and technicians, however, increasingly used computer-aided design instead of drawing instruments, sharply reducing the market for precise mathematical instruments and for those trained in their use. Representation of both three-dimensional surfaces and linkages relied extensively on virtual rather than tangible models.

As one might expect, the greatest changes occurred within electronic instruments. Teleconferencing, computer networking, and later the advent of the World Wide Web all offered new opportunities for diverse forms of distance learning, including home schooling and teacher training.[1] Within mathematics powerful handheld calculators reached the market in the mid-1980s. These tools not only performed numerical calculations but also graphed functions. Symbolic manipulation, integration, and differentiation on handheld calculators were not far behind. These capabilities, combined with concern about educating sufficient numbers of technically trained workers, prompted new discussion of the college curriculum, particularly in calculus. This same period also saw the widespread diffusion of integrated software packages with remarkable mathematical capabilities.[2]

Students, teachers, and research mathematicians reacted to the new tools with a mixture of awe, delight, and dismay. Between 1988 and 1994 the *Notices of the American Mathematical Society* published a column entitled "Computers and Mathematics," which served as a clearinghouse for information about how mathematicians were using computers in research, text processing, communication, and teaching. When Keith Devlin of Colby College took over the column in 1991, he expected that it would not last long. As he wrote: "This column is surely just a passing fad that will die away before long. Not because mathematics will cease to have much connection with computers, but rather quite the reverse: the use of computers by mathematicians will become so commonplace that no one thinks to mention it any more."[3]

Indeed, by the end of 1994 Devlin was happy to report that computers had become a fact of mathematical life. He had long looked forward to the day "when, for most of us, the computer settled back to become an accepted part of the mathematician's working environment, ranked alongside the blackboard, the pencil and scratch pad, the telephone, and the coffee pot." The computer was "simply one of the tools we use to get the job done." Recognizing that "the use of computer technology was now just one more aspect of mathematics," the *Notices* no longer singled it out for special attention. As Devlin put it, "The child has come of age."[4] Evidence of the spread of computers and calculators also appeared in journals for mathematics teachers who were less erudite than those who read the *Notices*. The *American Mathematical Monthly* published reviews of microcomputer software as early as 1982 and noted books received relating to computer science in its "Telegraphic Reviews" column. In April 1986 this column began listing software that the journal had received. The *Two-Year College Mathematics Journal* instituted a column on "Computers and Calculators" in 1983 and greatly expanded its discussion of computers in the early 1990s. As with the *Notices,* coverage tapered off in the late 1990s, although software reviews continued. *Mathematics Teacher,* which initially reviewed software in its "New Products" column, instituted a special column of "Courseware Reviews" as early as 1982. In 1991, after the advent of graphing calculators, the title was changed to "Technology Reviews." Reviews of new software continued to appear in *Mathematics Teacher* throughout the last decade of the twentieth century. Articles about how best to make use of "technology" in the classroom also appeared regularly. Of course, the widespread diffusion of computers and calculators in no way guaranteed that they were used in teaching. Nonetheless, calculators were also increasingly allowed—and sometimes required—on examinations administered by states and by national organizations such as the College Board. More and more, parents and teachers had grown up with electronic instruments and expected children to gain facility with them. This chapter describes two aspects of this diffusion of electronic technology, the graphing calculator and mathematical software packages.

Graphing Calculators

By the mid-1980s many mathematicians and mathematics educators had, as the British computer scientist Mike Beilby put it, "quite lost sight of the humble pocket calculator."[5] This situation soon changed. In November 1989 Beilby attended an international symposium on bridging mathematics between school and university at which he heard Bert K. Waits of Ohio State University lecture on work he was doing in this area. Beilby learned that recent technical developments had expanded the capability of calculators to include graphical presentation, algebraic manipulation, and matrix calculation. Moreover, calculators were widely accepted by students as affordable and accessible to them personally. Waits argued—and Beilby tended to agree—that the new calculators were well suited to classroom use, particularly in the last years of school and the early years of university.

The first commercial handheld graphing calculator was introduced by the Japanese firm of Casio Computer Company in 1985. Casio, founded in 1946, had sold electric desk calculators since the 1960s and introduced a transistorized form of the machine in 1965. In the 1970s and 1980s it released a variety of microprocessor-based consumer products, including handheld calculators, digital watches, electronic musical instruments, and televisions.[6] Its fx-7000G graphing calculator reportedly was the work of a team headed by engineer Hideshi Fukaya (Figure 18.1). Casio's new prod-

Figure 18.1. Three graphing calculators. The Casio fx-7000G (1986, gift of San Juan High School) is at the *front,* the Hewlett-Packard HP-28C (1987, gift of Norton S. Starr) is *behind* it, and the Texas Instruments TI-81 (1990, gift of Amherst College) is to the *right.* National Museum of American History collections. Smithsonian Negative no. 2006-21186.

uct seems to have received relatively little initial publicity among mathematicians in the United States. At a cost that settled around seventy-five dollars, these early graphing calculators may have been beyond the price range of many students, at least those in high school.[7] Some schools arranged to have calculators available for checkout from the library. By 1986 the device had been adopted in programs led by Waits and his colleague at Ohio State, Franklin Demana.

Ohio State had a tradition of interest in applications of electronic technology to mathematics teaching. In 1974 four-function calculators were introduced there as a requirement in remedial courses in elementary and intermediate algebra. By the late 1970s all freshmen taking mathematics courses, including those in calculus, were required to use a scientific calculator. Faculty also worked with high school and middle school teachers developing curricula that used calculators and microcomputers. Waits, Demana, and their colleagues were particularly interested in the use of graphing in instruction and began developing graphics software for microcomputers. In 1985 the group started the Calculator and Computer Precalculus Project, which involved students from Ohio high schools and colleges. The following year the project adopted the Casio fx-7000G for its courses.[8] Similarly, in the summer of 1987 John Kenelly of Clemson University used Casio graphing calculators in a training program for South Carolina teachers of Advanced Placement (AP) Calculus. Further interest in the Casio came from the state of Michigan, where a group of educators associated with the Michigan Council of Teachers of Mathematics prepared a user's guide to the Casio and similar calculators.[9]

Demana and Waits also wrote a series of textbooks on elementary algebra and trigonometry that included titles for high schools and for colleges. These books used both graphing software developed at Ohio State and graphing calculators.[10] At the same time, the authors began a series of annual conferences on technology in collegiate mathematics, the first held in the fall of 1988. These meetings brought together those interested in the use of microcomputers and graphing calculators at this more advanced level. Participants included college faculty, software and calculator developers and vendors, and representatives of interested organizations such as the National Science Foundation (NSF) and the Educational Testing Service.[11]

This focus on collegiate mathematics reflected a more general call for reform in calculus education. American manufacturers worried that Japanese firms such as Casio were commanding an increasing share of the market for electronic goods. Some business and professional societies were concerned about the training that American students received in mathematics and technical subjects. Some computer scientists thought that a course in discrete mathematics might be a more helpful introduction for their students than the usual calculus course. Mathematicians, engineers, and others were concerned that too many college students advanced no further than this level. In an oft-cited hydraulic metaphor Robert M. White, president of the National Academy of Engineering, argued that calculus should be "a pump instead of a filter" in the pipeline providing the nation with "technical manpower." Calculus should entice students toward further study of mathematics and related disciplines, not discourage them.[12] Conferences held in 1986 and 1987, as well as related publications, urged revision of calculus teaching, including increased use of electronic technology. NSF, which had largely eliminated its mathematics education programs in the early 1980s, provided funds for a variety of projects. These programs generally encouraged greater variety in the presentation of mathematical ideas; in addition to solving calculus problems algebraically, teachers were encouraged to place new emphasis on both numerical and graphical techniques.[13] Graphing calculators—as well as computer algebra systems—were called on to assist in the task.

Other calculator manufacturers took up the challenge of designing graphing calculators. In 1987, fifteen years after the introduction of the pioneering HP-35, Hewlett-Packard Corporation introduced its HP-28C calculator, which featured not only graphing but symbolic manipulation as well as limited integration and differentiation (see Figure 18.1). Much of the development was associated with William C. Wickes, a physicist by training who had been fascinated by the programming possibilities of the earlier HP-41C calculator. After Wickes published a book on the matter, Hewlett-Packard hired him away from the physics department of the University of Maryland, and soon he was working on the development of the next generation of Hewlett-Packard calculators.[14]

In addition to hiring academics, Hewlett-Packard paid careful attention to promoting its products. John Kenelly, that early user of the Casio fx-7000G, received an HP-28C on secret loan six months before the release of the instrument. He soon was planning how the new calculator might be incorporated into Clemson University's freshman and sophomore mathematics courses.[15] Thomas Tucker of Colgate University prepared a review on the new calculator published in the January 1987 issue of the newsletter of the Mathematical Association of America (MAA). Reworking the title of Wilf's earlier review of muMATH, Tucker entitled his article "Calculators with a College Education." Tucker discussed both the Casio fx-7000G and the HP-28C. He noted that graphing, symbolic differentiation, and symbolic integration would be of great benefit to students. He would have preferred to have such transcendental numbers as $\pi/4$ represented symbolically rather than as approximate quantities such as 0.785 and commented that entering the instructions for many manipulations required a considerable number of "button-pushes." Nonetheless, the performance of graphing calculators left Tucker asking when mathematicians would recognize the existence of the new instruments in their teaching.[16] A detailed account of the HP-28C soon appeared in the *American Mathematical Monthly.* Yves Nievergelt of the Graduate School of Business Administration at the University of Washington, again playing on Wilf, entitled his article "The Chip with the College Education: The HP-28C." He described the use of the new calculator in detail, concluding that it offered "awesome computing power at both a modest price and size, with admirable user-friendliness." At the same time, understanding what procedures to use and interpreting the results with lucidity would require knowledge of underlying concepts that no calculator could provide.[17]

Publicity for the HP-28C was not confined to MAA publications. A short article in the January 12, 1987, *Wall Street Journal* described what the newspaper called a "calculator for conceptual algebra."[18] Detailed advertisements appeared in *Science* magazine that same month and were published in the *Notices of the American Mathematical Society.*[19] Moreover, Hewlett-Packard was soon ready to launch a version of the HP-28C with expanded memory, known as the HP-28S. It chose to do so at the January 1988 centennial meeting of the American Mathematical Society. Those attending the annual banquet of the society traditionally received a useful trinket such as an alarm clock. At the centennial party the favor was an HP-28S, but it came with an extra charge of $60 (the banquet alone was $30). Considering that the list price of the calculator was $235, the fee was not unreasonable.[20]

Other calculator manufacturers soon offered graphing calculators. In the spring of 1987 Sharp Corporation introduced its EL-5200 graphics calculator at a price of $100, and John Kenelly planned to introduce it to South Carolina high school students.[21] Like Casio, Sharp was a Japanese firm that had manufactured both desktop and handheld electronic calculators as well as other products such as televisions and cash registers. Over the coming years it would produce a wide range of graphing calculators.

In the 1980s concern about appropriate mathematics curricula was not limited to calculus teaching. President Ronald Reagan strongly disapproved of federal action to influence school curricula, and during the early years of his administration federal funds for high school and elementary curriculum development were sharply curtailed. Nonetheless, the National Council of Teachers of Mathematics (NCTM) set out to expand the recommendations for mathematics teaching suggested in its *Agenda for Action* (1980). Largely using its own funds, the NCTM prepared a series of publications describing a vision of mathematics education. The first volume of its recommendations, *Curriculum and Evaluation Standards for School Mathematics*, appeared in 1989. It listed as an underlying assumption for the standards for grades 9 through 12 that "scientific calculators with graphing capabilities will be available to all students at all times."[22] The following year Texas Instruments introduced its TI-81 graphics calculator, the first graphing calculator made by the firm (see Figure 18.1). Advertisements for the device published in *Mathematics Teacher* claimed that the calculator was explicitly designed to support the NCTM standards. More generally, the TI-81 was "the first graphics calculator developed with leading mathematics educators and experienced classroom teachers specifically for the general needs of mathematics education."[23] Bert K. Waits served on the NCTM Curriculum and Evaluations Standards Working Group particularly concerned with the grades 9 through 12.[24] Demana and Waits contributed to the development of and documentation for the TI-81 and would be active in promoting its use.[25] The calculator and its successor, the TI-82 (1993), were aimed at the high school market, with no ability to integrate, differentiate, or manipulate symbols. Like the HP-28C and HP-28S, the calculators were programmable and could be used by students for playing games as well as solving problems in mathematics. The TI-81 had a list price of $110.[26]

The U.S. federal government might provide funds for the purchase of school equipment and for trials of its use, but decisions about whether to use the materials were generally left to states, local authorities, and the staff of private institutions. Some uniformity in practice, at least at the high school and university level, was provided by the policies of examining authorities such as the College Board. The Mathematical Sciences Advisory Committee of the College Board initially suggested permitting the use of calculators on national mathematics tests in 1976. The first of these tests to allow examinees to use calculators were AP Calculus examinations in May 1983 and 1984. Calculators were not required, and no brand was specified. A study of the results indicated that students who made considerable use of calculators did less well than those who did not. At the same time, a policy of allowing but not requiring these relatively expensive tools seemed inherently unfair. More generally, it was difficult for examination writers to predict what calculators might be able to do months, if not years, in the future. For all these reasons the use of calculators in national testing was suspended.[27]

On further reflection, in part in light of the NCTM standards, the College Board decided to try allowing calculator use once again.[28] A special form of the Scholastic Aptitude Test (SAT) was to require the use of calculators when it was given in June 1990. Calculators were to be allowed on the PSAT (Preliminary SAT) given to high school juniors in the fall of 1993 and to be required on the AP Calculus examination given in May of that year. All forms of the SAT were to allow calculators by 1994, with some requiring it.[29] By the school year 1998–99 the AP Calculus examination required a graphing calculator such as the HP-28, HP-48, or TI-81. Calculators in various forms have also come to be allowed on various state tests of mathematical learning, although the capabilities of the instruments are sometimes restricted. Thus, a calculator "flaw" can be the ability to perform automatically an operation that students are expected to do by hand.[30]

Software for Mathematics Teaching

In 1996 Texas Instruments released its TI-83 calculator, a successor to the TI-82. This graphing calculator was particularly oriented toward statistics, business, and finance. Reviewing the new product for the *American Statistician,* John C. Nash and Chris Olsen commented that the TI-83 was a creature of compromise. How users reacted to it would depend on previous experience. As they wrote, "College and university teachers who have used modern statistical and scientific software in the classroom may be concerned at how much is 'lost' in migrating to the calculator. High school teachers may be encouraged by what is 'gained' in the migration from a prestatistical scientific calculator."[31] Nash, a college professor, found the TI-83 a fascinating toy but a toy nonetheless. Olsen, a high school teacher who used the calculator with students in an AP Statistics class, reacted differently. For him the instrument offered an opportunity to teach statistics in an interesting way, freeing teacher and student from the tedium of both graph sketching and routine computation.

Major competition for the use of programmable calculators came from increasingly sophisticated microcomputer software. Four software packages for microcomputers that were introduced in the 1980s suggest this trend and merit discussion here. The first, Minitab, was designed and used for teaching statistics and had its roots in programs used for larger computers. The remaining three—DERIVE, Maple, and Mathematica—had their origins in symbolic manipulation programs but came to cover much wider areas of mathematics. All four programs had historic associations with academe but became, or even started as, commercial products. All four were used not only by students but for research in mathematics and other areas. They all went through several editions and required familiarity with programming standards and care in entering commands that were not uniformly welcomed by users.

In their *American Statistician* review Nash and Olsen specifically mentioned the statistics package Minitab. Statisticians had been interested in computing methods and machines long before the invention of electronic computers and made extensive use of calculating machines, tabulators, slide rules, and other aids to computation. Around 1960 a physicist and two computer scientists at the National Bureau of Standards (NBS) developed an omnibus table-making program written in FORTRAN that they dubbed "Omnitab." It was particularly designed for use by non–computer specialists, with commands written in familiar English words. Brian Joiner made extensive use of Omnitab when he worked at NBS from 1963 until 1971. When he left there to teach at Pennsylvania State University, Joiner tried to use Omnitab in teaching elementary statistics, but the program was too large to run in the memory allotted for student use in the school's mainframe computer. He and another Penn State faculty member, Thomas A. Ryan Jr., wrote a new program, which they named "Minitab." It was also written in FORTRAN and looked very much like a subset of Omnitab, except the new program offered better bookkeeping, efficiency, and portability between computers. By 1973 Joiner and Ryan could report that Minitab worked successfully teaching elementary statistics at Penn State. Copies of the program were available "for a nominal charge to cover costs," and a fifty-page reference manual was also ready. The program proved sufficiently powerful to be used in more advanced statistics courses and as a tool of the

Penn State Statistical Computing Service. It could tabulate and plot data in diverse ways, perform a variety of statistical tests, and even generate data for simulations.[32] In 1974 Joiner left Penn State for the University of Wisconsin, although he remained a consultant in the further development of Minitab. That same year Tom Ryan's wife, Barbara F. Ryan, also joined in the development.

By 1983 the Ryans were devoting their full time to the sale and development of Minitab. A version of the program for microcomputers was available as early as 1982 and was revised regularly thereafter. Several standard statistics textbooks were written specifically for use with the program. This alteration of textbooks to reflect available technologies is similar to what happened in the case of elementary algebra, trigonometry, and calculus instruction. Even liberal arts colleges that had no formal programs in statistics adopted software such as Minitab for introductory courses (Figure 18.2). In later years, as students trained in Minitab took jobs in business, commercial users formed a growing segment of the market for the software. The idea that simulation, statistical analysis, and data presentation should be taught using computer software—or calculators—endured.

Relatively few mathematics professors teach statistics. Software for other topics would have greater influence on their work. In 1988 the expansion of microcomputer memories, the transformation of operating systems, and the wider diffusion of microcomputers encouraged the introduction of several new symbolic manipulation packages. In the wake of the general call for reform in calculus teaching, several of these programs were put to educational use. At least three products introduced that year would prove sufficiently successful to influence both classroom practice and mathematical research. Like Minitab, all three of these programs would undergo substantial changes in both presentation and capabilities over the years. All three were sufficiently complex that it is difficult, if not impossible, to judge precisely how they were used in the classroom. Our goal is to describe the programs in general terms and to suggest broadly how they arose and how they were incorporated into instruction.

In the fall of 1988 Soft Warehouse introduced a successor to muMATH (discussed in chap. 17). The software was known as DERIVE: A Mathematical Assistant (Figure 18.3). In place of typed commands, the program offered a menu of choices, including entering new expressions, building new expressions from old ones, and performing operations such as integration, differentiation, arithmetic, and factoring. Unlike muMATH, it also was ca-

Figure 18.2. The Student Edition of MINITAB Statistical Software, 1989. National Museum of American History collections, gift of Amherst College. Smithsonian Negative no. 2001-9272.

pable of graphing in both two and three dimensions, in Cartesian, parametric, or polar coordinates. Writing in the *Notices,* reviewer Eric L. Grinberg dubbed it "The Menu with the College Education." The program also had the ability to handle at least some infinite series and to operate on small vectors and matrices. Early versions of the product offered only limited capabilities for incorporating results into handouts or text files or saving them for class demonstrations. Grinberg also regretted that the files that executed programs were much less accessible for student examination than in muMATH. Moreover, as Eugene Herman pointed out in a review of DERIVE in the *Monthly,* the package made no provision for writing additional programs. To use Grinberg's phrase, "DERIVE provides a very useful and competent mathematical assistant, albeit an underemployed one."[33] Later versions of the program attempted to incorporate additional features. Dedicated muMATH users also could continue to purchase muMATH for the Apple II from the University of South Carolina. A similar program, RIEMANN, which ran on the Atari ST, was developed in Germany.[34] DERIVE was aimed particularly at beginning calculus students. Grinberg had used muMATH for computations relating to his research on integral operators and thought that DERIVE might be useful in research as well as teaching. According to Soft Warehouse, the program proved helpful in calculating the sum of inverses of twin primes carried out by Thomas Nicely of Lynchburg College. These studies led Nicely to detect a flaw in the Pentium microprocessor manufactured by Intel Corporation and attracted considerable attention in the mid-1990s.[35]

A larger computer algebra package, first developed for minicomputers, was the Canadian product MAPLE. It was prepared by a team of computer scientists at the University of Waterloo, starting in 1980. MAPLE was envisioned as a tool for mathematicians, computer scientists, and others who would have use for symbolic manipulation. It was initially written in the programming language B and then in C. By 1983 there were about fifty minicomputer installations outside of Waterloo using the program. University of Waterloo staff arranged for MAPLE to be licensed and distributed by WATCOM Products, Inc. A user's manual had appeared as early as 1982, with a fourth edition distributed by WATCOM in 1985. By the end of 1988 a version of MAPLE was available for the Macintosh microcomputer. Keith O. Geddes, Gastón H. Gonnet, and their associates established the firm of Waterloo Maple Software (later Waterloo Maple, Inc.) to market the product. Purchasers could also consult the volume *First Leaves: A Tutorial Introduction to MAPLE.* A 1989 advertisement in the *American Mathematical Monthly* indicated that the software cost $395—not a product for introductory student use. In the early 1990s graphics and text processing capabilities were added to the product, and a student edition for both the Macintosh and IBM-compatible microcomputer was released at a price of only $99.[36]

The expansion of academic computer algebra systems into commercial productivity tools that one sees in the development of MAPLE also occurred in mathematical software associated with the youthful British-born physicist Stephen Wolfram. In 1980, the year he received his Ph.D. degree, Wol-

Figure 18.3. User manuals for DERIVE, 1989. National Museum of American History collections, gift of Amherst College. Smithsonian Negative no. 2001-9273.

fram and colleagues at the California Institute of Technology began developing what came to be called "Symbolic Manipulation Program" (SMP). The work was initially inspired by Wolfram's desire to carry out complicated algebraic computations in quantum field theory—calculations that overflowed the memory of programs available to him. SMP soon was available commercially at a price of $40,000. Yet a dispute over the appropriate role of Caltech faculty in commercial development of innovations such as SMP led Wolfram to resign from his teaching post there in 1982 and curtailed further development.[37] According to a 1986 article in the *College Mathematics Journal,* SMP could then be purchased from Los Angeles–based Inference Corporation for $2,000 for the VAX or $3,000 for IBM minicomputers (by contrast, Maple cost $700 at the time and muMATH $200). In addition to computer algebra, SMP offered graphics, numerical as well as symbolic integration and differentiation, number theoretic applications, and solutions to ordinary differential equations. It also included such relatively novel features as online help.[38]

Meanwhile, Wolfram moved to the Institute for Advanced Study at

Figure 18.4. Mathematica for the Macintosh, 1988. National Museum of American History collections, gift of Theodore W. Gray. Smithsonian Negative no. 2006-2060.

Princeton and then to the University of Illinois, embarking on development of another software package of even greater complexity (Figure 18.4). Wolfram entitled his introduction to the new software *Mathematica: A System for Doing Mathematics by Computer.*[39] It was a commercial product from the start. Introduced in June 1988, Mathematica was most immediately noted for its stunning graphical capabilities. It initially sold for the Apple Macintosh SE and the MAC II and was also distributed as part of the standard software of the short-lived NeXT workstation. Accounts of the new program appeared not only in mathematical and scientific journals but also in *Fortune* and the *New York Times.*[40] By 1989 versions of the program were available for IBM microcomputers. At a price that ranged upward from $495, with considerable expertise required, Mathematica was not designed for novices, although some colleges and universities found that a roomful of computers running Mathematica offered a useful resource for mathematics students. Graphics generated by Mathematica also appeared in the classroom. More generally, the package expanded to include numerous areas of mathematics. One 1999 review of the software was entitled "A Mathematical Swiss Army Knife."[41] The product remained closely associated with Stephen Wolfram, who also used his firm of Wolfram Research to spread his ideas about cellular automata.[42]

Minitab, DERIVE, MAPLE, and Mathematica represent only a handful of the mathematical software for microcomputers introduced in the last fifteen years of the twentieth century. Educational programs for very young children continued to sell well. Drill and practice software such as Math Blaster was used in many schools, and more sophisticated tools for teaching plane geometry such as the Geometer's Sketchpad and Cabri-Géomètre acquired devoted followers. Some of these packages, particularly muMATH and Cabri-Géomètre, were also incorporated into sophisticated graphing calculators.[43] At the same time, the diffusion and use of both graphing calculators and computers for educational use was quite spotty.[44] Acquiring and maintaining electronic technology, using the devices effectively, and judging what skills would be of long-term use to students remained a challenge.

In the second half of the twentieth century the capability of electronic devices vastly increased, while their size and cost declined dramatically. Many, though by no means all, of the educational applications of computers and calculators were in mathematics. Ready access to powerful electronic instruments led to questions about numerous aspects of the mathematics curriculum. Authors even conflated *technology* with *electronic technology.* To understand how objects have become standard components of American classrooms, however, one should consider not only the history of electronic devices but also the stories of older products—such as textbooks, blackboards, graph paper, and the overhead projector—which may have proved so practical and durable as to be taken for granted. Whether they served as tools of presentation, of calculation, or of measurement and representation, these instruments reflect new educational ideas and technical capabilities as well as changing views of mathematics and education throughout American culture.

Notes

Introduction

1. James Hall, *Statistics of the West, at the Close of the Year 1836* (Cincinnati: J. A. James, 1836), 213. Patricia Cline Cohen quotes Hall, and explores the use of numbers in early America more generally, in her book *A Calculating People: The Spread of Numeracy in Early America* (Chicago: University of Chicago Press, 1982).

2. Population figures are taken from the Web site of the U.S. Census Bureau at www.census.gov/.

3. Institutional histories are listed in individual chapters. For general introductions and useful bibliography relating to the organizations of American mathematics, see the opening chapters of George M. A. Stanic and Jeremy Kilpatrick, eds., *A History of School Mathematics*, vol. 1 (Reston, Va.: National Council of Teachers of Mathematics, 2003).

4. George Sarton, "Notes and Correspondence," *Isis* 30, no. 1 (Feb. 1939): 95.

CHAPTER ONE Textbooks

Epigraphs: Florian Cajori, *The Teaching and History of Mathematics in the United States* (Washington, D.C.: Government Printing Office, 1890), 63; Charles Davies, *The Logic and Utility of Mathematics, with the Best Methods of Instruction Explained and Illustrated* (New York: A. S. Barnes & Co., 1869), end matter, 11.

1. The evolution of cataloging efforts is represented by Henry Barnard, "American Textbooks," *American Journal of Education* 13 (1863): 625–40, 14 (1864): 753–77, 15 (1865): 539–75; Louis C. Karpinski, *Bibliography of Mathematical Works Printed in America through 1850* (Ann Arbor: University of Michigan Press, 1940); John A. Nietz, *The Evolution of American Secondary School Textbooks* (Rutland, Vt.: Charles E. Tuttle Co., 1966); *Early American Textbooks, 1775–1900: A Catalog of the Titles Held by the Educational Research Library* (Washington, D.C.: U.S. Department of Education, 1985); and Joe Albree, David C. Arney, and V. Frederick Rickey, *A Station Favorable to the Pursuits of Science: Primary Materials in the History of Mathematics at the United States Military Academy* (Providence, R.I.: American Mathematical Society, 2000).

2. Cajori, *Teaching and History,* 7–8, 45, 98–100. The theoretical "British, then French," pedagogical model was first questioned by Helena M. Pycior, "British Synthetic vs. French Analytic Styles of Algebra in the Early American Republic," in *The History of Modern Mathematics,* ed. David E. Rowe and John McCleary (San Diego: Academic Press, 1989), 3:125–54. Current conceptualizations of the history of the American mathematics textbook can be found in Karen D. Michalowicz and Arthur C. Howard, "Pedagogy in Text: An Analysis of Mathematics Textbooks from the Nineteenth Century"; and Eileen F. Donoghue, "Algebra and Geometry Textbooks in Twentieth-Century America," both in *A History of School Mathematics,* ed. George M. A. Stanic and Jeremy Kilpatrick (Reston, Va.: National Council of Teachers of Mathematics, 2003), 1:77–109, 329–98.

3. Typical of authors more interested in the various aspects of mathematics problems are Harold Don Allen, "The Verse Problems of Early American Arithmetics," *Journal of the Rutgers University Library* 33 (1970): 49–62; and Patricio G. Herbst, "Establishing a Custom of Proving in American School Geometry: Evolution of the Two-Column Proof in the Early Twentieth Century," *Educational Studies in Mathematics* 49 (2002): 283–312.

4. For the development of a specific subject, see George M. Rosenstein, "The Best Method: American Calculus Textbooks of the Nineteenth Century," in *A Century of Mathematics in America,* ed. Peter Duren (Providence, R.I.: American Mathematical Society, 1989), 3:77–109. Textbooks play a tangential role in the departmental histories and reminiscences in volume 2 of this work; as well as in Karen Hunger Parshall and David E. Rowe's groundbreaking *The Emergence of the American Mathematical Research Community, 1876–1900: J. J. Sylvester, Felix Klein, and E. H. Moore* (Providence, R.I.: American Mathematical Society, 1994).

5. P. Andrew Karam, "Mathematical Textbooks and Teaching during the 1700s," in *Science and Its Times: Understanding the Social Significance of Scientific Discovery,* ed. Neil Schlager (Farmington Hills, Mich.: Gale Research Group, 2000–2001), 4:238–40.

6. On the persistence of copybooks in academies, see Joel Silverberg, "Higher Mathematics Education in the United States: The Role of the Academy in the Years following the War for Independence," in *Proceedings of the Canadian Society for History and Philosophy of Mathematics,* ed. Antonella Cupillari, Twenty-ninth Annual Meeting, May 30–June 1, 2003, 16:234–49.

7. For general accounts of the nineteenth-century United States, consult textbooks such as Paul S. Boyer et al., *The Enduring Vision,* 5th ed. (Boston: Houghton Mifflin, 2004); Robert A. Divine et al., *America Past and Present,* 2nd ed. (Glenview, Ill.: Scott, Foresman & Co., 1987); or John Mack Faragher et al., *Out of Many,* 2nd ed. (Upper Saddle River, N.J.: Prentice Hall, 1997). See also Robert Freeman Butts and Lawrence A. Cremin, *A History of Education in American Culture* (New York: Holt, Rinehart, & Winston, 1953).

8. Compare "Of the Course of Academic Literature and Instruction in the College (1795)" and "Course of Instruction (1822–23)," reprinted in George W. Pierson,

A Yale Book of Numbers: Historical Statistics of the College and University, 1701–1976 (New Haven, Conn.: Yale University, 1983), 212–15. The full titles of the books in Yale's mathematics curriculum were: Jeremiah Day, *An Introduction to Algebra, Being the First Part of a Course of Mathematics, Adapted to the Method of Instruction in the American Colleges* (New Haven: Howe & Deforest, 1814); John Playfair, *Elements of Geometry, Containing the First Six Books of Euclid, with a Supplement on the Quadrature of the Circle and the Geometry of Solids*, 1st American ed. (Philadelphia: F. Nichols, 1806); Jeremiah Day, *A Treatise of Plain Trigonometry, To Which Is Prefixed, A Summary View of the Nature and Use of Logarithms, Being the Second Part of a Course of Mathematics, Adapted to the Method of Instruction in the American Colleges* (New Haven: Howe & Deforest, 1815); Jeremiah Day, *A Practical Application of the Principles of Geometry to the Mensuration of Superficies and Solids, Being the Third Part of a Course of Mathematics, Adapted to the Method of Instruction in the American Colleges* (New Haven: Oliver Steele, [1816]); Jeremiah Day, *The Mathematical Principles of Navigation and Surveying, Being the Fourth Part of a Course of Mathematics, Adapted to the Method of Instruction in the American Colleges* (New Haven: Steele & Gray, 1817); and Samuel Webber, "Conic Sections" and "Spheric Geometry," in *Mathematics, Compiled from the Best Authors and Intended to Be the Text-Book of the Course of Private Lectures on These Sciences in the University at Cambridge*, 2nd ed. (Cambridge, Mass.: William Hilliard, 1808), 2:271–335, 367–408.

9. Cajori, *Teaching and History*, 107.

10. Eugene Exman, *The Brothers Harper* (New York: Harper & Row, 1965); *The First One Hundred and Fifty Years: A History of John Wiley and Sons, Incorporated, 1807–1957* (New York: John Wiley & Sons, 1957).

11. Biographies of Davies include Amy Ackerberg-Hastings, "Mathematics Is a Gentleman's Art: Analysis and Synthesis in American College Geometry Teaching, 1790–1840" (Ph.D. diss., Iowa State University, 2000); George W. Cullum, *Biographical Register of the Officers and Graduates of the U.S. Military Academy at West Point, N.Y.*, 3rd ed. (Boston: Houghton, Mifflin, 1891), 1:151–55; and "Charles Davies," *National Cyclopedia of American Biography* (New York: James T. White, 1898–84), 3:26.

12. Stephen E. Ambrose, *Duty, Honor, Country: A History of West Point* (Baltimore: Johns Hopkins University Press, 1966), 62–86; Samuel E. Tillman, "The Academic History of the Military Academy, 1802–1902," in *The Centennial of the United States Military Academy: 1802–1902* (Washington, D.C.: Government Printing Office, 1904), 1:223–41.

13. Elizabeth Mansfield to Mary Ann Mansfield, May [n.d.], 1823, Charles Davies, Letters, Special Collections and Archives, U.S. Military Academy, West Point, N.Y.

14. Keith Hoskin, "Textbooks and the Mathematisation of American Reality: The Role of Charles Davies and the U.S. Military Academy at West Point," *Paradigm*, no. 13 (1994): 11–41.

15. Due to tendencies to put similar titles on different textbooks and to issue the

same textbooks with different titles, it is not possible to determine exactly how many textbooks Davies prepared.

16. "Sketches of the Publishers: A. S. Barnes & Co.," *Round Table* 3 (1866): 170–71, 186.

17. Ray's sales figures and content are explored in David E. Kullman, "Joseph Ray—The McGuffey of Mathematics," *Ohio Journal of School Mathematics* 38 (1998): 5–10.

18. Albree, *Station Favorable,* 14–15. For this subject the cadets likely had been studying manuscript notes prepared by professor of engineering Claudius Crozet. Crozet had left West Point in 1823 and was not available to publish his notes—or to object if the notes were printed without his permission.

19. Exman, *Brothers Harper,* 20–24.

20. Contract for *Elements of Surveying,* 1830; and contract for *Legendre's Geometry and Trigonometry,* Apr. 1, 1834, Archives of Harper & Bros., 1817–1914, reel 50, Correspondence Relating to Contracts, 1832–1914, Butler Library, Columbia University, New York. Exman, *Brothers Harper,* 29–31, 92.

21. Jennifer Monaghan, "The Textbook as a Commercial Enterprise: The Involvement of Noah Webster and William Holmes McGuffey in the Promotion of Their Reading Textbooks," *Paradigm,* no. 6 (1991), available online at www.ed.uiuc .edu/faculty/westbury/Paradigm/monaghan.html.

22. Hezekiah Howe to Jeremiah Day, Aug. 23, 1837; and Hezekiah Howe to Leavitt Lord & Co., Jan. 8, 1838, Hezekiah Howe & Co. Letter Book, 1833–38, Beinecke Rare Book and Manuscript Library, Yale University, New Haven, Conn.

23. Dennis Hart Mahan, *An Elementary Course of Civil Engineering for the Use of the Cadets of the United States Military Academy* (New York: Wiley & Putnam, 1837), vii.

24. *National Advocate,* Apr. 25, 1823; *New York Gazette and General Advertiser,* Apr. 25, 1823.

25. *In Memory of Alfred Smith Barnes* (N.p.: privately printed, 1889), 14.

26. Mary Ann Davies to Elizabeth Mansfield, Nov. 28, 1837, Davies Letters.

27. Michael V. Belok, *Forming the American Minds: Early School-Books and Their Compilers (1783–1873)* (Moti Katra, Agra-U.P., India: Satish Book Enterprise, 1973), 62–76; Butts and Cremin, *History of Education in American Culture,* 236–90.

28. *Connecticut Common School Journal* 2 (1839–40): 224.

29. "Sketches of the Publishers," 170–71; John Tebbel, *A History of Book Publishing in the United States* (New York: R. R. Bowker, 1972–81), 1:294–99.

30. Philip N. Schuyler, ed., *The Hundred Year Book: Being the Story of the Members of the Hundred Year Association of New York* (New York: A. S. Barnes & Co., 1942), 175.

31. "Presidents of New York State Teachers' Association," *American Journal of Education* 15 (1865): 477–87.

32. David F. Labaree, *The Making of an American High School: The Credentials Market and the Central High School of Philadelphia, 1838–1839* (New Haven, Conn.: Yale University Press, 1988), 1–8.

33. William J. Reese, *The Origins of the American High School* (New Haven, Conn.: Yale University Press, 1995), 80–122.

34. Margaret W. Rossiter, *Women Scientists in America: Struggles and Strategies to 1940* (Baltimore: Johns Hopkins University Press, 1982), 2–8; Edward W. Stevens Jr., *The Grammar of the Machine: Technical Literacy and Early Industrial Expansion in the United States* (New Haven, Conn.: Yale University Press, 1995), 133–47; Kim Tolley, "Science for Ladies, Classics for Gentlemen: A Comparative Analysis of Scientific Subjects in the Curricula of Boys' and Girls' Secondary Schools in the United States, 1794–1850," *History of Education Quarterly* 36 (1996): 129–53.

35. *Elements of Geometry and Trigonometry* was omitted, perhaps because Davies had only begun to list his name on the title page in 1834 or because the work was technically still under contract to the Harpers. See Albree, *Station Favorable*, 91, 161–62.

36. Charles Davies, *Practical Geometry: With Selected Applications in Mensuration, in Artificers' Work and Mechanics* (Philadelphia: A. S. Barnes & Co., 1842), 2.

37. Charles Davies, *Elementary Algebra* (New York: A. S. Barnes & Burr, 1852), 2. Davies had prepared four additional textbooks for the academic course: *Elementary Geometry and Trigonometry* (1841), *Elementary Algebra* (1842, with a separate key), *University Arithmetic* (1846, with a separate key), and *Practical Mathematics for Practical Men* (1852). *Elements of Surveying* was also placed in this section of the list because Davies had rewritten the textbook in 1841 to make it suitable for academies and schools and had added *navigation* to the content and title. *University Algebra* (1858, with separate key) appeared in later versions of the advertisement.

38. Charles Davies, *The Metric System, Explained and Adapted to the Systems of Instruction in the United States* (New York: A. S. Barnes & Co., 1867), 2. From 1870 the arithmetical course was renamed the "Common School Course." See Charles Davies, *Elements of Geometry and Trigonometry, from the Works of A. M. Legendre* (New York: A. S. Barnes & Co., 1870), 2. This rendition additionally reemphasized the original cachet of Davies's series by terming it "the West Point Course."

39. Charles Davies, *New University Arithmetic* (New York: A. S. Barnes & Burr, 1860), end matter, n.p.

40. Davies, *Logic and Utility of Mathematics*, end matter, 11.

41. Ibid. A set of testimonials from people who had a personal connection with Davies or with his cronies in the educational circles of New York and Connecticut and people from the lower tiers of postsecondary institutions—mainly from Barnes's generation rather than of Davies's—was added to the National Course advertisement in Charles Davies, *New Elementary Algebra* (New York: A. S. Barnes & Co., 1877), end matter, 9. Additionally, this advertisement for Davies's National Course organized the list of his textbooks by subject: arithmetic, algebra, geometry, mensuration, and "mathematical science" (works for teachers).

42. Davies and Barnes were in fact sued once for plagiarism, by Frederick Emerson, who charged that *First Lessons of Arithmetic* (1840) was copied from Emerson's 1830 *North American Arithmetic*, pt. 1. In 1845 Davies and Barnes paid an out-of-

court settlement to Emerson in exchange for his public announcement that any similarities between the two textbooks were "accidental." *Emerson v. Davies, et al.,* 8 F. Cas. 615 (D. Mass. 1845), available from LEXIS/NEXIS Print Delivery [Lexsee 8 FCAS 615, AT 619]; Charles Carpenter, *History of American Schoolbooks* (Philadelphia: University of Pennsylvania Press, 1963), 140–41.

43. [Charles Davies, ed.], *Elements of Geometry and Trigonometry; With Notes. Translated from the French of A. M. Legendre,* [trans. Thomas Carlyle], 3rd ed. (New York: N. & J. White; Collins & Hannary; Collins & Co.; and James Ryan, 1832), iv–v; Davies, *A Treatise on Shades and Shadows, and Linear Perspective,* 2nd ed. (New York: Wiley & Putnam et al., 1838), iii–iv; Davies, *Elements of Descriptive Geometry, with Their Application to Spherical Trigonometry, Spherical Projections, and Warped Surfaces* (New York: A. S. Barnes & Co., 1849), iii–iv.

44. Davies, *New Elementary Algebra,* iii–iv.

45. Charles Davies, *The Logic and Utility of Mathematics, with the Best Methods of Instruction Explained and Illustrated* (New York: A. S. Barnes & Co., 1850), 256–59, 341–45.

46. These statistics are based upon a Davies bibliography compiled by the authors from *The National Union Catalog,* 753 vols. (London: Mansell, 1976); and Henry Barnard, "American Textbooks," *American Journal of Education* 13 (1863): 625–40.

47. "The Book Trade: Publishers and Publishing in New York," *Norton's Literary Gazette,* n.s. 1 (1854): 164–67, 191–93.

48. These numbers are also compiled from the *National Union Catalog* as well as examinations of many of the textbooks prepared by Loomis, Olney, and Wells.

49. John Barnes Pratt, *Personal Recollections: Sixty Years of Book Publishing* (New York: A. S. Barnes & Co., 1942).

CHAPTER TWO The Blackboard

Epigraph: Franklin Sawyer Jr., Sept. 1842, quoted in Francis W. Shearman, *System of Public Instruction and Primary School Law of Michigan* (Lansing, Mich.: Ingals [*sic*], Hedges & Co., 1852), 453.

1. Peter Mackintosh to Boston School Committee, Oct. 17, 1823; and "Report of the Committee 'to Consider the Expediency of Making an Alteration in the Books Now Used . . . ,'" Boston School Committee Papers (BSCP), Boston Public Library, Boston.

2. Report of the Committee on Books, July 7, 1823, BSCP.

3. "Memorandum—Apparatus—Chairman of Subcommittee of Grammar Schools," 1831, BSCP. Those schools without blackboards were provided with one each at a cost of twelve dollars.

4. R. B. Parker to the Hon. Benjamin Rafael [Russell?], May 19, 1831, BSCP.

5. In 1868 the U.S. Government Printing Office published a small volume on school house architecture for use by those building schools for freed slaves. It repeated advice on the making of blackboards that had commonly been presented in

Connecticut earlier in the century. See C. Thurston Chase, *A Manual of School-houses and Cottages for the People of the South* (Washington, D.C.: Government Printing Office, 1868), 61–63.

6. Deborah Jean Warner, "Commodities for the Classroom: Apparatus for Science and Education in Antebellum America," *Annals of Science* 45 (1988): 387–97.

7. O. A. W. Dilke, *Mathematics and Measurement* (London: Trustees of the British Museum, 1987), 16.

8. Fletcher B. Dresslar, "Blackboards," *Cyclopedia of Education* (New York: Macmillan, 1911), 1:391–94.

9. John Taylor to John Bogart, July 2, 1779, published in John Bogart, *The John Bogart Letters* (New Brunswick: Rutgers College, 1914), 18.

10. Edward S. Holden, comp., *The Centennial of the United States Military Academy at West Point, New York* (Washington, D.C.: Government Printing Office, 1904), 1:218, 244.

11. Stephen E. Ambrose, *Duty, Honor, Country: A History of West Point* (Baltimore: Johns Hopkins Press, 1966), 19, 20.

12. Emmor Kimber, *Arithmetic Made Easy for Children . . .*, 4th ed. (Philadelphia: Kimber & Conrad, 1809), iv. Kimber's comments on the blackboard are mentioned in Clifton Johnson, *Old Time Schools and School-books* (1904; rpt., New York: Dover, 1964), 107.

13. On Lancaster, see Mora Dickson, *Teacher Extraordinary: Joseph Lancaster, 1778–1838* (Sussex: Book Guild Ltd., 1986). See also Carl F. Kaestle, *Joseph Lancaster and the Monitorial School Movement: A Documentary History* (New York: Columbia University Teachers College, 1973).

14. Emmor Kimber, *Arithmetic Made Easy for Children*, v–vi. On the monitorial schools in Philadelphia, see Charles C. Ellis, "Lancasterian Schools in Philadelphia" (thesis presented to the Faculty of the University of Pennsylvania, Philadelphia, 1909). For a later manual of the Lancasterian system, see *Manual of the Lancasterian System of Teaching Reading, Writing, Arithmetic, and Needle-work as Practiced in the Schools of the Free-School Society of New York* (New York: Free-School Society, 1820). Boards painted black (but not erasable) are mentioned on p. 13.

15. Henry Barnard, "Reverend Samuel J. May," *American Journal of Education* 16 (1866): 140–42. The quotation is from a talk May gave to the Normal Association in Bridgewater, Mass., on Aug. 8, 1855. See Florian Cajori, *The Teaching and History of Mathematics in the United States* (Washington, D.C.: Government Printing Office, 1890), 117.

16. Edward D. Mansfield, "The United States Military Academy at West Point," *American Journal of Education* 13 (Mar. 1863): 31. See V. Frederick Rickey, "The First Century of Mathematics at West Point," paper presented at a conference on the History of Undergraduate Mathematics in America, June 2001, U.S. Military Academy, West Point, N.Y.

17. Quoted in Rickey, "First Century of Mathematics at West Point." The quotation is from Francis H. Smith, *West Point Fifty Years Ago: An Address Delivered before*

the Association of Graduates of the U.S. Military Academy, West Point, at the Annual Reunion, June 12, 1879 (New York: Van Nostrand, 1879), 9.

18. Joseph Henry, "Journal of a Trip to West Point and New York," *Joseph Henry Papers* (Washington, D.C.: Smithsonian Institution, 1972), 1:157. The annotations to this journal provide a helpful introduction to several sources relating to the early history of the blackboard in the United States.

19. On Farrar, see Amy Ackerberg-Hastings, "Tragedy and Trigonometry: The Story of Harvard's John Farrar" (MS, 2001). An 1823 pamphlet signed by George E. Winthrop and containing the plates from Farrar's *Elements of Geometry, Trigonometry, Applications of Trigonometry, and Topography* is held by the Massachusetts Historical Society, Boston.

20. Cajori, quoting A. S. Packard, in *Teaching and History of Mathematics,* 170–71.

21. Yale College, class of 1832, *A Circular, Explanatory of the Recent Proceedings of the Sophomore Class in Yale College* (New Haven, Conn.: Yale College 1830), 1. Copies of this circular can be found at the Connecticut Historical Society in Hartford, Conn., and at Beinecke Library, Yale University. This was not the first protest by Yale students against alterations in the mathematics curriculum. During their freshman year sixty-four of sixty-six members of the class of 1830 had sent a petition to the faculty of Yale College protesting against the suggestion that they might be required not only to recite theorems from Euclid but also to draw the corresponding figures in public examinations. "The failure which must ensue," they wrote, "would in a measure unfit us for the more laborious studies of the ensuing yeare [*sic*]." See "To the Faculty of Yale College," [1826 or 1827], Student Petitions 1818–44, box 52, folder 498, Record Group 175, Series 3, Day Family Papers, Manuscripts and Archives, Yale University Library. Similarly, in the spring of 1830 over seventy members of the sophomore class had protested against the introduction of an evening class in mathematics. It was these sophomores who would cause such trouble that summer. See "Sophomore Petition," Mar., 1830, Student Petitions 1818–44, Day Family Papers.

22. [Yale College, Faculty], [letter dated Aug. 7, 1830], box 1, folder 8, Student Unrest, Record Group 41-C, Manuscripts and Archives, Yale University Library. A draft of this letter, in the hand of Yale College president Jeremiah Day, is in box 30, folder 353, Day Family Papers. See also Burton J. Bledstein, *The Culture of Professionalism: The Middle Class and the Development of Higher Education in America* (New York: W. W. Norton, 1976), 231; Henry D. Sheldon, *Student Life and Customs* (New York: Appleton & Co., 1901), 110; and Brooks Mather Kelley, *Yale: A History* (New Haven, Conn.: Yale University Press, 1974), 168–69.

23. Elizabeth Peabody to [Nathan Appleton], [1827], *Letters of Elizabeth Palmer Peabody: American Renaissance Woman,* ed. Bruce A. Ronda (Middletown, Conn.: Wesleyan University Press, 1984), 80.

24. Quoted in Edward Mayes, *History of Education in Mississippi* (Washington, D.C.: Government Printing Office, 1899), 40.

25. Blackboards were not confined to female academies in southern states. They

were used in geography teaching in the public schools of Washington, D.C., by 1837. See Samuel Yorke At Lee [*sic*], *History of the Public Schools of Washington City, D.C.* (Washington, D.C.: M'Gill [*sic*] & Witherow, 1876), 20. Caroline Howard Gilman described a boy in Charleston, S.C., who studied at a blackboard in *Recollections of a Southern Matron* (New York: G. P. Putnam & Co., 1852), 225. This fictionalized account was copyrighted in 1838.

26. Joseph Felt, *Annals of Salem* (Salem, Mass.: W. & S. B. Ives, 1845–49), 1:469.

27. Samuel R. Hall, *Lectures on School-Keeping* (Boston: Richardson, Lord & Holbrook, 1829), 169. Hall taught school in Rumford, Maine, from 1814 until sometime before 1822 and has been credited with making the first use of blackboards in the United States during his time there. See Frank Monaghan, "Samuel Read Hall," *Dictionary of American Biography* (New York: Scribner's, 1932), 8:142–43.

28. William J. Adams, "Lecture on the Construction and Furnishings of School Rooms and on School Apparatus," *Annual Meeting of the American Institute of Instruction; Proceedings, Constitution, List of Active Members, and Addresses* (Aug. 1830), 1:345–46.

29. Henry Barnard, "Report of the Secretary of the Board [of Commissioners of Common Schools]," *Connecticut Common School Journal* 1 (May 1839): 170.

30. New York State, *Statutes of the State of New York Relating to Common Schools . . .* (Albany: published by the Superintendent of Common Schools of New York, printed by Thurlow Weed, 1841), 167.

31. Josiah Bumstead, *The Blackboard in the Primary Schools* (Boston: Perkins & Marvin, 1841). Bumstead was active in the affairs of the Boston Primary School Committee, which, after some years of debate, recommended in 1840 that the city supply blackboards to primary as well as grammar schools. See Joseph M. Wightman, *Annals of the Boston Primary School Committee, from Its First Establishment in 1818, to Its Dissolution in 1855* (Boston: Geo. C. Rand & Avery, City Printers, 1860), esp. 192.

32. William A. Alcott, "Slate and Black Board Exercises for Common Schools," *Connecticut Common School Journal* 4, nos. 8–11 (Apr. 1, Apr. 15, May 1, and May 15, 1842): 70. William A. Alcott was a cousin and close friend of Bronson Alcott, the father of Louisa May Alcott. Bronson Alcott also was an advocate of blackboards, installing a slate board imported from Wales at a school at which he taught in Cheshire, Conn., from 1825 until 1827. See Odell Shepard, *Pedlar's Progress: The Life of Bronson Alcott* (Boston: Little, Brown, 1937), 75–101, esp. 77.

33. Thomas B. Stockwell, ed., *A History of Public Education in Rhode Island from 1636 to 1876* (Providence, R.I.: Providence Press Co., 1876), 324–26.

34. Personal Communication from Joan Swann, Westwood Historical Society, Westwood, Mass., Sept. 2002.

35. Henry Barnard, *School Architecture* (New York: A. S. Barnes & Co., 1848), 99.

36. Horace Greeley et al., *The Great Industries of the United States* (Hartford: J. B. Burr & Hyde, 1873), 213.

37. A. H. Andrews and Co., "Portable Blackboards," box 17: Schools, Warshaw

Collection of Business Americana, Archives Center, National Museum of American History (NMAH), Smithsonian Institution, Washington, D.C.

38. Deborah J. Warner, "Geography of Heaven and Earth, Part 4," *Rittenhouse* 2 (1988): esp. 110–12, 120, 127–29.

39. Michigan Department of Public Instruction, *School Funds and School Laws of Michigan: With Notes and Forms . . .* (Lansing, Mich.: Hosmet & Kerr, 1859), 418, describes the use of "Shepherd's patent slate globe" in geography teaching. See also Laura L. Runyon, "Elementary History Teaching in the Laboratory School," *Elementary School Teacher* 3 (June 1903): 704. On the use of a slated globe in high school astronomy teaching, see George W. Myers, "Experimental Course in Astronomy: Ninth and Tenth Grades," *Course of Study of the Chicago Institute* (later *The Elementary School Teacher*) 1, no. 6 (Feb. 1901): 670.

40. *American Educational Monthly* 1 (1864), n.p.

41. A. H. Andrews and Co., "Portable Blackboards," box 17: Schools, Warshaw Collection, Archives Center, NMAH, Smithsonian Institution.

42. A. H. Andrews & Co., *A. H. Andrews & Co.'s Illustrated Catalogue of School Merchandise* (Chicago: n.d.), 71; J. C. Brooks, *Church, School and Hall Furniture* (Cincinnati, ca. 1884). The second title is available on microfiche in *Trade Catalogues at Winterthur* (New York: Clearwater, 1995), no. 2343.

43. Edward E. Babb & Co., *Illustrated Catalogue of School Supplies* (Boston, 1897–98). Weber Costello Company of Illinois filed a trademark for the term *Hyloplate* as used for writing boards in 1947. In this application the first use of this sense of the term in commerce is given as 1895. The firm received the registration in 1948, and it was renewed in 1968. See U.S. Trademark 503,761, registered Nov. 9, 1948. This is available online at www.uspto.gov/index.html.

44. "Minutes of the Faculty of Centre College," Dec. 3, 1886, Special Collections, Centre College, Danville, Ky.

45. W. H. Beardsley, "Slate Blackboards for Schools, Colleges, Counting Rooms, Offices and Private Use" (Hamilton, Ohio: W. H. Beardsley, n.d.), box 17: Schools, Warshaw Collection, Archives Center, NMAH.

46. Samuel R. Hall, *Lectures on School-Keeping* (Boston: Richardson, Lord & Holbrook, 1829), 169; William A. Alcott, "Slate and Black Board Exercises for Common Schools," *Connecticut Common School Journal* 4, no. 8, Apr. 1, 1842, 70.

47. Henry Barnard, *School Architecture* (New York: A. S. Barnes & Co., 1848), 114–15, 208.

48. James Johonnot, *Country School-houses . . .* (New York: Ivison & Phinney, 1859), 189–92.

49. James Johonnot, *School-houses* (New York: J. W. Schermerhorn & Co., 1871), 206.

50. Joshua G. Fitch, *Notes on American Schools and Training Colleges* (London, 1890), 35–37.

51. Shortly after William Chauvenet began teaching midshipmen at the Philadelphia Naval Asylum in 1842, he introduced more extensive blackboards and the

practice of assigning a grade to each student each day. Chauvenet would transfer his teaching to Annapolis when the U.S. Naval Academy was established as an official institution, later in the 1840s. See Benjamin Park, *The United States Naval Academy* (New York: G. P. Putnam's Sons, 1900), 124. Photographs from the 1890s show students reciting at the blackboard. See also Ralph Earle, *Life at the U.S. Naval Academy: The Making of the American Naval Officer* (New York: G. P. Putnam's Sons, 1917), 106–7.

52. H. C. Wright, "Educational Equipment and Its Uses," *School Science and Mathematics* 15 (1915): 500–504.

CHAPTER THREE Standardized Tests

Epigraph: Agnes L. Rogers, "Tests of Mathematical Ability—Their Scope and Significance," *Mathematics Teacher* 11, no. 4 (June 1919): 162.

1. On the transformation of pen and paper manufacture in nineteenth-century Britain, see Asa Briggs, *Victorian Things* (London: B. T. Batsford Ltd., 1988), 182–87, 289–93, 322–26. On pencils, see Henry Petroski, *The Pencil: A History of Design and Circumstance* (New York: Knopf, 1990). On changes in papermaking in the United States, see David C. Smith, *History of Papermaking in the United States (1691–1969)* (New York: Lakewood Publishing, 1971), esp. 121–87.

2. On classroom use of pens, pencils, and inexpensive paper in the United States, see Charnel Anderson, *Technology in American Education* (Washington, D.C.: Government Printing Office, 1962), 34–37.

3. Clifford Upton, "Standardized Tests in Mathematics for Secondary Schools," in National Committee on Mathematical Requirements, *The Reorganization of Mathematics in Secondary Education . . .* ([Oberlin, Ohio]: Mathematical Association of America, 1923), 280–428.

4. Nancy Beadie, "From Student Markets to Credential Markets: The Creation of the Regents Examination System in New York State, 1864–1890," *History of Education Quarterly* 39, no. 1 (Spring 1999): 1–30.

5. Lawrence A. Cremin, *American Education: The Metropolitan Experience (1876–1980)* (New York: Harper & Row, 1988), 379.

6. "New Test System Gains in Colleges," *New York Times*, Nov. 7, 1925, 14.

7. John A. Valentine, *The College Board and the School Curriculum: A History of the College Board's Influence on the Substance and Standards of American Education, 1900–1980* (New York: College Entrance Examination Board, 1987).

8. Kenneth M. Ludmerer, *Learning to Heal: The Development of American Medical Education* (New York: Basic Books, 1985). On the history of American legal education, see Robert Stevens, *Law School: Legal Education in America from the 1850s to the 1980s* (Chapel Hill: University of North Carolina Press, 1983). On the U.S. Civil Service, see Paul Van Riper, *History of the United States Civil Service* (Evanston, Ill.: Row, Peterson, 1958). On municipal government reform (with brief mention of examinations), see Martin J. Schiesl, *The Politics of Efficiency: Municipal Administration and Reform in America 1800–1920* (Berkeley: University of California Press, 1977). On

actuarial examinations, see Preston C. Bassett, "To Become a Member," *Transactions of the Society of Actuaries* 37 (1985): 1–12.

9. Edward A. Krug, *The Shaping of the American High School* (New York: Harper & Row, 1964), 151–63.

10. Joseph M. Rice, "Our Public School System: Evils in Baltimore," *Forum* 14 (Oct. 1892): 147.

11. Joseph M. Rice, "The Futility of the Spelling Grind," *Forum* 23 (Apr. 1897): 163–72. The quote is from p. 171. Rice published a second article with the same title in the *Forum* 23 (June 1897): 409–19.

12. Joseph M. Rice, "Obstacles to Rational Educational Reform," *Forum* 22 (Dec. 1896): 389; and J. M. Rice, "Educational Research: A Test in Arithmetic," *Forum* 34 (Oct. 1902): 282.

13. Joseph M. Rice, "Educational Research," *Forum* 34 (July 1902): 118.

14. Joseph M. Rice, "Educational Research: A Test in Arithmetic," *Forum* 34 (Oct. 1902): 283 (the tests are printed on pp. 296–97).

15. Joseph M. Rice, "Educational Research: Causes of Success and Failure in Arithmetic," *Forum* 34 (Jan. 1903): 437–52.

16. Rice was hardly the only American to be concerned about efficiency around 1900. As Raymond E. Callahan has pointed out, Frederick W. Taylor's ideas about scientific management and their possible application in education were widely discussed from about 1910. Callahan mentions several educators who had discussed the efficiency of schools before Taylor but does not include Rice among them. See Raymond C. Callahan, *Education and the Cult of Efficiency* (Chicago: University of Chicago Press, 1962). For further sources, see Patricia Albjerg Graham, *Schooling America: How the Public Schools Meet the Nation's Changing Needs* (Oxford: Oxford University Press, 2005), 71–76.

17. For a general discussion of the development of mental tests in the United States, which does not discuss tests of mathematics, see Michael M. Sokal, ed., *Psychological Testing and American Society, 1890–1930* (New Brunswick, N.J.: Rutgers University Press, 1987).

18. Edward L. Thorndike, *Educational Psychology* (New York: Lemcke & Buecher, 1903), 164.

19. Geraldine Joncich traces the origins of this motto in *The Sane Positivist: A Biography of Edward L. Thorndike* (Middletown, Conn.: Wesleyan University Press, 1968), 283.

20. Edward L. Thorndike, *Introduction to the Theory of Mental and Social Measurements* (New York: Science Press, 1904). For a general overview of Columbia in Thorndike's years, see Michael Rosenthal, *Nicholas Miraculous: The Amazing Career of the Redoubtable Dr. Nicholas Murray Butler* (New York: Farrar, Strauss & Giroux, 2006).

21. David L. Roberts, "Mathematics and Pedagogy: Professional Mathematicians and American Educational Reform" (Ph.D. diss., Johns Hopkins University, 1997), esp. 57–83.

22. Joncich, *Sane Positivist*, 270–77.

23. Edward L. Thorndike, *Educational Psychology* (New York: Teachers College, 1910), 4–5. See also Thorndike, *Educational Psychology* (New York: Teachers College, 1914), 3:153.

24. W. A. Fox and Edward L. Thorndike, "The Relationships between the Different Abilities Involved in the Study of Arithmetic: Sex Differences in Arithmetical Ability," *Columbia University Contributions to Philosophy, Psychology and Education* 11, no. 2 (Feb. 1903): 32–40.

25. Cliff W. Stone, "Arithmetical Abilities and Some Factors Determining Them," *Teachers College Contributions to Education* 19 (1908).

26. J. F. Millis, "Review of C. W. Stone's 'Arithmetical Abilities and Some Factors Determining Them,'" *Elementary School Teacher* 9, no. 10 (June 1909): 526.

27. Stuart A. Courtis, "Measurement of Growth and Efficiency in Arithmetic," *Elementary School Teacher* 10, no. 2 (Oct. 1909): 58. The entire article is on pp. 58–74, with a second part in the Dec. 1909 issue on pp. 177–99. For an account of Courtis's ideas, see Edwin V. Johanningmeier, "Stuart Appleton Courtis: His Views and Contributions to Tests and Measurement" (MS). Some of the material has been published in Ervin V. Johanningmeier, "The Transformation of Stuart Appleton Courtis: Test Maker and Progressive," *American Educational History Journal* 31, no. 2 (2004): 201–10.

28. S. A. Courtis, "Our School as a Contributor to Educational Progress," *Rivista* (Detroit: Liggett School, 1910), n.p. We thank Jan Durecki of the University Liggett School for sending this reference.

29. S. A. Courtis, "Measurement of Growth and Efficiency in Arithmetic (Continued): A Year's Progress," *Elementary School Teacher* 11, no. 3 (Nov. 1910): 171–85. This was a continuation of the two articles cited previously.

30. "Editorial Notes," *Elementary School Teacher* 11, no. 8 (Apr. 1911): 450.

31. "Educational News and Editorial Comment," *Elementary School Teacher* 14, no. 4 (Dec. 1913): 145. A manual for the test was published in 1914.

32. "Educational News and Editorial Comment," *Elementary School Teacher* 14, no. 9 (May 1914): 403; "Knowledge of Pupils in Boston Schools Is Measured by Expert," *Christian Science Monitor,* Oct. 1, 1912, 4; "Schools Stand Low by Arithmetic Test," *New York Times,* Mar. 11, 1913, 7; "Foreign Children Make Good Showing," *New York Times,* Mar. 12, 1913, 8.

33. S. A. Courtis, "The Reliability of Single Measurements with Standard Tests," *Elementary School Teacher* 13, no. 7 (Mar. 1913): 327. Courtis discusses fluctuations in test scores on pp. 488–95.

34. Ervin V. Johanningmeier, "The Transformation of Stuart Appleton Courtis: Test Maker and Progressive," *American Educational History Journal* 31, no. 2 (Fall 2004): 202–10.

35. Ibid., 205.

36. Daniel Starch, *Educational Measurements* (New York: Macmillan, 1916), esp. 114–16; and "A Scale for Measuring Ability in Arithmetic," *Journal of Educational Psychology* 7 (1916): 213–22.

37. Clifford Woody, *Measurement of Some Achievements in Arithmetic* (New York: Teachers College Bureau of Publications, 1916). Woody ignored Courtis's finding that there was little correlation between the ability to perform arithmetic using single digits and skill in operations using larger numbers.

38. Clifford Woody and William A. McCall, *Woody-McCall Mixed Fundamentals* (New York: Teachers College Bureau of Publications, 1920).

39. The term *diagnostic test* would also be used to refer to tests of personality disorders. See David Rappaport, *Diagnostic Personality Testing* (Chicago: Year Book Publishers, 1945).

40. These are listed in detail in Walter S. Monroe, *Measuring the Results of Teaching* (Boston: Houghton Mifflin, 1918), 109–14.

41. Guy M. Wilson and Kremer J. Hoke, *How to Measure: Revised and Enlarged* (New York: Macmillan, 1929), 82–90.

42. S. A. Courtis, *Courtis Standard Practice Tests,* 2nd ed. (Yonkers: World Book Co., 1917).

43. Monroe, *Measuring the Results of Teaching,* 114.

44. John Ward Studebaker, *Studebaker Economy Practice Exercises* (New York: Scott, Foresman & Co., n.d.). These exercises are mentioned in Monroe, *Measuring the Results of Teaching,* 135. A manual for the exercises was published as late as 1930. See John W. Studebaker, Frederic B. Knight, George W. Myers, and Giles M. Ruch, *Manual of Directions for Economy Practice Exercises in Whole Numbers* (Chicago: Scott, Foresman & Co., 1930).

45. Walter S. Monroe, "A Test of the Attainment of First-Year High-School Students in Algebra," *School Review* 23, no. 3 (Mar. 1915): 159–71.

46. Edward L. Thorndike, "An Experiment in Grading Problems in Algebra," *Mathematics Teacher* 6, no. 3 (Mar. 1914): 123–34.

47. Harold O. Rugg, "The Experimental Determination of Standards in First-Year Algebra," *School Review* 24, no. 1 (Jan. 1916): 37–66.

48. Harold O. Rugg and John R. Clark, "Standardized Tests and the Improvement of Teaching in First-Year Algebra," *School Review* 25, no. 2 (Feb. 1917): 113–32; and 25, no. 3 (Mar. 1917): 196–213. Harold O. Rugg and John R. Clark, *Rugg-Clark Tests in First Year Algebra* (Chicago: University of Chicago Bookstore, 1918).

49. Harold O. Rugg and John R. Clark, "The Improvement of the Ability in the Use of the Formal Operations of Algebra by Means of Formal Practice Exercises," *School Review* 25, no. 8 (Oct. 1917): 546–54.

50. Henry G. Hotz, *Hotz First Year Algebra Scales* (New York: Teachers College Bureau of Publications, 1918).

51. John H. Minnick, "An Investigation of Certain Abilities Fundamental to the Study of Geometry" (Ph.D. diss., University of Pennsylvania, 1918).

52. Percival M. Symonds, *Measurement in Secondary Education* (New York: Macmillan, 1928), 126–27; Guy M. Wilson and Kremer J. Hoke, *How to Measure* (New York: Macmillan, 1929), 438.

53. Daniel Starch and Edward C. Elliott, "Reliability of Grading High-School

Work in English," *School Review* 20, no. 7 (Sept. 1912): 442–57; "Reliability of Grading Work in Mathematics," *School Review* 21, no. 4 (Apr. 1913): 254–59; and "Reliability of Grading Work in History," *School Review* 21, no. 10 (Dec. 1913): 676–81. A relevant figure from Starch and Elliott is reproduced in Monroe, *Measuring the Results of Teaching*, 9.

54. Daniel Starch, *Educational Measurements* (New York: Macmillan, 1916), 9.

55. Monroe, *Measuring the Results of Teaching*, 1–9.

56. Ben D. Wood, *The Reliability and Difficulty of the CEEB Examinations in Algebra and Geometry* (New York: College Entrance Examination Board [CEEB], 1921).

57. W. L. Crum, "Note on the Reliability of a Test, with Special Reference to the Examinations Set by the College Entrance Board," *American Mathematical Monthly* 30, no. 6 (Sept.–Oct. 1923): 296–301.

58. Ben D. Wood, *Measurement in Higher Education* (Yonkers-on-Hudson: World Book Co., 1923), esp. 124–28, 280–90.

59. Symonds, *Measurement in Secondary Education*, 13–14. Symonds was an associate professor of education at Teachers College. Giles M. Ruch, *The Objective or New-Type Examination* (Chicago: Scott, Foresman & Co., 1929), 82–85. Ruch was professor of education at the University of California at Berkeley. He had received his doctorate from Stanford in 1922.

60. Vera Sanford, "A New Type Final Geometry Examination," *Mathematics Teacher* 18, no. 1 (Jan. 1925): 22–36; Raleigh Schorling and Vera Sanford, *Schorling-Sanford Achievement Test in Plane Geometry* (New York: Teachers College Bureau of Publications, 1925–26).

61. Franz Samuelson, "Was Early Mental Testing (a) Racist Inspired, (b) Objective Science, (c) A Technology for Democracy, (d) The Origin of Multiple-Choice Exams, (e) None of the Above? (Mark the RIGHT Answer)," in *Psychological Testing and American Society 1890–1930*, ed. Michael M. Sokal (New Brunswick, N.J.: Rutgers University Press, 1987), 113–27.

62. Upton, "Standardized Tests in Mathematics," 390.

63. Herbert E. Hawkes and Ben D. Wood, *Columbia Research Bureau Plane Geometry Test* (Yonkers-on-Hudson: World Book Co., 1924–26). Hawkes and Wood also published *Hawkes-Wood Plane Geometry Examination* (Yonkers-on-Hudson: World Book Co., 1923); and *Placement Test in Plane Geometry* (Yonkers-on-Hudson: World Book Co., 1924).

64. G. M. Ruch, *The Objective or New-Type Examinations: An Introduction to Educational Measurement* (Chicago: Scott, Foresman & Co., 1929); and C. W. Odell, *Traditional Examinations and New-Type Tests* (New York: Century Co., 1928).

65. Valentine, *College Board and the School Curriculum.*

66. Matthew Downey, *Ben D. Wood, Educational Reformer* (Princeton: Educational Testing Service, 1965); Nicholas Leman, *The Big Test: The Secret History of the American Meritocracy* (New York: Farrar, Straus & Giroux, 1999), esp. 22–24, 34–39.

67. Charles A. Sproel, "The Actuarial Examinations," *Transactions of the Society of Actuaries* 1, no. 1 (Nov. 1949): 42–68.

68. On the early history of intelligence testing, see the articles and references in Michael M. Sokal, ed., *Psychological Testing and American Society, 1890–1930;* as well as Leila Zenderland, *Measuring Minds: Henry Herbert Goddard and the Origins of American Intelligence Testing* (Cambridge: Cambridge University Press, 1998).

69. Leonard Ayres, *Laggards in Our Schools* (New York: Charities Publication Committee, 1909). On Ayres, see Raymond E. Callahan, "Leonard Ayres and the Educational Balance Sheet," *History of Education Quarterly* 1 (1961): 5–13; and Callahan, *Education and the Cult of Efficiency*, esp. 15–18, 153–56, 165–69.

70. Agnes L. Rogers, "Tests of Mathematical Ability—Their Scope and Significance," *Mathematics Teacher* 11, no. 4 (June 1919): 156–57.

71. Agnes L. Rogers, *Rogers Test for Diagnosing Mathematical Ability: Grade IX* (New York: Teachers College Bureau of Publications, 1918).

72. Agnes L. Rogers, "Tests of Mathematical Ability and Educational Guidance," *Mathematics Teacher* 16, no. 4 (Apr. 1923): 193–203.

73. William D. Reeve, "A Better Use of Tests in Mathematics," *Mathematics Teacher* 17, no. 3 (Mar. 1924): 140–47.

74. David Eugene Smith, "On Improving Algebra Tests," *Teachers College Record* 25, no. 2 (Mar. 1923): 87–88.

CHAPTER FOUR The Overhead Projector

Epigraph: Text from an advertisement for the Beseler Porta-Scribe overhead projector entitled "Picture Story of How to Solve a Math Problem," *Mathematics Teacher* 58, no. 4 (Apr. 1965): 365.

1. Paul Saettler, *The Evolution of American Educational Technology* (Englewood, Colo.: Libraries Unlimited, 1990), esp. chaps. 4–5; Ann De Vaney and Rebecca P. Butler, "Voices of the Founders: Early Discourses in Educational Technology," in *Handbook of Research for Educational Communications and Technology: A Project of the Association for Educational Communications and Technology*, ed. David H. Jonassen (New York: Macmillan, 1996), 3–45.

2. John R. Miles and Charles R. Spain, *Audio-Visual Aids in the Armed Services: Implications for American Education* (Washington, D.C.: American Council on Education, 1947), 4.

3. For an early history of the discipline, see Paul Saetler, *A History of Instructional Technology* (New York: McGraw-Hill, 1968). Neither this volume nor a later edition of the book (*The Evolution of American Educational Technology* [Englewood, Colo.: Libraries, Unlimited, 1990]) discusses the overhead projector. They do offer valuable insight into the general development of audiovisual instruction, both before and after World War II.

4. In addition to trade literature, magazines on both mathematics education and audiovisual instruction, and Web sites of various manufacturers, we have consulted Pearl Mary Wilshaw, "The Overhead Projector: Its History, Development and Utilization in Education" (M.S. diss., Graduate Library School, C. W. Post College,

Brookville, N.Y., 1972). We thank Shannon Perich for bringing this work to our attention.

5. Deborah J. Warner, "Projection Apparatus for Science in Antebellum America," *Rittenhouse* 6 (May 1992): 87–94.

6. Debbie D. Griggs, "Projection Apparatus for Science in Late Nineteenth Century America," *Rittenhouse* 7 (Nov. 1992): 9–15. On Tyndall, see John Tyndall, "The Royal Institution Lecture by Professor Tyndall on 5 May, 1870," *Mechanic's Magazine* 92, May 13, 1870, 347–48.

7. Central Scientific Co., *Catalog M: Physical and Chemical Apparatus* (Chicago, 1909 and 1912). The quotation is given on p. 282 of the 1912 catalog.

8. Spencer Lens Co., *Spencer Delineascopes for Transparent, Opaque, Vertical, Microscopic Projection* (Buffalo, N.Y., 1912).

9. Spencer Lens Co., *Catalog of the More Popular Spencer Microscopes, Microtomes, Delineascopes, Optical Measuring Instruments, and Accessories* (Buffalo, N.Y., 1930), 94.

10. Spencer Lens Co., *The Only Lantern in the World Permitting the Lecturer to Face His Class and the Lantern* (Buffalo, N.Y., 1929). The first quotation is from p. 1 of this leaflet, the second from p. 2.

11. Bausch & Lomb Optical Co., *Projection Apparatus and Accessories* (Rochester, N.Y., 1908), esp. 9.

12. Bausch & Lomb Optical Co., *Bausch & Lomb Projection Apparatus* (Rochester, N.Y., ca. 1930). In 1966 Stephen Krulik and Irwin Kaufman of Lafayette High School in Brooklyn, N.Y., published a pamphlet for the National Council of Teachers of Mathematics (NCTM) on the use of the overhead projector. They state in a footnote that the overhead projector first was introduced for bed patients, with the ceiling serving as a screen. Although overhead projectors were used in hospitals in the 1950s, this was not the original use of the device. See Stephen Krulik and Irwin Kaufman, *How to Use the Overhead Projector in Mathematics Education* (Washington, D.C.: NCTM, 1966), 1. For an image of a projector designed for a hospital patient, see *Toomey j. [sic] Gazette,* no. 31 (Spring–Summer 1962): 30.

13. See the following publications of Bausch & Lomb Optical Company of Rochester, N.Y.: *Bausch & Lomb Balopticons,* 1933, 18–19; *Balopticons and Accessories,* 1934, 34–35; *Balopticons and Accessories,* 1939, 8–9; *Balopticons and Accessories,* 1948, 9.

14. Committee on Multi-Sensory Aids, NCTM, "Aids to Mathematics Teaching," A. Harry Wheeler Papers, Mathematics Collection, NMAH.

15. Committee on Multi-Sensory Aids, NCTM, *Multi-Sensory Aids in the Teaching of Mathematics, 18th Yearbook of the NCTM* (New York: Bureau of Publications, Teachers College, Columbia University, 1945).

16. "Tel-E-Score," registered as a trademark Mar. 20, 1939. For details, see the Trademark Electronic Search System of the U.S. Patent and Trademark Office.

17. Wilshaw, "Overhead Projector," 115–20.

18. John B. Coker and Harold G. Fitzgerald, "Score Projecting Apparatus," U.S. Patent 2,330,779, Oct. 5, 1943.

19. Brunswick-Balke-Collender Co., *Bowling-America's No. 1 Investment Opportunity* (Chicago: Brunswick-Balke-Collender, 1941). John B. Coker, "Bowling Score Projector," U.S. Patent 2,381,260, Aug. 7, 1945.

20. U.S. War Department, *Army Instruction* (Washington, D.C.: Government Printing Office, 1945). This is War Department Technical Manual TM 21-250.

21. Leo Shippay, "Leeside," *Los Angeles Times*, Nov. 14, 1944, A4.

22. James D. Finn, "The Emerging Technology of Education," *Extending Education through Technology* (Washington, D.C.: Association for Educational Communications and Technology, 1972), 277, uses the first phrase. John R. Miles and Charles R. Spain, for the Commission on Implications of Armed Services Educational Programs, *Audio-Visual Aids in the Armed Services, Implications for American Education* (Washington, D.C.: American Council on Education, 1947), 26–27, use the second and discuss post-war use of overheads. One should not overestimate the importance of overhead projectors as training aids in the World War II navy. In Feb., Mar., and Apr. 1944 Lt. Col. Francis W. Noel, officer-in-charge of the utilization and Evaluation Section of the Training Aids Division of the Bureau of Naval Personnel, published articles in *School Executive* on training aids, as used by the navy. His focus was on motion pictures, with no mention of overhead projectors. Noel did argue that techniques developed for training sailors and marines could be transferred to teaching children and called for expert instruction of and assistance for teachers in the use of new techniques. See Francis W. Noel, *The Navy Turns to Training Aids* (Camden, N.J.: RCA [1945?]). Similarly, a Mar. 1948 manual on shipboard training mentioned opaque projectors but not overheads. See Standards and Curriculum Division, Training, Bureau of Naval Personnel, *Shipboard Training Manual* (Washington, D.C.: Government Printing Office, 1948), 107–12.

23. V. Frederick Rickey, "The First Century of Mathematics at West Point," paper presented at a conference on the History of Undergraduate Mathematics in America, June 2001, U.S. Military Academy, West Point, N.Y.; and Chris Arney, personal communication, Apr. 27, 2005. See also the *Annual Report of the Superintendent*, U.S. Military Academy, 1949, 12; and later *Reports*.

24. This account is based on Wilshaw, "Overhead Projector," 143–56.

25. Ibid., 154, 156.

26. Charles Beseler Co., *Standard Vu-Graph by Beseler for Perfect Projection* (East Orange, N.J.: Beseler, ca. 1952).

27. Charles Beseler Co., *Beseler Vu-Graph Teaching Center* (East Orange, N.J.: Beseler, ca. 1957).

28. Central Scientific Co., *Instruments for the Laboratory Sciences, Catalog J-300* (Chicago: Central Scientific, ca. 1960), 519.

29. *Industrial Photography* 9, no. 12 (Dec. 1960): 10.

30. Bert Lefkowitz, "New Products," *Industrial Photography* 11, no. 9 (Sept. 1962): 16.

31. Roger H. Appeldorn, "Technology Transfer in a Diversified, Global Manufacturing Company," *Journal of Technology Transfer* 22, no. 3 (Fall 1997): 57–64.

32. Bert Lefkowitz, "New Products," *Industrial Photography* 12, no. 2 (Feb. 1963): 12.

33. Ibid., no. 6 (June 1963): 12; and no. 9 (Sept. 1963): 117.

34. Edmund Scientific Co., "Amazing Science Buys for Fun, Study or Profit," *Science* 134, no. 3489, Nov. 10, 1961, 1471.

35. Edmund Scientific Co., *Edmund Catalog Number 681* (Barrington, N.J.: Edmund Scientific, 1967), 5.

36. Anthony G. Oettinger, *Run, Computer, Run: The Mythology of Educational Innovation* (Cambridge: Harvard University Press, 1969), 115–16, 260–61.

37. Robert A. Wilkin, "Now-or-Never Instruction: Newton Schools 'for Depth,'" *Christian Science Monitor,* May 26, 1958, 1.

38. G. R. Anderson, J. D. Weaver, and C. T. Wolf, "Large Group Instruction in Elementary College Mathematics," *American Mathematical Monthly* 72, no. 2 (Feb. 1965): 179–82.

39. J. Sutherland Frame, *Buildings and Facilities for the Mathematical Sciences* (Washington, D.C.: Conference Board of the Mathematical Sciences, 1963), 14–17.

40. See, e.g., Viggo P. Hansen, "New Uses for the Overhead Projector," *Mathematics Teacher* 53, no. 6 (Oct. 1960): 469; Frances A. Cartier, ed., "Research Notes," *Journal of Communication* 10, no. 1 (Mar. 1960): 49–51; and Encyclopaedia Britannica Films, *Science: Catalog of Audio-Visual Materials* (Chicago: Encyclopaedia Britannica, 1959).

41. Eleanor P. Godfrey, *Audiovisual Equipment and Materials in U.S. Public School Districts—Spring, 1961* (Washington, D.C.: Bureau of Social Science Research, 1963); Eleanor P. Godfrey et al., *Audiovisual Programs in the Public Schools, Spring, 1962—Results of a Nationwide Survey* (Washington, D.C.: Bureau of Social Science Research, 1964). U.S. Department of Health, Education and Welfare, Office of Education, "The Cost of Audiovisual Instruction," *School Management* 8, no. 6 (June 1964): 81–94. This article noted that the greatest increase in the use of an audiovisual aid was with the overhead projector.

42. Horace C. Hartsell, "Arithmetic Seen Is Arithmetic Learned," *Audiovisual Instruction* 3, no. 8 (Nov. 1958): 238–39.

43. Calhoun C. Collier, "Blocks to Arithmetical Understanding," *Arithmetic Teacher* 6, no. 5 (Nov. 1959): 262–68. Collier quoted Hartsell on p. 267.

44. Horace C. Hartsell and Wilfred L. Veenendaal, *Overhead Projection* (Buffalo, N.Y.: H. Steward, 1960).

45. "Roger Appeldorn and the Creation of Microreplication Technology," *3M Worldwide,* www.3m.com/about3M/pioneers/appeldorn.jhtml.

46. Minnesota Mining and Manufacturing Co. (3M), "A New Dimension in Teaching," Visual Products Group, 3M, St. Paul, Minn., 1962.

47. "A Mobile Classroom," *Industrial Photography* 12, no. 7 (July 1963): 70.

48. "1.5 Million Grant to Schools Offers Visual Training Aids," *New York Times,* Aug. 14, 1963, 31.

49. Harold F. Bennett, "Visual Teaching Aids," *Christian Science Monitor,* May 18, 1964, 13.

50. Calhoun C. Collier, "Blocks to Arithmetical Understanding," *Arithmetic Teacher* 6, no. 5 (Nov. 1959): 262–68.

51. Viggo P. Hansen, "New Uses for the Overhead Projector," *Mathematics Teacher* 53, no. 6 (Oct. 1960): 467–69.

52. Alan R. Osborne, "Using the Overhead Projector in an Algebra Class," *Mathematics Teacher* 55, no. 2 (Feb. 1962): 135–39.

53. Frame, *Buildings and Facilities for the Mathematical Sciences,* esp. 137–44.

54. J. S. Frame, "Facilities for Secondary School Mathematics," *Mathematics Teacher* 57, no. 6 (Oct. 1964): 379–91.

55. School Mathematics Study Group, *A Brief Course in Mathematics for Junior High School Teachers [Supplement]* ([Palo Alto, Calif.: Stanford University, ca. 1965]).

56. Donovan R. Lichtenberg, "Projection Devices," *Instructional Aids in Mathematics, 34th Yearbook of the NCTM* (Washington, D.C.: NCTM, 1973), 217.

57. Paul D. Grimmer, "Slide Rule Demonstrator and Overhead Projector," U.S. Patent 2,841,888, July 8, 1958; Thomas C. Coale and Wesley F. Heyman, "Projectable Slide Rule," U.S. Patent 2,841,889, July 8, 1958.

58. Projection slide rules by Aristo, Beseler, Nestler, and Pickett are mentioned in Peter M. Hopp, *Slide Rules: Their History, Models, and Makers* (Mendham, N.J.: Astragal Press, 1999). On those of Faber-Castell and Graphoplex, see Dieter von Jezierski, trans. Rodger Shepherd, *Slide Rules: A Journey through Three Centuries* (Mendham, N.J.: Astragal Press, 2000). On those of Post, see an advertisement in *Mathematics Teacher* 58, no. 6 (Oct. 1965): 560. On one by Keuffel & Esser, see Keuffel & Esser Co., *Catalog 8, Slide Rules* (Morristown, N.J.: K&E, 1967), 24. This instrument is not listed in the K&E catalog for 1962 or in those for the 1970s. A projection slide rule also is shown in Welch Scientific Co., *Welch Mathematics Catalog: Instruments and Supplies for Teaching Mathematics* (Skokie, Ill.: Welch Scientific, 1965), 19.

59. Maria Kraus-Boelte and John Kraus, *The Kindergarten Guide and Illustrated Hand-book Designed for the Self-Instruction of Kindergartners, Mothers and Nurses* (New York: E. Steiger & Co., 1888), 1:47; Weber Costello Co., *School Supplies: Catalogue Number 12* (Chicago Heights, Ill.: Weber Costello, 1922), 9; Central Scientific Co., *Instruments for the Laboratory Sciences* (Chicago: Central Scientific, 1961), 504; Central Scientific Co., Physics, Chemistry (Chicago: Central Scientific, 1980), 552.

60. Alan R. Osborne, "Using the Overhead Projector in an Algebra Class," *Mathematics Teacher* 55, no. 2 (Feb. 1962): 135–39.

61. Welch Scientific Co., *Supplement to the Welch Science Catalog: New and Improved Science Teaching Apparatus* (Chicago: Welch Scientific, 1963), 18. See also *Welch Mathematics Catalog: Instruments and Supplies for Teaching Mathematics* (Skokie, Ill.: Welch Scientific, 1965), 48. In both of these catalogs Welch also offered slides marked with grids of lines for use with ordinary projectors.

62. One such device, sold by Stokes Publishing Co. of Palo Alto, Calif., was advertised in *American Mathematical Monthly* 84, no. 10 (Dec. 1977): U7.

63. *Mathematics Teacher* 82, no. 5 (May 1989): 348.

64. Ibid., 83, no. 5 (May 1990): 408–9.

65. Ibid., 82, no. 6 (Sept. 1989): 426.

66. Delta Education (Nashua, N.H.), *Hands-On Math* (Spring 1997): 52–53.

67. For a widely quoted comparison of overhead projector use in the classroom and the bowling alley, see George P. Landow, *Hypertext: The Convergence of Contemporary Critical Theory and Technology* (Baltimore: Johns Hopkins University Press, 1992), 161. Here Landow cites an unidentified participant at a 1988 Sloan-sponsored conference on educational hypermedia, held at Dartmouth College. Reportedly, the speaker said: "It took *only* twenty-five years for the overhead projector to make it from the bowling alley to the classroom. I'm optimistic about academic computing; I've begun to see computers in bowling alleys."

CHAPTER FIVE Teaching Machines and Programmed Instruction

Epigraph: Kenneth O. May, "Programming and Automation," *Mathematics Teacher* 59, no. 5 (May 1966): 445.

1. James T. Fey, Oral History Interview by David L. Roberts, Dec. 7, 2000, R. L. Moore Legacy Collection, 1890–1900, 1920–2003, Archives of American Mathematics, Center for American History, University of Texas at Austin.

2. For an indication of the range of programs written, see Charles I. Foltz, *The World of Teaching Machines: Programed Learning and Self-Instructional Devices* (Washington, D.C.: Electronic Teaching Laboratories, 1961). Further evidence of the importance of mathematics as a subject for programmed instruction appears in Center for Programed Instruction, *The Use of Programed Instruction in U.S. Schools . . .* (Washington, D.C.: Government Printing Office, 1963). This document presents the results of a survey carried out in the year 1961–62. During the early 1960s references such as this one used the spelling *programed*. We have tried to retain this spelling in citing sources.

3. For a general discussion of the history of programmed instruction, see Paul Saettler, *The Evolution of American Educational Technology* (Englewood, Colo.: Libraries Unlimited, 1990), esp. 69–72, 286–304.

4. Leonard Zusne, *Biographical Dictionary of Psychology* (Westport, Conn.: Greenwood Press, 1984), 346; Sidney L. Pressey, "A Simple Apparatus Which Gives Tests and Scores—and Teaches," *School and Society* 23 (1926): 373–76; Pressey, "A Machine for Automatic Teaching of Drill Material," *School and Society* 25 (1927): 549–52; and Pressey, "A Third and Fourth Contribution toward the Coming 'Industrial Revolution' in Education," *School and Society* 36 (1932): 668–72. For an overview of Pressey and Skinner's ideas, see Ludy T. Benjamin, "A History of Teaching Machines," *American Psychologist* 43, no. 9 (1988): 703–12. On Pressey, see Stephen Petrina, "Sidney Pressey and the Automation of Education, 1924–1934," *Technology and*

Culture 45 (Apr. 2004): 305–30. Pressey was granted U.S. Patent no. 1,670,480, issued on May 22, 1928.

5. Sidney L. Pressey, *Psychology and the New Education* (New York: Harper and Bros., 1933), 582.

6. Uta C. Merzbach, *Of Levers and Electrons and Learning and Enlightenment: Technological Augmentation of Cognition in the United States since 1776* (Washington, D.C.: Smithsonian Institution, n.d.), 33. For a more complete discussion of Pressey's work and his commercialization of his machine, see Stephen Petrina, "Psychology, Technology and Clinical Procedures in Education: The Cases of Luella W. Cole and Sidney L. Pressey, 1917–1934" (Ph.D. diss., University of Maryland, 1994).

7. For an account of this work, see E. A. Vargas and Julie A. Vargas, "B. F. Skinner and the Origins of Programmed Instruction," in *B. F. Skinner and Behaviorism in American Culture,* ed. Laurence D. Smith and William R. Woodward (Bethlehem, Pa.: Lehigh University Press, 1996), 237–53. Vargas and Vargas date the initial design of this machine to Apr. 1955 (242).

8. B. F. Skinner, "Teaching Machine," U.S. Patent 2,846,779, Aug. 12, 1958.

9. B. F. Skinner, "Teaching Machines," *Science* 128, Oct. 24, 1958, 969.

10. Foltz, *World of Teaching Machines,* 16–17, 50, 85, 115.

11. Mathematical titles by Nathan A. Crowder include *An Automatic Tutoring Book on Number Systems* (Arlington, Va.: Psychological Research Associates, 1958); *The Arithmetic of Computers: An Introduction to Binary and Octal Mathematics* (Garden City, N.J.: Doubleday, 1960); *Adventures in Algebra* (Garden City, N.J.: Doubleday, 1960); and *Trigonometry: A Practical Course* (Garden City, N.J.: Doubleday, 1962). The algebra and trigonometry books were coauthored by Grace C. Martin. The TutorText on bridge was officially authored by Charles H. Goren. For an advertisement, see *New York Times,* Nov. 6, 1960, BR46.

12. James G. Holland and B. F. Skinner, *The Analysis of Behavior: A Program for Self-Instruction* (New York: McGraw-Hill, 1961). On Skinner's difficulties with commercial manufacture of his teaching machines, see Daniel W. Bjork, *B. F. Skinner: A Life* (New York: Basic Books, 1993), 170–87.

13. "Machine Teaches as Socrates Did," *New York Times,* Mar. 4, 1959, 33; Foltz, *World of Teaching Machines,* 7, 9–11, 72, 80, 104.

14. John W. Blyth, "Teaching Machines and Logic," *American Mathematical Monthly* 67, no. 3 (Mar. 1960): 287.

15. Allen Calvin, *A Demonstration Class Using Teaching Machines for Ninth-Grade Algebra* (Roanoke, Va.: Hollins College, 1960). This publication is cited in Robert E. Silverman, "Auto-Instructional Devices: Some Theoretical and Practical Considerations," *Journal of Higher Education* 31 (1960): 481–86; reprinted in John P. De Cecco, ed., *Educational Technology: Readings in Programmed Instruction* (New York: Holt, Rinehart & Winston, 1964), 28–35.

16. Foltz, *World of Teaching Machines,* 82, 103, 105.

17. Ibid., 72.

18. Ibid., 91–92.

19. E. W. Rushton, *Programmed Learning: The Roanoke Experiment* (Chicago: Encyclopaedia Britannica Press, 1965).

20. Foltz, *World of Teaching Machines,* 72.

21. Evan R. Keislar, "The Development of Understanding in Arithmetic by a Teaching Machine," *Journal of Educational Psychology* 50 (1959): 247–53.

22. Foltz, *World of Teaching Machines,* 87, 96; Evan R. Keislar, *Abilities of First Grade Pupils to Learn Mathematics in Terms of Algebraic Structures by Teaching Machines,* Cooperative Research Program, Project SAE 8998, no. 1090 (Washington, D.C.: U.S. Office of Education, 1961).

23. Patrick Suppes and Newton Hawley, *Geometry for Primary Grades* (San Francisco: Holden-Day, 1960).

24. Adults also encountered programmed instruction on the job. A survey of federal agencies carried out in 1963 turned up 382 programs, most of them developed in-house and used for standard training. Mathematical topics covered included arithmetic, algebra, trigonometry, logarithms, Boolean algebra, and calculus. See Glenn L. Bryan and John A. Nagay, "Use of Programed Instructional Materials in Federal Government Agencies," in *Teaching Machines and Programed Learning, II: Data and Directions,* ed. Robert Glaser (Washington, D.C.: Department of Audiovisual Instruction, National Education Association of the United States, 1965), 743–67. Several industrial firms polled that spring also reported offering programmed courses on basic mathematics for their employees. See H. A. Shoemaker and H. O. Holt, "The Use of Programed Instruction in Industry," in Glaser, *Teaching Machines and Programed Learning,* 2:685–742.

25. Ralph T. Heimer, "Some Implications of Programmed Learning for the Teaching of Mathematics," *Mathematics Teacher* 54, no. 5 (May 1961): 333.

26. J. F. Clark, "Programmed Learning: My First Six Months," *Mathematics Teacher* 55, no. 7 (Nov. 1962): 579.

27. Ibid.

28. Paul McGarvey, "Programmed Instruction in Ninth Grade Algebra," *Mathematics Teacher* 55, no. 7 (Nov. 1962): 576–79.

29. David L. Roberts and Angela L. E. Walmsley, "The Original New Math: Storytelling versus History," *Mathematics Teacher* 96, no. 7 (Oct. 2003): 468–73.

30. Herbert Wills, "The UICSM Programmed Instruction Project," *American Mathematical Monthly* 69, no. 8 (Oct. 1962): 804–6; O. R. Brown, "Using Programmed Instruction in Mathematics Curriculum Development: The UICSM Project," *Programed Instruction* 4, no. 3 (Dec. 1964): 1, 3–5, 9–10.

31. School Mathematics Study Group, *Programed First Course in Algebra Revised Form H* (Palo Alto, Calif.: Stanford University, 1963–65). The paperback text was in four parts, which differed in copyright date. William Wooten, *SMSG: The Making of a Curriculum* (New Haven, Conn.: Yale University Press, 1965); Leander W. Smith, "The School Mathematics Study Group Programed Learning Project," *Programed Instruction* 4, no. 3 (Dec. 1964): 2, 4.

32. "Meeting of the Board of Governors," *American Mathematical Monthly* 69, no. 4 (Apr. 1962): 327–32.

33. "Officers and Committees," *American Mathematical Monthly* 70, no. 4 (Apr. 1963): 477; "Course Content Work at Stanford," *American Mathematical Monthly* 71, no. 7 (Aug.–Sept. 1964): 792.

34. Kenneth O. May, "Programming and Automation," *Mathematics Teacher* 59, no. 5 (May 1966): 444–54. This paper was based on K. O. May, *Programed Learning and Mathematical Education: A CEM Study* (San Francisco: Committee on Educational Media, 1965).

35. "Mathematical Education Notes," *American Mathematical Monthly* 72, no. 8 (Oct. 1965): 900–901.

36. Mathematical Association of America, Committee on Educational Media, *A Programed Course in Calculus* (New York: W. A. Benjamin, 1968); "Telegraphic Reviews," *American Mathematical Monthly* 75, no. 10 (Dec. 1968): 1137; R. G. Tross, "A Programmed Text in Calculus," *American Mathematical Monthly* 77, no. 2 (Feb. 1970): 201–2; "Telegraphic Reviews," *American Mathematical Monthly* 77, no. 5 (May 1970): 551.

37. F. Joe Crosswhite, Oral History Interview by David L. Roberts, July 29–30, 2003, National Council of Teachers of Mathematics Oral History Project Records, 1992–1993, 2002–2003, Archives of American Mathematics, Center for American History, University of Texas at Austin.

38. Shirley A. Hill, Oral History Interview by David L. Roberts, Sept. 13, 2002, National Council of Teachers of Mathematics Oral History Project Records, 1992–1993, 2002–2003, Archives of American Mathematics, Center for American History.

39. James T. Fey, Oral History Interview by David L. Roberts, Dec. 7, 2000, R. L. Moore Legacy Collection, 1890–1900, 1920–2003, Archives of American Mathematics, Center for American History.

40. Allen Calvin, *A Demonstration Class Using Teaching Machines for Ninth-Grade Algebra* (Roanoke, Va.: Hollins College, 1960); cited in Robert E. Silverman, "Auto-Instructional Devices: Some Theoretical and Practical Considerations," *Journal of Higher Education* 31 (1960): 481–86; and reprinted in John P. De Cecco, ed., *Educational Technology: Readings in Programmed Instruction* (New York: Holt, Rinehart & Winston, 1964), 28–35. Herbert Wills, "The UICSM Programmed Instruction Project," *American Mathematical Monthly* 69, no. 8 (Oct. 1962): 804–6.

41. George L. Henderson, "An Independent Classroom Experiment Using Teaching Machine Programmed Materials," *Mathematics Teacher* 56, no. 4 (Apr. 1963): 251.

42. O. Robert Brown Jr., "Using a Programmed Text to Provide an Efficient and Thorough Treatment of Solid Geometry under Flexible Classroom Procedures," *Mathematics Teacher* 60, no. 5 (May 1967): 492–503; Edmund C. Lasswell, "An Open Letter to Authors and Publishers," *Mathematics Teacher* 59, no. 8 (Dec. 1966): 739–40.

43. Ruth Onetta Lane, "Uses of Cardboard," *Mathematics Teacher* 18, no. 4 (Apr. 1925): 238–41; Lane, Frederic Butterfield Knight, and Giles Murrel Ruch, *Manual of Directions for Geometry Rapid Drill Cards* (Chicago: Scott Foresman, 1929).

44. Edmund Scientific advertised its "Trig and Calculus Cards" as their "newest teaching aid" in *Mathematics Teacher* 52, no. 7 (Nov. 1959): 528. The cards were advertised in at least three publications of Edmund Scientific Co.: *Edmund Catalog 606* (Barrington, N.J., ca. 1960), 81; *Edmund Catalog 625* (1962), 97; and *Edmund Catalog 651* (1964), 29. They were not included in *Edmund Catalog 681* (1968). The flash cards for teaching arithmetic were shown only in *Edmund Catalog 606* (ca. 1960), 78. There they were associated with research by Garry Cleveland Myers. Myers wrote about flash cards in the 1920s.

45. "How to Use the New Math Flash Cards," *New Math Addition Flash Cards* (Long Island City, N.Y.: Ed-U-Cards Manufacturing Corp., ca. 1966), card no. 1A. An example of this set of flash cards is in the collections of the National Museum of American History.

CHAPTER SIX The Abacus

Epigraph: Samuel Wilderspin, *Infant Education; or Remarks on the Importance of Educating the Infant Poor,* 3rd ed. (London: W. Simpkin & R. Marshall, 1825), 94.

1. Computations on a counting table are shown in a few early European printed arithmetics. Illustrations from several of them are reproduced in David Eugene Smith, *Rara arithmetica* (Boston: Ginn & Co., 1908). We have found no evidence that these early modern devices were used in instruction.

2. For references on the history of the abacus in the West, see Charles Burnet and W. F. Ryan, "Abacus (Western)," in *Garland Encyclopedia of Scientific Instruments,* ed. Deborah J. Warner and Robert Bud (Hamden, Conn.: Garland, 1998), 5–7.

3. Michel Chasles, "Histoire de l'arithmétique. Développements et détails historiques sur divers points du système de l'abacus," *Comptes Rendus des Séances de l'Académie des Sciences* 16 (1843): 1409. For an overview of Poncelet's career, see Morris Kline, *Mathematical Thought from Ancient to Modern Times* (New York: Oxford University Press, 1972), 834–50. For a general overview of the French mathematical community of his time, see Ivor Grattan-Guiness, *Convolutions in French Mathematics, 1800–1840,* 3 vols. (Basel: Birkhäuser Verlag, 1990). On Poncelet and the abacus, see Irina Gouzévitch and Dmitri Gouzévitch, "La guerre, la captivité et les mathématiques," *SABIX: Bulletin de la Société des Amis de la Bibliothèque de l'École Polytechnique* 19 (1998): 31–68. On later use of a teaching abacus in France—without reference to Poncelet—see Jean-Claude Régnier, "Le Boulier-Numérateur de Marie-Pape Carpantier," *Bulletin de l'Association des Professeurs de Mathématiques de l'Enseignement Public,* no. 447 (Sept.–Oct. 2003): 457–71.

4. Most accounts of Robert Owen and the infant school movement draw heavily on Owen's own published works. See Owen, *The Life of Robert Owen Written by Himself* (London: Effingham Wilson, 1857), 1:60–62, 134–35, 152–54, 174–79.

5. Samuel Wilderspin, *Infant Education; or Remarks on the Importance of Educating the Infant Poor* (London: W. Simpkin & R. Marshall, 1825), 90–94. There is a diagram of the "transposition frame" on p. 90. The quotations are from p. 94.

6. Samuel Wilderspin, *On the Importance of Educating the Infant Children of the*

Poor . . ., 2nd ed. (1823; rpt., London: W. Simpkin & R. Marshall, 1824), 48–50. This book mentions cubical blocks used in teaching arithmetic but not the abacus. *The Infant System, for Developing the Intellectual and Moral Powers of All Children, from One to Seven Years of Age* (London: J. S. Hodson, 1852), 14, 138–42, 150–52. This book is available online at the *Gutenberg* Web site. It is the eighth edition of the book Wilderspin first published in 1823. On Wilderspin, see Phillip McCann and Francis A. Young, *Samuel Wilderspin and the Infant School Movement* (London: Croom Helm, 1982). In *Infant System* Wilderspin refers to a depot he had established for the distribution of the arithmeticon and other infant school apparatus (151). McCann and Young do not discuss this aspect of his life, nor is it mentioned extensively in his surviving papers.

7. A Friend of the Poor, *Infant Education, or, Remarks on the Importance of Educating the Infant Poor: From the Age of Eighteen Months to Seven Years: With an Account of Some of the Infant Schools in England and the System of Education There Adopted: Selected and Abridged from the Works of Wilderspin, Goyder, and Others: And Adapted to the Use of Infant Schools in America* (New York: printed by J. Seymour, 1827).

8. For an overview of the history of infant schools, see Barbara Beatty, *Preschool Education in America: The Culture of Young Children from the Colonial Era to the Present* (New Haven, Conn.: Yale University Press, 1995), esp. 1–37. On the decline of the infant school movement in the United States, see Caroline Winterer, "Avoiding a 'Hothouse System of Education': Nineteenth-Century Early Childhood Education from the Infant Schools to the Kindergartens," *History of Education Quarterly* 32, no. 3 (Fall 1992): 289–314.

9. William Wilson, *The System of Infants' Schools*, 2nd ed. (London: for G. Wilson, 1825). The third edition, with the same imprint, appeared the next year. The sections of this book concerning the abacus were reprinted in "Infant Schools," *American Sunday School Magazine* 3 (Sept. 1826): 257–61. See also William Wilson, *A Manual of Instruction for Infants' Schools* (New York: G. & C. & H. Carvill, 1830).

10. "Infant Schools," 260. Early Sunday schools taught not only religious topics but also basic skills such as reading and writing. See Anne M. Boylan, *Sunday School: The Formation of an American Institution, 1790–1880* (New Haven, Conn.: Yale University Press, 1988).

11. Wilson, *Manual of Instruction for Infants' Schools*, 23, 60–68.

12. On Neef, see Gerald Lee Gutek, *Joseph Neef: The Americanization of Pestalozzianism* (Tuscaloosa: University of Alabama Press, 1978).

13. For biographical information about Phiquepal, see John S. Doskey, ed., *The European Journals of William Maclure* (Philadelphia: American Philosophical Society, 1988), xxxviii, 685, 687.

14. Johann Henrich Pestalozzi, *How Gertrude Teaches Her Children*, trans. Lucy Holland and Frances C. Turner (Syracuse, N.Y.: C. W. Bardeen, 1892), 214–18; Joseph Neef, *Sketch of a Plan and Method of Education, Founded on an Analysis of the Human Faculties and Natural Reason, Suitable for the Offspring of a Free People and for All Rational Beings* (Philadelphia: for the author, 1808); Warren Colburn, *First Lessons*

in Arithmetic on the Plan of Pestalozzi with Some Improvements (Boston: Hilliard & Metcalf, 1821). On Pestalozzi, Colburn, and the new arithmetic, see Joel Silverberg, "The Teaching and Study of Mercantile Mathematics in New England during the Colonial and Early Federal Periods: Sources, Content and Evolution," *Proceedings of the Canadian Society for the History and Philosophy of Mathematics* 17 (2004): 219–40, esp. 232–36.

15. "Miscellaneous Remarks on the Comparative Situation of the Old and New World, and on the Progressive Improvements in Education," *American Journal of Science,* ser. 1, 8 (1824): 189.

16. William Maclure, "An Epitome of the Improved Pestalozzian System of Education, as Practised by William Phiquepal and Madam Fretageot, Formerly in Paris, and Now in Philadelphia; Communicated at the Request of the Editor," *American Journal of Science,* ser. 1, 10 (1826): 145–50.

17. William Maclure to Marie D. Fretageot, July 15, 1825. This letter is published in Arthur E. Bestor Jr., ed., *Education and Reform at New Harmony: Correspondence of William Maclure and Marie Duclos Fretageot, 1820–1833* (Clifton, N.J.: Augustus M. Kelley, 1973), 322. Other letters in this volume trace the history of the schools at New Harmony. For more on Maclure as an educator, see John J. Lucas, "William Maclure: Visionary and Innovative Pioneer of American Science and Education," *American Educational History Journal* 31, no. 2 (2004): 171–77.

18. Sarah Cox Thrall is quoted in Will S. Monroe, *History of the Pestalozzian Movement in the United States* (Syracuse, N.Y.: C. W. Bardeen, 1907), 121. Monroe, in turn, refers to George Browning Lockwood, *New Harmony Communities* (Marion, Ind.: n.p., 1902), 282.

19. On the history of New Harmony and the later fortunes of those who lived there, see the works of Doskey, Gutek, and Bestor already cited; as well as Arthur Bestor, *Backwoods Utopias: The Sectarian Origins and the Owenite Phase of Communitarian Socialism in America* (Philadelphia: University of Pennsylvania Press, 1970). See also Celia Morris, *Fanny Wright: Rebel in America* (Urbana: University of Illinois Press, 1992). The June 1998 issue of the *Indiana Magazine of History* (94, no. 2) is devoted to the subject of William Maclure and the New Harmony experiment, although there is no mention of the abacus.

20. Wilson would not describe a 10 × 10 abacus until 1830.

21. We thank Frank Smith of the Working Men's Institute Library in New Harmony for information about these objects.

22. For biographical information about Holbrook, see "Josiah Holbrook," *American Journal of Education* 8 (1860): 229–56. According to this source, Holbrook attended Silliman's lectures while running his school in Derby. Carl Bode reports, however, that Holbrook served as Silliman's lab assistant during his last two years as an undergraduate. See Carl Bode, *The American Lyceum: Town Meeting of the Mind* (New York: Oxford University Press, 1956), 8. For a more recent discussion of Holbrook, see Angela G. Ray, *The Lyceum and Public Culture in the Nineteenth-Century United States* (East Lansing: Michigan State University Press, 2004).

23. Josiah Holbrook, "Associations of Adults for Mutual Instruction," *American Journal of Education* 1 (Oct. 1826): 594–97. This article was reprinted in the previously cited 1860 biographical article on Holbrook.

24. Josiah Holbrook, *American Lyceum of Science and the Arts: Composed of Associations for Mutual Instruction, and Designed for the General Diffusion of Useful and Practical Knowledge* (Worcester, Mass.: Samuel H. Colton & Co., 1827). On lyceums and mechanics' institutes, see Bode, *American Lyceum,* 3–18.

25. Doskey, *European Journals of William Maclure,* 734.

26. Josiah Holbrook, *Easy Lessons in Geometry for Use in Infant and Primary Schools, as Well as Academies, Lyceums and Families* (Boston: Carter, Hendee & Babcock, 1832).

27. Josiah Holbrook, *American Lyceum, or Society for the Improvement of Schools and Diffusion of Useful Knowledge* (Boston: T. Marvin, 1829), esp. 15–16; Holbrook, *American Lyceum, or Society for the Improvement of Schools and Diffusion of Useful Knowledge* (Boston: Perkins & Marvin, 1829), esp. 23–24.

28. See, e.g., "Apparatus for Schools," *Workingman's Advocate,* Feb. 5, 1831. This notice in a New York newspaper was reprinted from the *Philadelphia Gazette.*

29. "Apparatus," *Family Lyceum: Designed for Instruction and Entertainment and Adapted to Families, Schools and Lyceums* 1, Aug. 10, 1833, 52.

30. Josiah Holbrook, *Apparatus Designed for Families, Schools, Lyceums and Academies* (Boston: Allen & Ticknor, 1833), 4–5. Here Holbrook described an instrument he called the "arithmeticon" (5–6). It was not an abacus but a sheet of paper with a 12 × 12 grid, with a dot in each square. Covering up the squares in different ways suggested laws of arithmetic.

31. Josiah Holbrook, ed., *Self Instructor* 1, no. 3 (Mar. 1841). On the Lyceum Village at Berea, see H. O. Sheldon, *A Lecture Delivered before the Society for the Promotion of Useful Knowledge upon the Lyceum System of Education . . .* (Cincinnati: Ephraim Morgan & Co., 1842). Sheldon indicated that both boys and girls and young men and young ladies were enrolled at Berea, but he did not specifically state the ages of the students. Both students and instructors were expected to spend six hours a day in manual labor (9).

32. In the preface to *First Lessons: Intellectual Arithmetic, upon the Inductive Method of Instruction* Colburn said that he had not had the opportunity to see Pestalozzi's own work on the subject "but only a brief outline of it by another." He does not specify whose outline he had read. See Colburn, *Colburn's First Lessons . . .* (Baltimore: Cushing & Sons; Boston: Hilliard, Gray, Little & Wilkins, 1828), ix. This book is a later edition of Colburn's *An Arithmetic.*

33. Colburn, *An Arithmetic* (1821); Neef, *Sketch of a Plan;* Pestalozzi, *How Gertrude Teaches Her Children,* 214–18.

34. Peter J. MacKintosh to Boston School Committee, Oct. 17, 1823, Papers of the Boston School Committee, Boston Public Library.

35. Report of the Committee "to Consider the Expediency of Making an Alter-

ation in the Books Now Used in the Public Schools for Teaching Arithmetic," May 9, 1826, Papers of the Boston School Committee.

36. Return of the Hancock School, Jan. 1827, Papers of the Boston School Committee.

37. Schoolmasters to H. T. Otis, June 18, 1829, Papers of the Boston School Committee; Frederick Emerson, *Primary Lessons in Arithmetick,* 3rd ed. (Boston: Lincoln & Edmunds, 1827). We have found no evidence of surviving copies of earlier editions.

38. Frederick Emerson, *Arithmetic: Part First, for Young Learners* (Boston: Lincoln & Edmunds, 1829).

39. The Hawes School, which taught African-American girls and boys, was not included in the scheme. School Committee Report, May 1830, Papers of the Boston School Committee.

40. School Committee of the City of Boston, Regulations, Nov. 3, 1831; and Jonathan Snelling to the School Committee, Jan. 21, 1833, Papers of the Boston School Committee.

41. Holbrook, *Easy Lessons in Geometry,* back cover.

42. Joseph M. Wightman, *Annals of the Boston Primary School Committee . . .* (Boston: George C. Rand & Avery, 1860), 184.

43. An example of Emerson's set of cards is in the collections of the National Museum of American History. Cards were mentioned as a teaching device in the "Rules and Regulations" of the Boston Primary School Committee, which operated from 1818 to 1825. See *American Journal of Education* 1 (July 1826): 385–91.

44. Hubbard Winslow, *On the Dangerous Tendency to Innovations and Extremes in Education* (Boston: Tuttle & Weeks, 1835), 5. A copy of this work is in vol. 16 of the Benjamin Silliman Miscellaneous Pamphlets, Beinecke Library, Yale University, New Haven, Conn.

45. Ibid., 12.

46. Ibid., 14.

47. Wightman, *Annals of the Boston Primary School Committee,* 123.

48. Ibid., 124.

49. Minutes of Subcommittee of School Committee Relating to Apparatus, Sept. 2, 1831, Papers of the Boston School Committee.

50. William C. Woodbridge, "Palpable Arithmetic," *American Annals of Education* 1, pt. 2 (1831): 384–87.

51. Thomas H. Palmer, *Prize Essay: The Teacher's Manual: Being an Exposition of an Efficient and Economical System of Education, Suited to the Wants of a Free People* (Boston: Marsh, Capen, Lyon & Webb, 1840), esp. 141–43.

52. Benjamin A. Gould for the Committee on Books, Feb. 14, 1833, Masters to the School Committee, Nov. 6, 1834; and Recommendation of the Committee on Books, Nov. 18, 1834, Papers of the Boston School Committee.

53. H. K. Bugbee, *A Revised Key to Gould's Patent Arithmetical Frame* (New York:

John Gould, 1881). Bugbee first patented an arithmetical frame in 1864. Gould modified the instrument slightly but kept Bugbee's key.

CHAPTER SEVEN The Slide Rule

Epigraph: Timothy Claxton, *Memoir of a Mechanic: Being a Sketch of the Life of Timothy Claxton . . . Written by Himself . . .* (Boston: George W. Light, 1839), 55.

1. See, e.g., Ezekiel Little, *The Usher: Comprising Arithmetic in Whole Numbers; Federal Money; Decimal and Vulgar Fractions; a Description and Use of Coggeshall's Sliding Rule; Some Uses of Gunter's Scale; Superficial and Solid Measuring; Geometrical Definitions and Problems; Surveying; The Surveyor's Pocket Companion, or Trigonometry Made Easy; A Table of Sines; A Table of Tangents; Miscellany, Tables of the Weight and Value of Gold Coins. Calculated and Designed for Youth* (Exeter: H. Ranlet, 1799), 153–54; James Thompson, *A Complete Treatise on the Mensuration of Timber* (Troy, N.Y.: Wright, Wilbur & Stockwell, 1805), 99; and Charles Hutton, *The Compendious Measurer, Being a Brief, yet Comprehensive, Treatise on Mensuration and Practical Geometry. With an Introduction to Duodecimal Arithmetick . . .* (Philadelphia: Hugh Maxwell, 1807), 99–110, 163–75.

2. Useful histories of the slide rule include the classic Florian Cajori, *History of the Logarithmic Slide Rule* (New York: Engineering News Publishing Co., 1909); and Dieter von Jezierski, *Slide Rules: A Journey through Three Centuries,* trans. Rodger Shepherd (Mendham, N.J.: Astragal Press, 2000).

3. Henry Coggeshall, *Timber-Measure by a Line of More Ease, Dispatch and Exactness, Then [*sic*] Any Other Way Now in Use, by a Double Scale . . .* (London: for the author, 1677).

4. Edmund Stone's 1758 English translation of Nicholas Bion's treatise on mathematical instruments reports that Coggeshall's rule was made in three forms: "for some have the two Rulers composing them sliding by one another, like Glazier's Rules; and sometimes there is a Groove made in one Side of a Two-Foot Joint-Rule, in which a thin sliding Piece being put, the Lines put upon this Rule, are placed upon the said Side. And lastly, one Part sliding in a Groove made along the middle of the other, the Length of each of which is a Foot." Stone illustrated the third form of the rule. The second form is what we have called a "carpenter's rule." See Edmund Stone, trans., *The Construction and Principal Uses of Mathematical Instruments. Translated from the French of M. Bion . . .* (London: J. Richardson, 1758), 26. This volume has been reprinted by Holland Press and by the Astragal Press.

5. John Farey, *A Treatise on the Steam Engine: Historical, Practical, and Descriptive* (London: Longman, Rees, Orme, Brown & Green, 1827), 531–40; J. Wess, "The Soho Rule," *Journal of the Oughtred Society* 6, no. 2 (1997): 23–26; Paul Zoller, "The Soho Slide Rule: Genesis and Archeology," *Bulletin of the Scientific Instrument Society* no. 57 (1998): 5–12.

6. Edmé-François Jomard, "Description d'une règle à calculer, employée en Angleterre et appelée sliding rule; précédée de quelques réflexions sur l'état de l'in-

dustrie anglaise, en avril 1815," *Bulletin de la Société d'Encouragement pour l'Industrie Nationale* 14 (1815): 179–90.

7. On the emigration of steam engineers to the United States, see Carroll W. Pursell Jr., *Early Stationary Steam Engines in America: A Study in the Migration of a Technology* (Washington, D.C.: Smithsonian Institution Press, 1969).

8. On the Routledge and carpenter's rules, see Kenneth D. Roberts, "Carpenters' and Engineers Slide Rules (Part 1, History)," *Chronicle of the Early American Industries Association* 36, no. 1 (Mar. 1983): 1–5; and Philip E. Stanley, "Carpenters' and Engineers' Slide Rules: Routledge's Rule," *Chronicle of the Early American Industries Association* 37, no. 2 (June 1984): 25–27. For Routledge's original account, see J. Routledge, *Instructions for Engineers' Improved Slide Rules* (London: for John Jones, 1823). This work was republished in 1983 in Fitzwilliam, N.H., by K. Roberts Publishing Co.

9. On the slide rule, see L. C. Karpinski, *A Bibliography of Mathematical Works Printed in America through 1850* (1940; rpt., New York: Arno Press, 1980); and Peter M. Hopp, "A Slide Rule Bibliography," *Journal of the Oughtred Society* 3, no. 2b (Sept. 1994).

10. John W. Nystrom, *A Treatise on Screw-Propellers and Their Steam-Engines, with Practical Rules and Examples How to Calculate and Construct the Same . . . Also, a Full Description of a Calculating Machine* (Philadelphia: H. C. Baird, 1852).

11. For further information about Nystrom and his rules, see Peggy A. Kidwell, "Nystrom's Calculating Rule," *Rittenhouse* 1 (1987): 102–5; and R. C. Miller, "Nystrom's Calculator," *Journal of the Oughtred Society* 4, no. 2 (1995): 7–11. For an advertisement for the rule, see, e.g., John W. Nystrom, *Pocket-Book of Mechanics and Engineering* (Philadelphia: Lippincott, 1855), 55.

12. On Fuller, see Bob Otnes, "George Fuller and His Slide Rule," *Journal of the Oughtred Society* 8, no. 2 (Fall 1999): 4.

13. Keuffel & Esser, *Catalogue and Price-List . . .* (New York: Keuffel & Esser, 1880), 71.

14. William Franklin, ed., *Partners in Creating: The First Century of K+E, 1867–1967* (N.p.: Keuffel & Esser, 1967), 11.

15. Robert Riddell, *The Slide Rule Simplified, Explained, and Illustrated. Showing Its Wonderful Powers of Calculation and Its Rapidity in Solving All Questions Relative to Mechanical Trades. Designed Especially for the Instruction of Young Carpenters and Joiners . . .* (Philadelphia: J. B. Lippincott & Co., 1881). Kathryn Coyle of the Winterthur Library has kindly examined a copy of Riddell's *Carpenter and Joiner Modernized* (1880) and reports that it contains no illustrations of slide rules.

16. On Thacher's cylindrical slide rule, see Wayne E. Feeley, "Thacher Cylindrical Slide Rules," *Chronicle of the Early American Industries Association* 50, no. 4 (Dec. 1997): 125–29. For an early account of the instrument, see "Thacher's Calculating Instrument or Cylindrical Slide Rule," *Engineering News* 16, Dec. 18, 1886, 410.

17. Francis Wells and Tom Wyman, "La Règle à Calcul: Lenoir, Gravet-Lenoir and

Tavernier-Gravet Slide Rules," *Journal of the Oughtred Society* 11, no. 1 (Spring 2002): 23–27.

18. Amadée Mannheim, "Règle à calculs modifiée," *Nouvelles Annales de Mathématiques*, ser. 1, 12 (1853): 327–29.

19. Léon Lalanne, *Instruction sur les règles à calcul et particulèrement sur la nouvelle règle à enveloppe de verre* (Paris: Hachette, 1854), x.

20. C. V. Boys, "The Slide Rule," *Van Nostrand's Engineering Magazine* 33 (1885): 512.

21. "Topical Discussions and Interchange of Data—No. 259-42," *Transactions of the American Society of Mechanical Engineers* 8 (1886): 707–10. The quotation from Emery is from p. 709; that from Scott is on p. 708.

22. John Butler Johnson, *The Theory and Practice of Surveying*, 16th ed. (1900; rpt., New York: John Wiley, 1908), iv. Johnson explicitly mentions a one-dollar student rule recently introduced by Keuffel & Esser on p. 171 of the text. He prepared a leaflet of instructions for this instrument entitled "Simple Rules for Using the Pocket Slide Rule with Transparent Sliding Indicator," which was distributed by Keuffel & Esser. A copy of this leaflet is in the Thayer School Papers, Collection DA-4, box 3, Rauner Special Collections Library, Dartmouth College.

23. For an overview of American engineering education in this period, see James Gregory McGivern, *First Hundred Years of Engineering Education in the United States (1807–1907)* (Spokane, Wash.: Gonzaga University Press, 1960). On the conflict between the cultures of the shop and the classroom, see Monte A. Calvert, *The Mechanical Engineer in America, 1830–1910: Professional Cultures in Conflict* (Baltimore: Johns Hopkins Press, 1967).

24. Walter S. Church to Robert Fletcher, Feb. 3, 1887, Robert Fletcher Correspondence, Thayer School Papers, Collection DA-4, box 1, Rauner Special Collections Library, Dartmouth College. We thank Sarah Hartwell, of this library, for her remarkable skill and persistence in uncovering information about early use of the slide rule at Dartmouth.

25. Minutes of the Dartmouth Scientific Association, Feb. 16, 1887, Collection DO-6, box 1, Rauner Special Collections Library, Dartmouth College.

26. We thank David Pantalony for this information.

27. Minutes of the Dartmouth Scientific Association, Dec. 18, 1889, Collection DO-6, box 1, Rauner Special Collections Library, Dartmouth College.

28. An 1894 list of books to be procured by students during their first year at the Thayer School includes as "very desirable" a manual on the use of the Mannheim slide rule by Charles W. Crockett of Rensselaer Polytechnic Institute. See Aug. 1, 1894, Receipts, 1894, Robert Fletcher—Finance, Thayer School, Thayer School Papers, Collection DA-4, box 5, Rauner Library Special Collections, Dartmouth College. See also Charles W. Crockett, *An Explanation of the Principles and Operation of the Mannheim Slide Rule* (Troy, N.Y.: E. H. Lisk, 1891).

29. This photograph has been published in Lawrence P. Grayson, *The Making of*

an Engineer: An Illustrated History of Engineering Education in the United States and Canada (New York: Wiley, 1993).

30. For general accounts of Keuffel & Esser slide rules, see Wayne E. Feely, "K&E Slide Rules," *Chronicle of the Early American Industries Association* 49, no. 3 (June 1996): 51–52; and Robert K. Otnes, "Keuffel & Esser—1880 to 1899," *Journal of the Oughtred Society* 10, no. 1 (Spring 2001): 18–28.

31. Dartmouth College, *Catalogue,* 1898–99, 181.

32. Aug. 4, 1898, Receipts, 1898, Robert Fletcher—Finance, Thayer School, Thayer School Papers, Collection DA-4, box 5, Rauner Library Special Collections, Dartmouth College.

33. Crockett, *Explanation of the Principles and Operation of the Mannheim Slide Rule.*

34. Amy P. Rupert of the Rensselaer Polytechnic Institute Archives and Special Collections kindly provided this information.

35. University of Minnesota, College of Engineering, *Catalog,* 1907–15.

36. Stevens Institute of Technology, *Annual Announcement,* 1875–1921.

37. U.S. Naval Academy, *Annual Register,* 1890–1940.

38. Maryland Agricultural College, *Catalogue,* 1889–1912. We thank Heather Huntington for her assistance in searching for these and other college catalogs.

39. Georgia School of Technology, *Catalog,* Atlanta.

40. V. Frederick Rickey and Amy Shell-Gellasch, "201 Years of Mathematics at West Point," in *West Point: Two Centuries and Beyond,* ed. Lance Betros (Abilene, Tex.: McWhiney Foundation Press, 2004), 601.

41. These comments are based on slide rules in the collections of the National Museum of American History, Smithsonian Institution, Washington, D.C.

42. Theodore Lindquist, "Mathematics for Freshmen Students of Engineering" (Ph.D. diss., University of Chicago, 1911). Lindquist describes the courses in computation on pp. 61–63. He gives general information about his questionnaire on pp. 67–70 and specific data on the question on a course in computation on pp. 85–86. For those who like precise numbers, Lindquist sent out 651 questionnaires and received 237 replies. Of the respondents, 124 favored a course in computation, while 49 were opposed.

43. Massachusetts Institute of Technology, *Annual Catalogue,* 1914–15, 397. Lipka published the notes from this course as *Graphical and Mechanical Computation* (New York: Wiley & Sons, 1918).

44. Massachusetts Institute of Technology (MIT), *Annual Catalogue,* 1917–18, 348.

45. Instruction in the use of adding machines also was offered as part of courses in bookkeeping, but they generally were offered entirely outside of mathematics departments.

46. George W. Myers, "The Laboratory Method in the Secondary School," *School Review* 11 (1903): 737, 738.

47. C. E. Comstock, "The Mathematical Library," *Mathematical Supplement of School Science* 1, no. 1 (Apr. 1903): 18.

48. Mathematical Association of America, National Committee on Mathematical Requirements, *The Reorganization of Mathematics in Secondary Education* (N.p.: Mathematical Association of America, 1923), 27. David Lindsay Roberts has described the origins and workings of this committee in "Mathematics and Pedagogy: Professional Mathematicians and American Educational Reform, 1893–1923" (Ph.D. diss., Johns Hopkins University, 1997), 367–91.

49. The first advertisement for a slide rule we found in *Mathematics Teacher* was for the Richardson direct reading slide rule, an all-metal instrument sold out of Chicago that offered scales for addition and subtraction as well as multiplication and division. See *Mathematics Teacher* 6 (Dec. 1913): 122; 8 (Sept. 1915): 64; and 8 (Dec. 1915): 111. Keuffel & Esser ran an advertisement for slide rules that appealed to "Progressive Teachers of Mathematics" in Sept. and Dec. 1916 (*Mathematics Teacher* 9:70 and 137). The following year the firm prepared an ad touting the advantages of teaching the slide rule to junior high school students (*Mathematics Teacher* 9:177). In a 1920 blurb K&E emphasized that "Engineering and Trade Schools no longer monopolize courses in Slide Rule Practice" (*Mathematics Teacher* 12:136). A 1923 advertisement stressed the role of the slide rule in trigonometry teaching (*Mathematics Teacher* 16:192). Keuffel & Esser ads, which often stressed the role of the slide rule as a check in trigonometry teaching, ran in the *American Mathematical Monthly* almost monthly from 1923 through at least 1930.

50. For a brief account of demonstration slide rules, see Jezierski, *Slide Rules*, 85–86.

CHAPTER EIGHT The Cube Root Block

Epigraph: The Teacher's Guide to Illustration: A Manual to Accompany Holbrook's School Apparatus (Hartford, Conn., 1856), 34.

1. James H. Porter, *A New System of Arithmetic and Mathematics* (New York: Piercy & Reed, Printers, 1845), 175–76.

2. Ibid., 176.

3. For a short survey, see David L. Roberts, "Romancing the Root: An Episode in the Promotion of Geometric Aids for Arithmetic Instruction," *Rittenhouse* 7 (1993): 33–36.

4. Martin A. Nordgaard, "The Origin and Development of Our Present Method of Extracting the Square and Cube Roots of Numbers," *Mathematics Teacher* 17 (1924): 223–38.

5. David Eugene Smith, *History of Mathematics* (1925; rpt., New York: Dover Publications, 1953), 2:148.

6. See, e.g., Morris Kline, *Mathematical Thought from Ancient to Modern Times* (New York: Oxford University Press, 1972), 256–58.

7. Herman Goldstine, *A History of Numerical Analysis from the 16th through the 19th Century* (New York: Springer-Verlag, 1977), 16.

8. Oliver Byrne, *The Pocket Companion for Machinists, Mechanics, and Engineers* (New York: Dewitt & Davenport, 1851), 30.

9. Nick Kollerstrom, "Thomas Simpson and 'Newton's Method of Approximation': An Enduring Myth," *British Journal for the History of Science* 25 (1992): 347–54.

10. Goldstine, *History of Numerical Analysis,* 64–68, 278–80.

11. Specifically, x_{k+1} is generated by intersecting the tangent to the function at x_k with the horizontal axis. See, e.g., Kaiser S. Kunz, *Numerical Analysis* (New York: McGraw-Hill, 1957), 11.

12. F. B. Knight, J. W. Studebaker, and G. M. Ruch, *Standard Service Arithmetics, Grade Eight* (Chicago: Scott, Foresman & Co., 1928), 343.

13. Roswell C. Smith, *Practical and Mental Arithmetic on a New Plan* (Philadelphia: Marshall Clarke, & Co., 1833), vi. The preface is dated Jan. 1829.

14. Oliver A. Shaw, *A Description of the Visible Numerator, with Instructions for Its Use* (Boston: T. R. Marvin, Printer, 1832), 85–93. Shaw was promoting the use of a set of solid blocks of his invention, one configuration of which was essentially equivalent to the cube root blocks shown in Figures 8.3–5.

15. Benjamin Greenleaf, *The National Arithmetic* (Boston: Robert S. Davis & Co., 1856), 242–50.

16. Charles Davies, *New University Arithmetic* (New York: A. S. Barnes & Burr, 1860), 300–317, explains the binomial method for square and cube root but uses the diagram for square root only. Horatio N. Robinson, *The Progressive Arithmetic* (New York: Ivison, Phinney, Blakeman & Co., 1868), 381–405, explains the binomial method for square and cube root, with no diagrams. Daniel W. Fish, *Robinson's Progressive Practical Arithmetic* (New York: Ivison, Blakeman, Taylor, & Co., 1871), 315–28, explains the binomial method for square and cube root, using diagrams for both. *Robinson's New Higher Arithmetic* (New York: American Book Co., 1895), 429–47, is virtually identical with the preceding reference.

17. Brownell, *Teacher's Guide to Illustration,* 34.

18. See Lawrence Cremin, *The Transformation of the School* (New York: Alfred A. Knopf, 1961), 134.

19. Advertising supplement to James Johonnot, *School Houses* (New York: J. W. Schermerhorn & Co., 1871), 84–85.

20. See Cremin, *Transformation of the School,* 26–34, 128–42.

21. One of the earliest uses of the term *mathematics laboratory* seems to have been Truman H. Safford, *Mathematics Teaching and Its Modern Methods* (Boston: D. C. Heath & Co., 1887), 43. The idea received a strong boost in Eliakim Hastings Moore, "On the Foundations of Mathematics," *Science,* n.s. 17, Mar. 13, 1903, 401–16, and further amplification soon after by a number of others. See, e.g., G. W. Myers, "The Laboratory Method in the Secondary School," *School Review* 11 (1903): 727–41; and J. W. A. Young, *The Teaching of Mathematics in the Elementary and Secondary Schools* (New York: Longman, Green & Co., 1907).

22. *Priced and Illustrated Catalogue and Descriptive Manual of Meteorological and Philosophical Instruments and School Apparatus* (Philadelphia: James W. Queen & Co.,

1867), 6. The "trinomial cube" was a cube dissected into twenty-seven pieces to illustrate the expansion of $(a + b + c)^3$. *Catalogue of Physical Instruments,* 29th ed. (Philadelphia: James W. Queen & Co., 1882), 3. *Catalogue of Physical Apparatus* (Philadelphia: Queen & Co., 1898), 7–8.

23. *A Catalogue of Scientific Instruments* (Boston: L. E. Knott Apparatus Co., 1916), 68; *Physics Equipment* (Cambridge, Mass.: L. E. Knott Apparatus Co., 1925), 39–40; *Catalog "G," Laboratory Apparatus and Supplies for Physics and Chemistry* (Chicago: W. M. Welch Scientific Co., 1922), 52. See Figure 8.4; *Catalog G, Scientific Instruments, Laboratory Apparatus and Supplies for Physics, Chemistry and General Science* (Chicago: W. M. Welch Scientific Co., 1929), 28–29.

24. See, respectively, Glenn Hobbs et al., *Practical Mathematics, Part III* (Chicago: American School of Correspondence, 1910), 86–87; George H. Van Tuyl, *Complete Business Arithmetic* (New York: American Book Co., 1911), 150–54; and Claude Irwin Palmer, *Practical Mathematics for Home Study* (New York: McGraw-Hill, 1919), 105.

25. James K. Bidwell and Robert G. Clason, *Readings in the History of Mathematics Education* (Washington, D.C.: National Council of Teachers of Mathematics [NCTM], 1970), 131.

26. See, e.g., National Committee on Mathematics Requirements, *The Reorganization of Mathematics in Secondary Education* (Boston: Houghton Mifflin, 1923), 27.

27. Knight, Studebaker, and Ruch, *Standard Service Arithmetics,* 338.

28. See, e.g., Robert E. Kohler, "The Ph.D. Machine: Building on the Collegiate Base," *Isis* 81 (Dec. 1990): 638–62.

29. G. M. Wilson, *A Survey of the Social and Business Usage of Arithmetic* (New York: Bureau of Publications, Teachers College, Columbia University, 1919), 52–53.

30. On the rise of algebra in the schools, see Philip S. Jones, ed., *A History of Mathematics in the United States and Canada* (Washington, D.C.: NCTM, 1970), 27–28; and David Lindsay Roberts, "Mathematics and Pedagogy: Professional Mathematicians and American Educational Reform, 1893–1923" (Ph.D. diss., Johns Hopkins University, 1997), 78–83.

31. See Warren Colburn, *An Introduction to Algebra upon the Inductive Method of Instruction* (1825; rpt., Boston: Jordan & Wiley, 1846), 133–41, 151–61. For later texts, see, e.g., Charles W. Hackley, *A Treatise on Algebra* (New York: Harper & Bros., 1854), 84–87; Webster Wells, *College Algebra* (Boston: D. C. Heath & Co., 1890), 136–63; Albert Harry Wheeler, *First Course in Algebra* (Boston: Little, Brown, 1907), 381–90; and William J. Milne, *First Year Algebra* (New York: American Book Co., 1911), 209–11.

32. David Eugene Smith, *The Teaching of Elementary Mathematics* (New York: Macmillan, 1902), 132.

33. Patricia Cline Cohen, *A Calculating People: The Spread of Numeracy in Early America* (Chicago: University of Chicago Press, 1982), 130–33.

34. Greenleaf, *National Arithmetic,* 3.

35. Porter, *New System of Arithmetic and Mathematics,* 178.

36. From Frederick Winsor's *The Space Child's Mother Goose* (New York: Simon & Schuster, 1958), as reprinted in Clifton Fadiman, ed., *The Mathematical Magpie* (New York: Simon & Schuster, 1962), 263–64.

37. Laurence R. Veysey, *The Emergence of the American University* (Chicago: University of Chicago Press, 1965), 55–56.

38. Cremin, *Transformation of the School,* 113.

39. David Eugene Smith, *The Teaching of Arithmetic* (Boston: Ginn & Co., 1909), 186.

40. See chapter 7.

41. Marilyn N. Suydam, ed., *Developing Computational Skills* (Washington, D.C.: NCTM, 1978), 5. See also pp. 220–22.

42. Moore, "On the Foundations of Mathematics."

43. Ibid., 410.

44. See, e.g., Ernest Breslich, *Senior Mathematics,* bk. 1 (Chicago: University of Chicago Press, 1928), 164–66. Moore coauthored the preface.

45. See David Lindsay Roberts, "E. H. Moore's Early Twentieth-Century Program for Reform in Mathematics Education," *American Mathematical Monthly* 108 (Oct. 2001): 689–96.

CHAPTER NINE Blocks, Beads, and Bars

Epigraphs: John Holt, *How Children Fail* (New York: Pitman Publishing, 1964), 78, 80.

1. A Friend of the Poor, *Infant Education, or, Remarks on the Importance of Educating the Infant Poor: From the Age of Eighteen Months to Seven Years: With an Account of Some of the Infant Schools in England and the System of Education There Adopted: Selected and Abridged from the Works of Wilderspin, Goyder, and Others: And Adapted to the Use of Infant Schools in America* (New York: printed by J. Seymour, 1827), 21–22.

2. There is an extensive literature on Friedrich Froebel and the dissemination of the kindergarten in Europe and the United States. See Barbara Beatty, *Preschool Education in America* (New Haven, Conn.: Yale University Press, 1995); Norman Brosterman, *Inventing Kindergarten* (New York: Harry N. Abrams, 1997); Catherine A. Cosgrove, "A History of the American Kindergarten Movement from 1860 to 1916" (Ed.D. diss., Northern Illinois University, 1989); and Ann Taylor Allen, "'Let Us Live with Our Children': Kindergarten Movements in Germany and the United States, 1840–1914," *History of Education Quarterly* 28 (Spring 1988): 23–48.

3. Froebel was not the first to propose using blocks to assist in learning basic arithmetic. Not only Wilderspin but Ernst Tillich, a German admirer of Pestalozzi, made suggestions concerning this topic. Tillich described his methods for teaching arithmetic in *Allgemeines Lehrbuch der Arithmetik; oder, Anleitung zur Rechenkunst für Jedermann,* originally published in 1806. A second edition, revised by Friedrich Wilhelm Lindner, appeared in 1821, in two volumes, published in Leipzig by Gräffschen Buchhandlung.

4. This account is based on Edward Wiebé, *The Paradise of Childhood: A Manual*

of Self-Instruction in Friedrich Froebel's Educational Principles and a Practical Guide to Kindergartners (Springfield, Mass.: Milton Bradley & Co., 1869).

5. Ibid., 26–27 and pl. 16.

6. Ibid. On Wiebé and Bradley, see the centennial history of Milton Bradley, James J. Shea as told to Charles Mercer, *It's All in the Game* (New York: G. P. Putnam's Sons, 1960). J. W. Schemerhorn & Co., *School Material* (New York: Schemerhorn, 1871) does not include kindergarten apparatus but mentions that the company soon planned to publish a guide that would contain these materials. They are listed in J. W. Schemerhorn & Co., *School Material* (New York: Schemerhorn, 1874), 116–21. Maria Kraus-Boelte and John Kraus, *The Kindergarten Guide: An Illustrated Handbook Designed for the Self-Instruction of Kindergartners, Mothers and Nurses* (New York: E. Steiger, 1892). By 1900 *Steiger's Kindergarten Catalogue* had reached its eighth edition and ran to sixty-five pages (a copy of this catalog is bound with the preceding work in the Smithsonian Institution Libraries collections). In addition to these materials, Steiger published "kindergarten occupations for the family," which were variations on Froebel's gifts and occupations designed for home use. He also distributed a set of pattern books for use with Froebelian materials that were arranged by Kraus-Boelte and Kraus and published under the general title *Kindergarten Gifts and Occupation Material.*

7. E. Seguin, "Report on Education," in *Reports of the Commissioners of the United States to the International Exposition Held at Vienna,* ed. Robert H. Thurston (Washington, D.C.: Government Printing Office, 1876), 2:17.

8. Examples of Glidden's proposed gift and Bradley's manufactured "curvilinear gift for the kindergarten" are in the Association of Childhood Education International Archives, Archives and Manuscripts Department, University of Maryland Libraries, College Park. See also Minnie M. Glidden, "Educational Appliance," U.S. Patent 595,455, Dec. 14, 1897. For further information about mathematical uses of kindergarten gifts, see Minnie M. Glidden, *A Mathematical Study: Froebel's Building Gifts Seventh and Eighth* (New York: Pratt Institute, 1906); and Kristina Leeb-Lundberg, "Friedrich Froebel's Mathematics for the Kindergarten: Philosophy, Program, and Implementation in the United States" (Ph.D. diss., School of Education, New York University, 1971). Leeb-Lundberg discusses ideas of Glidden, of the philosopher Denton J. Snider of Chicago, and of the kindergarten teacher Florence Lawson of the Los Angeles State Normal School.

9. On kindergarten exhibits at the 1876 and 1893 fairs and on the rift between Coe and Peabody, see Cosgrove, "History of the American Kindergarten Movement." See also Elizabeth Palmer Peabody, *Letters of Elizabeth Palmer Peabody: American Renaissance Woman,* ed. Bruce A. Ronda (Middletown, Conn.: Wesleyan University Press, 1984), 356, 364. On Peabody and her family, see Megan Marshall, *The Peabody Sisters: Three Women Who Ignited American Romanticism* (Boston: Houghton Mifflin, 2005).

10. On kindergartens and primary schools, see David Tyack and Larry Cuban,

Tinkering toward Utopia: A Century of Public School Reform (Cambridge: Harvard University Press, 1995), 64–69. Barbara Beatty and Catherine Cosgrove also discuss the growth of kindergartens in the public schools in the works cited previously. Brosterman argues that Froebel's gifts did have considerable influence on abstract art, both in the United States and elsewhere; see *Inventing Kindergarten.*

11. Unless otherwise noted, information about Montessori is taken from Rita Kramer, *Maria Montessori: A Biography* (New York: G. P. Putnam's Sons, 1976).

12. For accounts of Montessori's apparatus, see Maria Montessori, *The Montessori Method,* trans. Anne E. George (New York: Frederick A. Stokes Co., 1912); and the House of Childhood, *The Montessori Didactic Apparatus* (New York: House of Childhood, 1913). A copy of the latter catalog is found in the Association of Childhood Education International Archives, Archives and Manuscripts Department, University of Maryland Libraries, College Park.

13. Maria Montessori, "Educational Device," U.S. Patent 1,103,369, July 14, 1914.

14. Maria Montessori, "Cut-Out Geometrical Figure for Didactical Purposes," U.S. Patent 1,173,298, Feb. 29, 1916.

15. Advertisements for the House of Childhood, Inc., appeared in the *New York Times,* Dec. 30, 1911, 4, and Dec. 31, 1911, X13. On Byoir, see Gerald Carson, "Carl Robert Byoir," *Dictionary of American Biography,* supp. 6 (New York: Scribner's Sons, 1980), 89–91.

16. Kramer, *Maria Montessori,* 221. Here Kramer states that the Montessori class received both of the only two gold medals awarded in the field of education. This claim is highly dubious. At the Panama-Pacific International Exposition the highest level of award was the grand prize, the next the medal of honor, and the third the gold medal. See Frank M. Todd, *The Story of the Exposition,* vol. 5 (New York: G. P. Putnam Sons, 1921), app., 131. Awardees in the Department of Education from the state of Massachusetts alone included medal of honor winners Ginn & Co., Milton Bradley Co. (for its kindergarten and school supplies), and Standard Electric Time Co. Gold medal winners from the state included Harvard University, Massachusetts Agricultural College, Smith College, and Springfield High Schools. It is possible that Montessori won the grand prize, but only one such prize was awarded in any department. See Massachusetts, Board of Panama-Pacific Managers, *Massachusetts at the Panama-Pacific International Exposition* (Boston: Wright & Potter Printing Co., 1915[?]), 213.

17. Maria Montessori, *L'autoeducazione nelle scuole elementari* (Roma: P. Maglione & C. Strini, 1916).

18. Maria Montessori, *The Montessori Elementary Material* (New York: F. A. Stokes, 1917). This is the second volume of *The Advanced Montessori Method.*

19. Eric Pace, "Nancy Rambusch, 67, Educator Who Backed Montessori Schools," *New York Times,* Oct. 30, 1964, 49.

20. "Montessori Form of Teaching Gains," *New York Times,* Oct. 4, 1964, 64.

21. Deborah J. Warner, personal communication, July 15, 2005.

22. Nancy Shute, "Madam Montessori," *Smithsonian* 33, no. 6 (Sept. 2002): 70–74.

23. "Catherine Stern," *National Cyclopedia of American Biography* (Clifton, N.J.: James T. White, 1977), 57:660–61; and Richard D. Troxel, "Catherine Brieger Stern," *Notable American Women* (Cambridge, Mass.: Belknap Press, 1971), 3:659–60.

24. Catherine Stern, *Look and See! Touch and Feel! Training the Senses of Children from Six to Fourteen Years of Age with New Play Materials,* Mathematics Collections, National Museum of American History.

25. Montessori, *Montessori Method,* 330–32.

26. Catherine Stern, *Children Discover Arithmetic: An Introduction to Structural Arithmetic* (New York: Harper & Row, 1949), 263–69. See also Stern, "Concrete Representation of Geometric Progression (with Illustrations from the Decimal and the Binary Number Systems)," *Mathematics Teacher* 33 (Mar. 1951): 170–76.

27. Delta Education, *Hands-On Math, K–8* (Nashua, N.H., Spring 1997): 20–21; ETA/Cuisenaire, *Math K–12 2004 Catalog* (Vernon Hills, Ill.: ETA/Cuisenaire, 2004), 99–102; Learning Resources, *2004 Educational Catalog Grades Pre K–8* (Vernon Hills, Ill.: Learning Resources, 2004), 97.

28. John E. Pfeiffer, "Math Can Be Child's Play," *New York Times,* June 5, 1949, BR12; Maxine Dunfee, "Review of *Children Discover Arithmetic,*" *Childhood Education* 27 (Sept. 1950): 44; Ida Mae Heard, "Review of *Children Discover Arithmetic,*" *Mathematics Teacher* 44 (May 1951): 349.

29. Catherine Stern, *Experimenting with Numbers: Teacher's Manual for Use with Beginners* (Boston: Houghton Mifflin, 1950).

30. This series was advertised in *Arithmetic Teacher* 3 (Apr. 1956): 126.

31. Troxel, "Catherine Brieger Stern," 660.

32. Stern and her daughter-in-law critically examined many assumptions of arithmetic teaching associated with the New Math, in Catherine Stern and Margaret B. Stern, *Children Discover Arithmetic: An Introduction to Structural Arithmetic* (New York: Harper & Row, 1971). This is a "revised and enlarged edition" of Stern's 1949 book.

33. From 1948 through 1953 Henry W. Syer of the School of Education of Boston University and Donovan A. Johnson of the College of Education of the University of Minnesota edited a regular column on aids to teaching, including booklets, charts, equipment, films, filmstrips, mathematical instruments, and geometric and scientific models. This column discussed several sets of blocks. On the "Teach-a-Number" blocks manufactured in New Orleans, see comments in *Mathematics Teacher* 42 (Nov. 1949): 369. On the walnut cubes manufactured by the Mathaids Company of Syracuse, N.Y., see vol. 43 (Jan. 1950): 43. On the Math-o-Blocks made by the D. T. Davis Co. of Lexington, Ky., see vol. 45 (Mar. 1952): 209.

34. For an account of Baratta-Lorton and her encounter with mathematical manipulatives, see Bob Baratta-Lorton, "A Tribute to Mary," *Math Their Way Summary*

Newsletter (Saratoga, Calif.: Center for Innovation in Education, 1997); available online at www.center.edu/NEWSLETTER/newsletter.shtml. For fragmentary information about Kathy Richardson, see the Web site of *Math Perspectives Teacher Development Center* at www.mathperspectives.com/. On the number of Unifix cubes sold, see the press release "Didax Turns Twenty-Five," Didax, Inc., Dec. 31, 2001; available online at www.didax.com/pressroom/pressdetails.cfm/PressID/2.cfm.

35. Louis Jeronnez, "Hommage à Georges Cuisenaire," *Mathématique et Pédagogie* 6 (Mar.–Apr. 1976): 75–81.

36. Rosemary March, "Georges Cuisenaire and His Rainbow Rods," *Learning* 6 (Nov. 1977): 81–88.

37. ETA/Cuisenaire, *Celebrating 75 Years of Cuisenaire Rods, 1931–2006* (Vernon Hills, Ill.: ETA/Cuisenaire, [2006]), 2.

38. Caleb Gattegno, "Numbers in Colour," *Bulletin of the Association for Teaching Aids in Mathematics* 2 (1953): 2–3; and "Arithmetic with Coloured Rods," *Times Educational Supplement,* Nov. 19, 1954.

39. Georges Cuisenaire and Caleb Gattegno, *Numbers in Colour: A New Method of Teaching the Processes of Arithmetic to All Levels of the Primary School* (Melbourne: Heinemann, 1954).

40. Dale Seymour and Patricia S. Davidson, "A History of Nontextbook Materials," in *A History of School Mathematics,* ed. George M. A. Stanic and Jeremy Kilpatrick (Reston, Va.: NCTM, 2003), 989–1035.

41. Cuisenaire and Gattegno, *Numbers in Colour.*

42. Caleb Gattegno, "New Developments in Arithmetic Teaching in Britain: In Britain: Introducing the Concept of 'Set'," *Arithmetic Teacher* 3 (Apr. 1956): 85.

43. Ibid., 87.

44. Charles F. Howard, "British Teachers' Reactions to the Cuisenaire-Gattegno Materials: The Color-Rod Approach to Arithmetic," *Arithmetic Teacher* 4 (Nov. 1957): 191–95.

45. James Carberry, "British Expert on Math Unveils Novel Teaching Methods Here," *Washington Post,* Nov. 8, 1958, C1; and Carolyn F. Hummel, "Colorful Rods 'Teach' Arithmetic," *Christian Science Monitor,* May 25, 1960, 2.

46. *Arithmetic Teacher* 6 (Oct. 1959): 240. The same advertisement also ran in the Nov. and Dec. issues of the magazine.

47. William P. Hull, *An Introduction to the Cuisenaire Rods: An Approach to the Teaching of Elementary Mathematics* (Boston: National Council of Independent Schools, 1961). This work was noted in *American Mathematical Monthly* 69 (Apr. 1962): 311.

48. Holt, *How Children Fail,* 78–129.

49. Martin P. Mayer, *The Schools* (New York: Harper & Bros., 1961), 168, 234–59.

50. See, e.g., Negative no. 4937, coll. 2–13 (Committee on School Mathematics), box 99, Photographic Subject File, R.S. no. 39/2/20, University Archives, University of Illinois at Urbana-Champaign. This is one of a group of photographs.

51. "Captivating Key to Math," *Life,* Aug. 10, 1961. The article in *PTA Magazine* is reported in Kenneth G. Gehret, "As Teacher Finds Herself Famous," *Christian Science Monitor,* Dec. 30, 1961, 7.

52. Joseph Turner, "Model Teaching," *Science* 134, no. 3491, Nov. 24, 1961, 1661.

53. Glenn Fowler, "Caleb Gattegno, 76, a Proponent of Novel Learning Theories, Dies," *New York Times,* Aug. 4, 1988, D21. For further information on Gattegno's later enterprises, see the trademarks registered by Educational Solutions.

54. ETA/Cuisenaire, *Celebrating 75 Years of Cuisenaire Rods,* 1–7; Cathy L. Seeley, "Mathematics Textbook Adoption in the United States," and Dale Seymour and Patricia S. Davidson, "A History of Nontextbook Materials," both in *A History of School Mathematics,* ed. G. M. A. Stanic and J. Kilpatrick (Reston, Va.: NCTM, 2003), 2:957–88 and 989–1035, respectively.

CHAPTER TEN The Protractor

Epigraph: Webster Wells and Walter W. Hart, *Modern Plane Geometry: A Graded Course* (Boston: D. C. Heath & Co., 1926), 16, 18.

1. Amy Ackerberg-Hastings, "Protractors in the Classroom: An Historical Perspective," in *From Calculus to Computers: Using the Last 200 Years of Mathematical History in the Classroom,* ed. Richard Jardine and Amy Shell-Gellasch, Mathematical Association of America (MAA) Notes no. 68 (Washington, D.C.: MAA, 2005), 217–28; and Ackerberg-Hastings, "Rectangular Protractors and the Mathematics Classroom," in *Hands on History: A Resource for Teaching Mathematics,* ed. Amy Shell-Gellasch (Washington, D.C.: MAA, 2007), 35–40.

2. Thomas Blundeville, *A Brief Description of Universal Maps and Cards and of Their Use: Necessary for Those That Delight in Reading of Histories; and Also for Travellers by Land or Sea,* ed. and trans. Harold M. Otness (1589; rpt., Ashland, Ore.: Detu Press, 1977), 13. See also David W. Waters, *The Art of Navigation in England in Elizabethan and Early Stuart Times* (London: Hollis & Carter, 1958), 212, 347–50.

3. Maurice Daumas, *Scientific Instruments of the Seventeenth and Eighteenth Centuries and Their Makers,* ed. and trans. Mary Holbrook (London: B. T. Batsford, 1972), 16; Maya Hambly, *Drawing Instruments, 1580–1980* (London: Sotheby's Publications, 1988), 120.

4. E. G. R. Taylor, *The Mathematical Practitioners of Tudor and Stuart England* (Cambridge: [published] for the Institute of Navigation at the University Press, 1970), 173.

5. Nicolas Bion, *Traité de la construction et des principaux usages des instruments de mathematique* (Paris, 1709), 25–28; Edmond Stone, trans., *The Construction and Principal Uses of Mathematical Instruments, Translated from the French of M. [Nicolas] Bion,* 2nd ed. (London, 1758), 12–14. A condensed version of Bion's treatise is [Henry Kemble Oliver], *An Elementary Treatise on the Construction and Use of the Mathematical Instruments Usually Put into Portable Cases* (Boston, 1830).

6. Edmond R. Kiely, *Surveying Instruments: Their History and Classroom Use,* Na-

tional Council of Teachers of Mathematics 19th Yearbook (New York: Bureau of Publications, Teachers College, Columbia University, 1947), 241.

7. Alexis-Claude Clairaut, *Éléments de géométrie* (Paris: Lambert & Durand, 1741), 1–72. See also Jean Itard, "Clairaut, Alexis-Claude," in *Dictionary of Scientific Biography,* ed. Charles Coulston Gillispie (New York: Charles Scribner's Sons, 1970), 3:281–86; and Alexis-Claude Clairaut to Unknown, Feb. 4, 1751, Dibner Library [MSS000348A]. Some French Euclidean geometry textbooks included a protractor in the plates but did not discuss this instrument in the text: Olry Terquem, *Manuel de géométrie. Ou exposition élémentaire des principes de cette science,* 2nd ed. (Paris: Roret, 1835).

8. T. Drummond, *The Young Ladies' and Gentlemen's Auxiliary in Taking Heights and Distances, Containing the Use of the Small Pocket Case of Mathematical Instruments, Illustrated by Practical Geometry,* 2nd ed. (Norwich, Eng.: Bacon, Kinnebrook, & Co., [1814]), 1:18–26. Similar textbooks include Thomas Kentish, *A Treatise on a Box of Instruments and the Slide-Rule, for the Use of Excisement, Engineers, Seamen, and Schools,* 2nd ed. (London, 1847), 21–26; and Henry Angel, *Practical Plane Geometry and Projection, for Science Classes, Schools, and Colleges* (London: William Collins, Sons, & Co., Ltd., n.d.), 20–27.

9. The prolific Charles Davies published textbooks on practical geometry that contained training in protractors, including *Practical Geometry: With Selected Applications in Mensuration, in Artificers' Work and Mechanics* (Philadelphia: A. S. Barnes & Co., 1839), 87–115; and *Elements of Drawing and Mensuration Applied to the Mechanic Arts* (New York: A. S. Barnes & Co., 1846), 38–52.

10. *Encyclopaedia Britannica; Or, A Dictionary of Arts and Sciences, Compiled upon a New Plan* (Edinburgh: Bell and Macfarquhar, 1771), 1:684–710; George Gregory, *A Dictionary of Arts and Sciences* (London: Richard Phillips, 1806); George Gregory, *A New and Complete Dictionary of Arts and Sciences: Including the Latest Improvement and Discovery and the Present State of Every Branch of Human Knowledge,* 1st American, from the 2nd London, ed., vol. 1 (Philadelphia: Isaac Peirce, 1815–16).

11. The many editions of Abel Flint's *System of Geometry and Trigonometry* never addressed how to make surveying drawings with protractors: *A System of Geometry and Trigonometry: Together with a Treatise on Surveying; Teaching Various Ways of Taking the Survey of a Field; Also to Protract the Same and Find the Area. Likewise, Rectangular Surveying; Or, An Accurate Method of Calculating the Area of Any Field Arithmetically, without the Necessity of Plotting It. To the Whole Are Added Several Mathematical Tables, Necessary for Solving Questions in Trigonometry and Surveying; with a Particular Explanation of Those Tables, and the Manner of Using Them,* 3rd ed. (Hartford, Conn.: Oliver D. Cooke, 1813); *A System of Geometry and Trigonometry,* 4th ed. (Hartford, Conn.: Cooke & Hale, 1818); *A System of Geometry and Trigonometry,* 5th ed. (Hartford, Conn.: Oliver D. Cooke & Co., 1825); *A System of Geometry and Trigonometry,* 7th ed. (Hartford, Conn.: Cooke & Co., 1833). Two surveying textbooks that did mention protractors were Samuel Alsop, *A Treatise on Surveying,* 3rd ed.

(Philadelphia: E. C. & J. Biddle & Co., 1860); and T. G. Bunt, *Crocker's Elements of Land Surveying,* new ed. (London: Longman, 1864).

12. Allan Chapman, *Dividing the Circle: The Development of Critical Angular Measurement in Astronomy, 1500–1850,* 2nd ed. (Chichester: John Wiley & Sons, 1995), 16–65.

13. Ibid., 123–37; J. Elfreth Watkins, "The Ramsden Dividing Engine," in *Annual Report of the Board of Regents of the Smithsonian Institution . . . to July, 1890* (Washington, D.C.: Government Printing Office, 1891), 721–39; Anthony J. Leiserowitz, "The Dividing Engine in History," *Virtual Museum of Surveying,* www.surveyhistory.org/the-dividing-engine.htm.

14. William Y. McAllister, *A Priced and Illustrated Catalogue of Mathematical Instruments* (Philadelphia: McAllister, 1867), 6, 22.

15. Amy Ackerberg-Hastings, "Jeremiah Day and Navigation Instruction at Yale," in *Proceedings of the Canadian Society for History and Philosophy of Mathematics,* ed. Antonella Cupillari, Twenty-ninth Annual Meeting, May 30–June 1, 2003, 16:4–13.

16. Edward W. Stevens Jr., *The Grammar of the Machine: Technical Literacy and Early Industrial Expansion in the United States* (New Haven, Conn.: Yale University Press, 1995), 156; Lawrence P. Grayson, "Civil Engineering Education: An Historical Perspective," in *Civil Engineering History: Engineers Make History,* ed. Jerry R. Rogers et al. (New York: American Society of Civil Engineers, 1996), 44–52.

17. Robert Coleman Jr., *The Development of Informal Geometry* (New York: Bureau of Publications, Teachers College, Columbia University, 1942). As Coleman notes, the subject existed before it had a name. The phrase *informal geometry* appeared at least as early as 1930; see W. D. Reeve, ed., *The Teaching of Geometry,* National Council of Teachers of Mathematics 5th Yearbook (New York: Columbia University Bureau of Publications, 1930), vii, 1–13.

18. Thomas Hill, *A Second Book in Geometry* (Boston: Brewer & Tileston, 1863), 11–54.

19. Ibid., 62–65.

20. The American edition of this book was William George Spencer, *Inventional Geometry: A Series of Problems, Intended to Familiarize the Pupil with Geometrical Conceptions, and to Exercise His Inventive Faculty* (New York: D. Appleton and Co., 1877).

21. Typical examples included Charles W. Hackley, *Elementary Course of Geometry, for the Use of Schools and Colleges* (New York: Harper & Bros., 1847); and Samuel Edward Warren, *A Primary Geometry, with Simple and Practical Exercises in Plane and Projection Drawing, and Suited to All Beginners* (New York: John Wiley & Sons, 1887). In contrast, there were no protractors in Walter Smith, *The Child's Practical Geometry; Being a Series of Elementary Problems in Drawing Plane Geometrical Figures, as Given in the Course of Lessons in Public Schools* (Boston: James R. Osgood & Co., 1872).

22. See, e.g., D. M. Knapen, *The Mechanic's Assistant: A Thorough Practical Treatise on Mensuration and the Sliding Rule* (New York: D. Appleton & Co., 1849); and John Gadsby Chapman, *The American Drawing-Book: A Manual for the Amateur, and*

Basis of Study for the Professional Artist: Especially Adapted to the Use of Public and Private Schools, as Well as Home Instruction (New York: J. S. Redfield, 1849).

23. Charles Davies, *Elements of Geometry and Trigonometry from the Works of A. M. Legendre, Adapted to the Course of Mathematics Instruction in the United States,* ed. J. Howard Van Amringe (New York: American Book Co., 1885); George Albert Wentworth, *Elements of Plane and Solid Geometry* (Boston: Ginn & Co., 1877).

24. Charles Davies, *The Logic and Utility of Mathematics, with the Best Methods of Instruction Explained and Illustrated* (New York: A. S. Barnes & Co., 1850), 223–59.

25. Trigonometry textbooks that described protractors and were reprinted numerous times between 1860 and 1890 included Charles Davies, *Elements of Geometry and Trigonometry* (New York: A. S. Barnes & Burr, 1859), B16; and Elias Loomis, *Elements of Plane and Spherical Trigonometry, with Their Applications to Mensuration, Surveying, and Navigation* (New York: Harper & Bros., 1848), 34–35. Less popular but making more extensive use of protractors was Charles W. Hackley, *A Treatise on Trigonometry, Plane and Spherical, with Its Application to Navigation and Surveying, Nautical and Practical Astronomy and Geodesy, with Logarithmic, Trigonometrical, and Nautical Tables. For the Use of Schools and Colleges,* 2nd ed. (New York: George P. Putnam, 1851), iv–vii, 7–10, 121–22, 233.

26. David Lindsay Roberts, "Mathematics and Pedagogy: Professional Mathematicians and American Educational Reform, 1893–1923" (Ph.D. diss., Johns Hopkins University, 1997). See also Phillip S. Jones and Arthur F. Coxford Jr., eds., *A History of Mathematics Education in the United States and Canada,* 32nd Yearbook (Washington, D.C.: NCTM, 1970); Ernst R. Breslich, "Mathematics," in *A Half Century of Science and Mathematics Teaching* (Oak Park, Ill.: Central Association of Science and Mathematics Teachers, 1950), 39–79; and George M. A. Stanic and Jeremy Kilpatrick, eds., *A History of School Mathematics,* 2 vols. (Washington, D.C.: NCTM, 2003).

27. On the end of college geometry instruction, including the decline of Euclid's *Elements of Geometry* as a textbook, see also Joan L. Richards, *Mathematical Visions: The Pursuit of Geometry in Victorian England* (Boston: Academic Press, 1988); W. H. Brock, "Geometry and the Universities: Euclid and His Modern Rivals, 1860–1901," *History of Education* 4, no. 2 (1975): 21–35; Florian Cajori, "Attempts Made during the Eighteenth and Nineteenth Centuries to Reform the Teaching of Geometry," *American Mathematical Monthly* 17 (1910): 181–201; and John Donald Wilson, "An Analysis of the Plane Geometry Content of Geometry Textbooks Published in the United States before 1900" (Ed.D. diss., University of Pittsburgh, 1959).

28. F. L. Wren and H. B. McDonough, "Development of Mathematics in Secondary Schools of the United States, Part III," *Mathematics Teacher* 27 (1934): 218. On the broader framework of defining the audience and levels of American education, see also Laurence R. Veysey, *The Emergence of the American University* (Chicago: University of Chicago Press, 1965); Frederick Rudolph, *The American College and University: A History* (New York: Alfred A. Knopf, 1962); Lawrence A. Cremin, *American Education: The National Experience 1783–1876* (New York: Harper & Row, 1980);

Cremin, *American Education: The Metropolitan Experience 1876–1980* (New York: Harper & Row, 1988); and Edward A. Krug, *The Shaping of the American High School, 1880–1920* (New York: Harper & Row, 1964).

29. The mathematics portion of the Committee of Ten Report, along with other significant documents, was printed with the entire report but can also be found in James K. Bidwell and Robert G. Clason, eds., *Readings in the History of Mathematics Education* (Washington, D.C.: NCTM, 1970), 129–41.

30. National Educational Association of the United States, *Report of the Committee of Ten on Secondary School Studies, with the Reports of the Conferences Arranged by Committee* (New York: American Book Co., 1894), 110.

31. Bidwell and Clason, *Readings*, 189–209; Breslich, "Mathematics." See also David E. Smith, *The Teaching of Elementary Mathematics* (New York: Macmillan, 1900).

32. David Lindsay Roberts, "E. H. Moore's Early Twentieth-Century Program for Reform in Mathematics Education," *American Mathematical Monthly* 108 (2001): 689–96. The facets of Moore's overall career are analyzed in Karen Hunger Parshall and David E. Rowe, *The Emergence of the American Mathematical Research Community, 1876–1900: J. J. Sylvester, Felix Klein, and E. H. Moore,* History of Mathematics vol. 8 (Providence, R.I.: American Mathematical Society, 1994).

33. See, e.g., Bidwell and Clason, *Readings*, 220–45, esp. 241–42.

34. J. Fred Smith, *School Geometry: Inductive in Plan, Containing the Elements of Plane Geometry and Selections from Solid Geometry for Use in Schools, High Schools, and Academies* (Chicago: Scott, Foresman & Co., 1897), 3–4, 7. Another school textbook that was prepared as an explicit response to the Committee of Ten was Adelia Roberts Hornbrook, *Concrete Geometry for Beginners* (New York: American Book Co., 1895), 3–4.

35. Smith, *School Geometry*, 37–40, 78.

36. *Catalogue of Keuffel & Esser Co.* (New York: Keuffel & Esser, 1909), 172, 174–75, 214–15.

37. J. C. Packard, *School Review* (1903), quoted in Byron Cosby, "Efficiency in Geometry Teaching: A Study—An Experiment—A Result," *School Science and Mathematics* 12 (1912): 409.

38. Edith Long and W. C. Brenke, *Plane Geometry* (New York: Century Co., 1916), 15–28.

39. Webster Wells and Walter W. Hart, *Plane and Solid Geometry* (Boston: D. C. Heath & Co., 1915), 18–22.

40. Ibid., iii.

41. Wells and Hart, *Modern Plane Geometry*, 3, 16. See also Walter Wilson Hart, *Progressive Plane and Solid Geometry* (Boston: D. C. Heath & Co., 1936), iii–vii, 16; Claude H. Ewing and Walter W. Hart, *Essential Vocational Mathematics* (Boston: D. C. Heath & Co., 1945), 99; and Walter Wilson Hart, Veryl Schult, and Henry Swain, *Plane Geometry and Supplements* (Boston: D. C. Heath & Co., 1959), 24.

42. William Charles Brenke, *A Text-Book on Advanced Algebra and Trigonometry*

with Tables (New York: Century Co., 1910), 94–96; Robert E. Moritz, *Plane and Spherical Trigonometry* (New York: John Wiley & Sons, 1913), 1. See also W. A. Granville, "On the Teaching of the Elements of Plane Trigonometry," *American Mathematical Monthly* 17 (1910): 25–31; and F. L. Griffin, "An Experiment in Correlating [College] Freshman Mathematics," *American Mathematical Monthly* 22 (1915): 325–30.

43. M. J. Newell and G. A. Harper, "First Lessons in Demonstrative Geometry," *Mathematics Teacher* 14 (1921): 43.

44. National Committee on Mathematical Requirements, under the auspices of the Mathematical Association of America, *The Reorganization of Mathematics in Secondary Education* ([Oberlin, Ohio]: Mathematical Association of America, 1923), 22.

45. John Charles Stone, *The New Mathematics,* 3 vols. (Chicago: Benj. H. Sanborn & Co., 1926), 1:121–33; 2:59–120. Stone was employed by the state normal school in New Jersey.

46. Leo John Brueckner, Laura Farnham, and Edith Woolsey, *Mathematics for Junior High Schools: Book Three, Modern Algebra* (Chicago: John C. Winston Co., 1931), 53–78. Mathematics educators classified seventh, eighth, and ninth grade as the "junior high" years by 1927. For instance, see the table of contents in *Curriculum Problems in Teaching Mathematics,* National Council of Teachers of Mathematics Second Yearbook (New York: Teachers College, Columbia University, 1927).

47. There were a few nontraditional students who also were instructed in protractor use. During World War II, e.g., GIs learned that rulers and protractors were for drawing, measuring, and checking one's work, thereby connecting geometrical drawings to navigational applications while also mastering the preparation of abstract, indirect proofs. Paul Harold Daus, John M. Gleason, and William Marvin Whyburn, *Basic Mathematics for War and Industry* (New York: Macmillan, 1944); War Department Education Manual, *Plane Geometry: Course One, a Self-Teaching Course Based on Modern School Geometry by John R. Clark and Rolland R. Smith* (Madison, Wis.: United States Armed Forces Institute and World Book Co., 1944). There were also some textbooks that contained protractors but never explained in the text why these objects were useful or appropriate; see, e.g., Claude Irwin Palmer and Daniel Pomeroy Taylor, *Solid Geometry,* ed. George William Myers (Chicago: Scott, Foresman & Co., 1918).

48. *Principles and Standards for School Mathematics* (Reston, Va.: NCTM, 2000), 242–43; available online as "Measurement Standard for Grades 6–8," http://standards.nctm.org/document/chapter6/meas.htm.

CHAPTER ELEVEN Metric Teaching Apparatus

Epigraph: Birdsey G. Northrop, "The Metric System," "Report of the Secretary of the Connecticut Board of Education for 1877," Hartford, Conn., [1877], 45.

1. Ira Mayhew, *School Funds and School Laws of Michigan . . .* (Lansing, Mich.: Hosner & Kerr, 1856), 421–22.

2. Henry Barnard, *Object Teaching and Oral Lessons on Social Science and Common Things . . .* (New York: F. C. Brownell, 1860), 451.

3. Chicago Board of Education, *Graded Course of Instruction for the Public Schools of Chicago,* 3rd ed. (Chicago: Church, Goodman and Donnelley, 1869), 66. This book suggested making apparatus. Alfred Holbrook, *School Management* (Lebanon, Ohio: Josiah Holbrook, 1871), 258. Holbrook suggested that measures could be borrowed or purchased.

4. Thomas Jefferson, *Papers,* ed. J. P. Boyd (Princeton: Princeton University Press, 1961), 16:484–675. Correspondence relating to this report, as well as the report itself, are scattered through these pages.

5. John L. Heilbron, "The Measure of Enlightenment," in *The Quantifying Spirit in the Eighteenth Century,* ed. T. Frängsmyr, J. L. Heilborn, and R. E. Rider (Berkeley: University of California Press, 1990); Witold Kula, *Measures and Men,* trans. R. Szreter (Princeton: Princeton University Press, 1986); and Charles C. Gillispie, *Science and Polity in France: The Revolutionary and Napoleonic Years* (Princeton: Princeton University Press, 2004), 223–85, 458–94. The term *mètre* for unit of length was introduced in 1792. On Apr. 7, 1795 (18 germinal, an III), the French Convention adopted a set of interrelated "republican" measures. No later than 1799 the new units were referred to as the "metric system."

6. Republic of France, "Letter from the Minister of the French Republic to the Secretary of State of the United States, with Decrees of the National Assembly of France Establishing a New System of Weights and Measures," Philadelphia, 1795.

7. *Gloria Greenleaf's New-York, Connecticut, & New Jersey Almanack, or Diary, with an Ephemeris for the Year of Our Lord 1796 . . .* (New York: Thomas Greenleaf, 1795); and "Weights and Measures," *Supplement to the Encyclopaedia, or Dictionary of Arts, Sciences and Miscellaneous Literature in Three Volumes* (Philadelphia: Thomas Dobson, 1803), 3:543–44. Dobson's *Encyclopaedia* was a reprint of the third edition of the *Encyclopaedia Britannica,* which he had published between 1789 and 1798. His three-volume *Supplement* appeared between 1800 and 1803. See Robert D. Arner, *Dobson's Encyclopaedia: The Publisher, Text and Publication of America's First Britannica, 1789–1803* (Philadelphia: University of Pennsylvania Press, 1991).

8. C. F. Treat, *A History of the Metric Controversy in the United States* (Washington, D.C.: National Bureau of Standards, 1971), 20.

9. H. L. Mason, ed., *Catalog of Artifacts on Display in the NBS Museum* (Washington, D.C.: National Bureau of Standards, 1977).

10. Arthur H. Frazier, *United States Standards of Weights and Measures: Their Creation and Creators* (Washington, D.C.: Smithsonian Institution Press, 1978).

11. Edward Younger, *John A. Kasson: Politics and Diplomacy from Lincoln to McKinley* (Iowa City: State Historical Society of Iowa, 1955), 141–52.

12. Quoted in Northrop, "Metric System," 78.

13. Hubert A. Newton, *The Metrical System of Weights and Measures* (New Haven, Conn.: E. Hayes, 1864).

14. Charles Davies, *The Metric System Explained and Adapted to the Systems of Instruction in the United States* (New York: A. S. Barnes & Co., 1867), 5–6.

15. Frederick W. True, *A History of the First Half-Century of the National Academy of Sciences* (Washington, D.C.: National Academy of Sciences, 1913), 206–13.

16. Treat, *History of the Metric Controversy,* 39–46. U.S. Coast and Geodetic Survey, *Papers Relating to Metric Standards . . .* (Washington, D.C.: Government Printing Office, 1876).

17. Davies, *Metric System Explained.* On the metric system in American arithmetics, see Peggy A. Kidwell, "The Metric System Enters the American Classroom, 1790–1890," in *From Calculus to Computers: Using the Last 200 Years of Mathematical History in the Classroom,* ed. Amy Shell-Gellasch and Richard Jardine (Washington, D.C.: Mathematical Association of America, 2005), 229–36.

18. W. and L. E. Gurley, *A Manual of the Principal Instruments Used in American Engineering and Surveying* (Troy, N.Y.: Gurley, 1871), 150.

19. J. W. Queen & Co., *Priced and Illustrated Catalogue and Descriptive Manual of Mathematical Instruments and Materials . . .* (Philadelphia: Queen & Co., 1874), inside back cover.

20. Northrop, "Metric System," 45.

21. Ibid., 45.

22. A. and T. W. Stanley, *The Metric System* (New Britain, Conn.: Stanley, 1876). A copy of this publication is in "Tools," Warshaw Collection of Business Americana, NMAH Archives, Smithsonian Institution, Washington, D.C.

23. Northrop, "Metric System," 46.

24. Ibid., 72, 28; and Francis A. Walker, ed., *International Exhibition, 1876. Reports and Awards* (Washington, D.C.: Government Printing Office, 1880), 8:82.

25. Northrop, "Metric System," 79.

26. For biographical information about Dewey, we have relied principally on Wayne A. Wiegand, *Irrepressible Reformer: A Biography of Melvil Dewey* (Chicago: American Library Association, 1996). Wiegand mentions these particular firms on p. 35. An earlier biographer of Dewey, Grosvenor Dawe, quoted Dewey's own 1931 autobiographical notes on the topic, which were written in the system of reformed spelling Dewey advocated and read as follows: "I arranjd with Fairbanks & Co in skales, with the Standard [*sic*] Rule & Level Co, & Keuffel & Esser for length mezures, & with George M. Eddy & Co for tape mezures & with the Dover Stamping Co for capasiti mezures & we suplyed some of the state departments & 100s of skools with standard metrik weits & mezures." Dawe, *Melvil Dewey Seer: Inspirer: Doer, 1851–1931* (Lake Placid, N.Y.: Lake Placid Club, 1932), 278.

27. *Metric Bulletin,* 1, no. 5 (Nov. 1877).

28. For a description of the terms under which the Metric Bureau distributed apparatus, see Baker, Pratt & Co., *Illustrated Catalogue of School Merchandise* (New York: Baker, Pratt & Co., 1879), 82, 168–69.

29. *Metric Bulletin,* no. 25 (July 1878): 391–400.

30. James W. Queen & Co., *Priced and Illustrated Catalogue of Physical Instruments . . .* (Philadelphia: Queen, 1881), 2.

31. Primary schools offered basic instruction in reading and numbers. They came to include what would now be called grades 1–3 and, beginning in 1870 in Boston, kindergarten. Grammar school offered intermediate instruction and high schools more advanced courses in what might now be called grades 9–12. Boston, *School Document #12—Expenditures for the Public Schools—Report of the Committee on Accounts,* Boston School Committee Papers (Boston: Rockwell & Churchill, 1877), 23; Charles K. Dillaway, "Education, Past and Present: The Rise of Free Education and Educational Institutions," in *The Memorial History of Boston,* ed. Justin Winsor (Boston: Osgood, 1881), 4:235–78.

32. Boston, *A Uniform Course of Study for Three Years for the High Schools,* Boston School Committee Papers (Boston: 1877). This printed broadside is in the Boston School Committee Papers.

33. Boston, *School Document No. 13—Majority and Minority Report of Committee on Text—Books on the Metric System* (Boston: Rockwell & Churchill, 1877), 3. A copy of this document is in the Boston School Committee Papers.

34. Ibid., 4.

35. On the meaning of measurement in late-nineteenth-century British electromagnetism, see Graeme J. N. Gooday, *The Morals of Measurement: Accuracy, Irony, and Trust in Late Victorian Electrical Practice* (Cambridge: Cambridge University Press, 2004).

36. Bruce Sinclair, *A Centennial History of the ASME* (Toronto: for ASME by the University of Toronto Press, 1980).

37. Edward F. Cox, "The International Institute: First Organized Opposition to the Metric System," *Ohio Historical Quarterly* 68 (1959): 54–83.

38. "Editor's Table," *Philadelphia Photographer* 19 (1882): 191.

39. Francis A. Walker, *Arithmetic in the Boston Schools,* reprinted from the *Academy* of Jan. 1887, n.p., n.d. A copy of this document is in the Yale University Libraries.

40. L. E. Knott Apparatus Co., *A Catalogue of Physical Instruments, Catalogue 17* (Boston: Knott, 1912), 32.

41. W. M. Welch Scientific Co., *Catalog G* (Chicago: Welch, 1931), 5.

42. Cambosco Scientific Co., *Order Book 1956–1957 for Science Supplies* (Boston: Cambosco, 1956), 63.

43. Norman A. Calkins, *Primary Object Lessons . . .* (New York: Harper & Row, 1861), 160–81.

44. J. W. Schermerhorn & Co., *School Material* (New York: Schermerhorn, 1873), 134–36.

45. Union School Furniture Co., *Illustrated Catalogue* (Battle Creek, Mich.: Union School Furniture Co., 1889), 21.

46. R. O. Evans, *Evans' Arithmetical Study Applying the Object Method to the Entire Subject of Practical Arithmetic . . .* (Chicago: Caxton Publishing Co., 1897).

47. *Supplement to the Metric Bulletin* (Dec. 1876): 82.

48. J. W. Schermerhorn & Co., *School Material* (New York: Schermerhorn, 1873), 156.

49. Baker, Pratt & Co., *Illustrated Catalogue of School Merchandise* (New York: Baker, Pratt & Co., 1879), 168–69.

50. Ibid., 67.

51. J. L. Hammett, *Catalogue* (Boston: Hammett, 1895), 68.

52. Milton Bradley Co., *Bradley's Souvenir Catalogue for Graduates of Teaching and Normal Schools* (Philadelphia: Milton Bradley, 1909), 67.

CHAPTER TWELVE Graph Paper

Epigraph: Eliakim Hastings Moore, "The Cross-Section Paper as a Mathematical Instrument," *School Review* 14 (May 1906): 338.

1. William H. Brock and Michael H. Price, "Squared Paper in the Nineteenth Century: Instrument of Science and Engineering, and Symbol of Reform in Mathematical Education," *Educational Studies in Mathematics* 11 (1980): 365–66. Much older isolated instances have been claimed. David Eugene Smith asserts, e.g., that "use of a kind of coordinate paper for the graphic representation of the course of the planets" can be found in "a manuscript of the 10th century." Smith, *History of Mathematics* (New York: Dover, 1958), 2:320.

2. William Smyth, *Elements of Plane Trigonometry, with Its Application to Mensuration of Heights and Distances, Surveying and Navigation* (Portland, Maine: Sanborn & Carter, 1852), 132–33. In chapter 2 we encountered Smyth as a younger man, championing the blackboard.

3. *Appleton's Cyclopedia of Drawing, Designed as a Text-Book for the Mechanic, Architect, Engineer, and Surveyor* (New York: Appleton, 1857), 365.

4. William Johnson, *The Practical Draughtsman's Book of Industrial Design* (New York: Stringer & Townsend, 1854), 135.

5. Truman Henry Safford, *A Catalogue of Standard Polar and Clock Stars* (Cambridge, Mass.: Welch, Bigelow, & Co., 1863), 121.

6. See H. D. Monachesi and Albert B. Yohn, *The Stationers' Handbook* (New York: Office of the Publishers' Weekly, 1876), 41. See also the advertisement for G. S. Woolman in John Woodbridge Davis, *Formulae for the Calculation of Railroad Excavation and Embankment* (New York: J. Dickson & Bro., 1876), two pages after p. 88. The title page of Davis's book notes that it was "used as a text-book in the School of Mines, Columbia College."

7. W. M. Gillespie, *A Treatise on Levelling Topography and Higher Surveying* (New York: D. Appleton & Co., 1870), 42.

8. Ellwood Morris, *Easy Rules for the Measurement of Earthworks, by Means of the Prismoidal Formula* (Philadelphia: T. R. Callender & Co., 1872), 48.

9. Arthur Wellington, *Methods for the Computation from Diagrams of Preliminary and Final Estimates of Railway Earthwork* (New York: D. Appleton & Co., 1875), 3.

10. Mark Twain and Charles Dudley Warner, *The Gilded Age: A Tale of To-Day*

(1873), *The Gilded Age and Later Novels* (New York: Library of America, 2002), 100. This passage was part of a chapter written by Warner, according to Twain's later account (1025). As a young man, Warner spent two years as a railroad surveyor. Justin Kaplan, *Mr. Clemens and Mark Twain* (New York: Simon and Schuster, 1966), 159.

11. Board of Water Commissioners, *The Brooklyn Water Works and Sewers: A Descriptive Memoir* (New York: D. Van Nostrand, 1867), 136.

12. George Sarton, "Notes and Correspondence," *Isis* 30 (Feb. 1939): 95–96.

13. W. F. Durand, "The Uses of Logarithmic Cross-Section Paper," *Engineering News*, Sept. 28, 1893, 248–51.

14. Sarton, "Notes and Correspondence." See also W. F. Durand, "Notes and Correspondence," *Isis* 32 (Jul. 1940): 117–18. Logarithmic paper was still being sold under Durand's name by K&E as late as 1949. Keuffel & Esser Co., "K&E Price List Applying to the 41st Edition Catalog," Oct. 15, 1949, 9.

15. North Carolina Senate, *Report of the Committee on Claims on the Western Turnpike Vouchers* (Raleigh: Thos. J. Lemay, 1850), 378; U.S. War Department, Exec. Doc. No. 155, "Letter from the Secretary of War, in Answer to *A Resolution of the House of June 4, Relative to Railroad Property in the Possession of the Government*" (Washington, D.C., 1866), 479.

16. Monachesi and Yohn, *Stationers' Handbook,* 41. See also advertisement for G. S. Roberts, in Davis, *Formulae,* one page after p. 88.

17. See James W. Queen & Co., *Catalogue of Mathematical and Engineering Instruments and Materials* (Philadelphia: Queen, 1887), 98–101.

18. Davis R. Dewey, "The Study of Statistics," *Publications of the American Economic Association* 4 (Sept. 1889): 48–49.

19. Franklin W. Barrows, "The New York State Teachers Association," *Science,* n.s. 5, Mar. 19, 1897, 466.

20. *Keuffel & Esser Catalogue,* 1906, 39.

21. *Atlas School Supply Company, Catalog No. 22,* 1909, 88.

22. Brock and Price, "Squared Paper in the Nineteenth Century," 367.

23. David Tyack and Larry Cuban, *Tinkering toward Utopia: A Century of Public School Reform* (Cambridge: Harvard University Press, 1995), 65–66.

24. Elizabeth P. Peabody, *Guide to the Kindergarten and Intermediate Class;* and Mary Mann, *Moral Culture of Infancy* (New York: E. Steiger, 1877), 50. Peabody's contribution to her sister's compilation (Mary Mann was married to Horace Mann) was written in 1869 (iv, 21).

25. Ibid., 21.

26. Eliakim Hastings Moore, "On the Foundations of Mathematics," *Science,* n.s. 17, Mar. 13, 1903, 410.

27. See, e.g., a nineteenth-century best seller by Joseph Ray, *Primary Elements of Algebra, for Schools and Academies* (Cincinnati: Wilson, Hinkle & Co., 1866). Other examples include George R. Perkins, *A Treatise on Algebra: Embracing besides the Elementary Principles, All the Higher Parts Usually Taught in Colleges* (New York: D. Appleton, 1857); and D. H. Hill, *Elements of Algebra* (Philadelphia: J. B. Lippincott &

Co., 1861). In these books no appeal whatsoever is made to geometry in introducing numbers.

28. See G. A. Wentworth, *Elements of Algebra* (Boston: Ginn, Heath, & Co., 1881), 5–7.

29. Simon Newcomb, *Algebra for Schools and Colleges,* 2nd ed. (New York: Holt, 1881), 5.

30. See, e.g., "Geometrical Illustration of the Rule of Signs," in ibid., 42–43.

31. Newcomb, *Algebra,* iii.

32. On the resulting development of the theory of complex variables, or complex function theory, see Morris Kline, *Mathematical Thought from Ancient to Modern Times* (New York: Oxford University Press, 1972), 628–32.

33. See *Report of the Committee of Ten on Secondary School Studies,* reprinted in Theodore R. Sizer, *Secondary Schools at the Turn of the Century* (New Haven, Conn.: Yale University Press, 1964), 209–71.

34. See *University of Chicago General Register of the Officers and Alumni, 1892–1902* (Chicago: University of Chicago Press, 1903), 73.

35. On this merger, see Richard J. Storr, *Harper's University: The Beginnings* (Chicago: University of Chicago Press, 1966), 300–302. Myers held the title of "professor of the teaching of mathematics and astronomy" and was sometimes listed in both the Department of Mathematics and the School of Education. See *University of Chicago General Register, 1892–1902,* 47; and *Circular of the Departments of Mathematics, Astronomy and Astrophysics, Physics, Chemistry,* University of Chicago, 1908, 7.

36. George W. Myers, "Mathematics in the High and Pedagogic Schools," *Course of Study: A Monthly Publication for Teachers and Parents Devoted to the Work of the Chicago Institute* 1 (Oct. 1900): 133.

37. On Moore's pedagogical ideas, see David Lindsay Roberts, "E. H. Moore's Early Twentieth-Century Program for Reform in Mathematics Education," *American Mathematical Monthly* 108 (Oct. 2001): 689–96.

38. Moore, "On the Foundations of Mathematics," 401–16.

39. Ibid., 407.

40. John Perry, "The Teaching of Mathematics," *Educational Review* 23 (Feb. 1902): 158–81.

41. On Perry, see Graeme J. N. Gooday, "Perry, John," *Oxford Dictionary of National Biography* (Oxford: Oxford University Press, 2004), 43:832–33; and Brock and Price, "Squared Paper in the Nineteenth Century," 373–76. On the earlier discontent with Euclid in Britain, see W. H. Brock, "Geometry and the Universities: Euclid and His Modern Rivals 1860–1901," *History of Education* 4 (1975): 21–29.

42. John Perry, ed., *Discussion on the Teaching of Mathematics* (London: Macmillan, 1902), 2.

43. Brock and Price, "Squared Paper in the Nineteenth Century," 372–75.

44. Brock, "Geometry and the Universities," 30.

45. Perry to D. E. Smith, Oct. 10, 1901, David Eugene Smith Papers, Columbia University, Rare Books and Manuscripts Library.

46. Ibid., Nov. 30, 1901. Smith agreed with Perry in general, offering the supportive comment that from his observation of education in England, France, Germany, and the United States, he judged England to be doing the worst job of teaching mathematics (he rated Germany best). Perry, *Teaching of Mathematics,* 89.

47. John Perry, "Address to the Engineering Section of the British Association," *Science* 16, Nov. 14, 1902, 761–82.

48. Moore, "Foundations of Mathematics," 406–7.

49. See H. E. Cobb, "The Need of a Perry Movement in Mathematical Teaching in America," *Mathematical Supplement of School Science* 1 (Oct. 1903): 121–24; and Victor C. Alderson, "Five Cardinal Points in the Perry Movement," *School Mathematics* 2 (1904): 193–95.

50. Moore, "Foundations of Mathematics," 407.

51. Ibid., 407–8.

52. Ibid., 409.

53. *Steiger's Kindergarten Catalogue,* 8th ed. (New York: E. Steiger & Co., 1900), 45–46. It is possible that Moore may have had a child in kindergarten prior to 1902. He was married in 1892 and had two children. *Who Was Who in America,* vol. 1: *1897–1942* (Chicago: Marquis, 1943), 359.

54. Moore, "Foundations of Mathematics," 412–13.

55. H. W. Tyler et al., "Report of the Committee of the American Mathematical Society on Definitions of College Entrance Requirements in Mathematics," *Bulletin of the American Mathematical Society* 10 (Nov. 1903): 74–77.

56. Moore originally presented this paper at a meeting of the Junior Mathematical Club and the Mathematical Club of the University High School of the University of Chicago, Mar. 5, 1906. It was subsequently published in *School Review* 14 (May 1906): 317–38; and in *School Science and Mathematics* 6 (June 1906): 429–50. References in the present chapter are to the *School Review* version.

57. Moore, "Cross-Section Paper," 317.

58. Ibid., 318.

59. Ibid., 319.

60. Ibid., 322–37.

61. See Thomas L. Hankins, "Blood, Dirt, and Nomograms: A Particular History of Graphs," *Isis* 90 (Mar. 1999): 50–80; and Ivor Grattan-Guinness, *The Norton History of the Mathematical Sciences* (New York: W. W. Norton, 1998), 514–17. On Ocagne, see Jean Itard, "Ocagne, Philbert Maurice d'," *Dictionary of Scientific Biography,* 10:170. Moore referred to him on p. 325 of "Cross-Section Paper."

62. For mathematical details, see Václav Pleskot, "Nomography and Graphical Analysis," in *Survey of Applicable Mathematics,* ed. Karel Rektorys, trans. Rektorys et al. (Cambridge, Mass.: MIT Press, 1969), 1183–1219.

63. N. J. Lennes, "The Graph in High-School Mathematics," *School Review* 14 (May 1905): 339–49.

64. G. W. Myers, "A Class of Content Problems for High School Algebra," *School Review* 14 (Oct. 1906): 565–66.

65. "Notes and News," *American Mathematical Monthly* 14 (1907): 62.

66. J. W. A. Young, *The Teaching of Mathematics in the Elementary and the Secondary School* (New York: Longmans, Green & Co., 1907), 111–15.

67. "The Course of Study in the Elementary School," *Elementary School Teacher* 8 (May 1908): 537–39. This publication was the successor to *The Course of Study*, for which Myers wrote in 1900 (see n. 36).

68. See Roberts, "E. H. Moore's Early Twentieth-Century Program."

69. Arthur Whipple Smith, "What Results Are We Getting from Graphic Algebra?" *Mathematics Teacher* 4 (Sept. 1911): 13.

70. Emily G. Palmer, "History of the Graph in Elementary Algebra in the United States," *School Science and Mathematics* 12 (1912): 692–93.

71. Percey F. Smith and Arthur Sullivan Gale, *New Analytic Geometry* (Boston: Ginn & Co., 1912), 10.

72. F. L. Griffin, "An Experiment in Correlating Freshman Mathematics," *American Mathematical Monthly* 22 (Dec. 1915): 329.

73. R. H. Henderson, "Recent Advances in the Teaching of Mathematics," *Mathematics Teacher* 9 (Mar. 1917): 143.

74. D. C. Heath & Co. advertisement for Karpinski, Benedict, and Calhoun's *Unified Mathematics*, in *American Mathematical Monthly* 26 (May 1919). See also Agnes E. Wells, review of *Plane and Solid Geometry* by Walter Burton Ford and Charles Ammerman, *American Mathematical Monthly* 21 (Sept. 1914): 222; F. L. Griffin, "An Experiment in Correlating Freshman Mathematics," *American Mathematical Monthly* 22 (Dec. 1915): 329; and H. J. Ettlinger, "An Introduction to Plane Trigonometry by Graphical Methods," *American Mathematical Monthly* 27 (Feb. 1920): 64.

75. Farrington Daniels, "Mathematics for Students of Chemistry," *American Mathematical Monthly* 35 (Jan. 1928): 7.

76. Phillip S. Jones, ed., *A History of Mathematics Education in the United States and Canada*, National Council of Teachers of Mathematics 32nd Yearbook (Washington, D.C.: NCTM, 1970), 158–59.

77. *Price-List Applying to General Catalogue*, 37th ed. (New York: Keuffel & Esser Co., 1928), 11–13.

78. *Price-List Applying to General Catalogue*, 38th ed. (New York: Keuffel & Esser Co., 1937), 16–18.

79. Cf. *Catalog of Eugene Dietzgen Co.*, 14th ed. (Chicago: Dietzgen, 1931), 48–49; and *Catalog of Eugene Dietzgen Co.*, 15th ed. (Chicago: Dietzgen, 1938), 70–71.

80. *K&E Graph Sheets: Coordinate Papers & Cloths with Illustrations of Their Use* (New York: Keuffel & Esser Co., 1941). The Codex Book Co., focusing on the business market, in 1925 sold "chart sheets," of which two types were "cross-section sheets" and "quadrille ruled sheets." See *List Giving Catalogue Numbers and Prices of Codex Data Paper* (New York: Codex Book Co., 1925), 1. In 1937 the same company was using "graphic chart sheets" as a generic designator. See cover of *Codex, 1937 Edition* (Norwood, Mass.: Codex Book Co.).

81. Norma Sleight, "An Experiment in the Use of Graph Papers," *Mathematics Teacher* 35 (Feb. 1942): 84.

82. The following books have sections or chapters on graphing, with pictures of squared grids: William J. Milne and Walter F. Downey, *New Second Course in Algebra* (New York: American Book Co., 1945), "squared paper" mentioned on p. 239; Isidore Dressler, *Reviewing Elementary Algebra* (New York: Amsco School Publications, 1949), "graph paper" mentioned on p. 142; Julius Freilich, Simon L. Berman, and Elsie Parker Johnson, *Algebra for Problem Solving,* bk. 1 (Boston: Houghton Mifflin, 1952), "paper ruled in squares" mentioned on p. 86; Edward I. Edgerton and Perry A. Carpenter, *Intermediate Algebra* (Boston: Allyn and Bacon, 1953), "squared paper" mentioned on p. 292; A. M. Welchons, W. R. Krickenberger, and Helen R. Pearson, *Algebra,* bk. 1 (Boston: Ginn & Co., 1956), "graph paper" mentioned on p. 297; Max Beberman and Herbert E. Vaughan, *High School Mathematics* (Boston: D. C. Heath & Co., 1964), "cross-section paper [that is: graph paper]" mentioned on p. 462; and Mary P. Dolciani, Simon L. Berman, and Julius Freilich, *Modern Algebra: Structure and Method,* bk. 1 (Boston: Houghton Mifflin, 1965), "graph paper" mentioned on p. 358.

83. Herbert E. Hawkes, William A. Luby, and Frank C. Touton, *First Course in Algebra* (Boston: Ginn & Co., 1910), 194. The same paragraph was included in the 1926 edition of this text, on p. 278.

84. Joseph Lipka, "Alignment Charts" *Mathematics Teacher* 14 (Apr. 1921): 171. Lipka was the author of *Graphical and Mechanical Computation* (New York: Wiley, 1918).

85. J. A. Roberts, "How to Figure Averages with the Top of a Shoe Box; Or the Use of an Alignment Chart in Averaging," *Mathematics Teacher* 17 (Dec. 1924). "For those who have never heard of alignment charts I recommend Wentworth-Smith-Schlauch's 'Commercial Algebra' (Ginn), or Mr. Ralph Beatley's course 'The Teaching of Senior High School Mathematics' at a Harvard Summer School session" (471).

86. Douglas P. Adams, "The Preparation and Use of Nomographic Charts in High School Mathematics," in *Multisensory Aids in the Teaching of Mathematics,* ed. W. D. Reeve, NCTM 18th Yearbook (New York: Teachers College Press, 1945), 181.

87. Hankins, "Blood, Dirt, and Nomograms," 71.

88. On this period, see David L. Roberts and Angela L. E. Walmsley, "The Original 'New Math': Storytelling versus History," *Mathematics Teacher* 96 (Oct. 2003): 468–73.

89. Dolciani, Berman, and Freilich, *Modern Algebra,* 1, 180, 356, 420.

90. "MINNEMAST Final Report to the National Science Foundation," 9, University of Minnesota Archives, Collection 35, Minnesota School Mathematics and Science Teaching project, box 22, folder "Minnemast Project Folder #1."

91. *Curriculum and Evaluation Standards for School Mathematics* (Reston, Va.: NCTM, 1989), 34, 80; *Principles and Standards for School Mathematics* (Reston, Va.: NCTM, 2000), 46, 165, 228, 235, 253, 260, 314. Thanks to the fully searchable elec-

tronic version of the latter document, it can be asserted that none of the older designations for graph paper (cross-section, squared, etc.) is used, nor is there any reference to nomography.

92. *Curriculum and Evaluation Standards*, 106, 168; *Principles and Standards*, 163, 252.

93. *Curriculum and Evaluation Standards*, 152; *Principles and Standards*, 70.

CHAPTER THIRTEEN Geometric Models

Epigraphs: Josiah Holbrook, *Apparatus Designed for Families, Schools, Lyceums and Academies* (Boston: Allen & Ticknor, 1833), 6; Eliakim Hastings Moore, "On the Foundations of Mathematics," *Science*, n.s. 17 Mar. 13, 1903, 412.

1. Soraya de Chadarevian and Nick Hopwood, eds., *Models: The Third Dimension of Science* (Stanford: Stanford University Press, 2004).

2. Specific references to topics touched on in this chapter are given in later notes. See also Gerd Fischer, *Mathematische Modelle / Mathematical Models* (Braunschweig: Friedr. Vieweg & Sohn, 1986); Peter R. Cromwell, *Polyhedra* (Cambridge: Cambridge University Press, 1999); Magnus J. Wenninger, *Polyhedron Models* (Cambridge: Cambridge University Press, 1974); and Florence Fasanelli, "Three Dimensional Links between the History of Mathematics and the History of Art," *Newsletter of the British Society for the History of Mathematics* 45 (Spring 2002): 7–19.

3. Holbrook, *Apparatus Designed for Families, Schools, Lyceums and Academies*, 6.

4. For a more detailed overview of geometrical models used in the United States, see P. A. Kidwell, "American Mathematics Viewed Objectively—the Case of Geometric Models," in *Vita Mathematica: Historical Research and Integration with Teaching*, ed. Ronald Calinger (Washington, D.C.: Mathematical Association of America, 1996), 197–208. Frederick Rickey and Amy Shell-Gellasch have studied the Olivier models in some detail. See Amy Shell-Gellasch, "The Olivier String Models at West Point," *Rittenhouse* 17 (2003): 71–84.

5. "Christian Gottlieb Ferdinand Engel," *American Journal of Science* 45 (1868): 282–84; C. G. F. Engel, *Axonometrical Projections of the Most Important Geometrical Surfaces . . .* (New York: H. Goebler, 1855). Examples of Engel's optical surfaces survive in the collections of the University Museums at the University of Mississippi, Oxford.

6. See Karen V. H. Parshall and David E. Rowe, "Embedded in the Culture: Mathematics at the World's Columbian Exposition of 1893," *Mathematical Intelligencer* 15, no. 2 (1993): 40–45. See also Ulf Hashagen, *Walther von Dyck (1856–1934): Mathematik, Technik und Wissenschaftsorganisation an der TH München* (Stuttgart: Steiner-Verlag, 2003); and Herbert Mehrtens, "Mathematical Models," in Chadarevian and Hopwood, *Models*, 276–306.

7. Amy Ackerberg-Hastings, "Mathematics Is a Gentleman's Art: Analysis and Synthesis in American College Geometry Teaching, 1790–1840" (Ph.D. diss., Iowa State University, 2000); Joan L. Richards, *Mathematical Visions: The Pursuit of Geometry in Victorian England* (Boston: Academic Press, 1988); Florian Cajori, "Attempts

Made during the Eighteenth and Nineteenth Centuries to Reform the Teaching of Geometry," *American Mathematical Monthly* 17 (1910): 181–201.

8. Cf. such books as Charles Davies, *Elements of Geometry and Trigonometry from the Works of A. M. Legendre* (New York: A. S. Barnes & Co., 1853); Thomas Hill, *A Second Book on Geometry* (Boston: Brewer & Tileston, 1863); George A. Wentworth, *Solid Geometry* (Boston: Ginn & Co., 1899); and Webster Wells and Walter W. Hart, *Solid Geometry* (Boston: D. C. Heath & Co., 1916).

9. Hill, *Second Book of Geometry*, 106.

10. I. Harrington, "Improvement in Apparatus for Teaching Mensuration," U.S. Patent 137,075, Mar. 25, 1873; C. Anderson, *Technology in American Education, 1650–1900* (Washington, D.C.: U.S. Department of Health, Education and Welfare, 1962), 45.

11. On Kennedy, see "Paintings Presented to Local Schools," *Rockport Journal*, May 15, 1964. For catalogs of improved versions of his models, see S. M. Stancliff & Co., *Arithmetic of Practical Measurements for Teachers' Instruction and Class Work in Mensuration*, 19th ed. (Des Moines, Iowa: S. M. Stancliff & Co., 1893). See also *Arithmetic of Practical Measurements for Teacher's Instruction and Class Work in Mensuration* (Chicago: A. Cowles & Co., 1897).

12. W. W. Ross, *Mensuration Taught Objectively with Lessons on Form* (Fremont, Ohio: n.p., ca. 1891), 3.

13. W. D. Ross, "The Ross Mensuration Blocks," box 14, Richard P. Baker Collection, University Archives, Department of Special Collections, University of Iowa Libraries, Iowa City. Hereafter this collection will be cited as Baker Collection. Ross's undated brochure was sent to the address in Iowa City where Baker moved in 1905. On Kennedy's patented apparatus, see A. H. Kennedy, "Educational Appliance," U.S. Patent 296,018, Apr. 1, 1884.

14. Moore, "On the Foundations of Mathematics," 401–16; David Lindsay Roberts, "E. H. Moore's Early Twentieth-Century Program for Reform of Mathematics Education," *American Mathematical Monthly* 108 (2001): 689–96.

15. Quoted in "Prof. Baker of Math Faculty Dies at Home," *Daily Iowan*, Aug. 14, 1937.

16. For a description of Baker's early schooling, see Richard P. Baker, *The Problem of the Angle Bisectors* (Ph.D. diss.) (Chicago: University of Chicago Press, 1911), 99.

17. See "Construction note," box 1; and "The Angle Bisector Equation," box 10, Baker Collection.

18. Virgil Snyder, "On the Quintic Scroll of Three Double Conics," *Bulletin of the American Mathematical Society* 9 (1903): 236–46.

19. See Virgil Snyder to Richard P. Baker, Apr. 14 and 29, May 5, Oct. 6, and June 3, 1904, and Jan. 16, 1905, box 25, Baker Collection.

20. Marin Liljeblad to Richard P. Baker, Aug. 31, 1904, and Apr. 27, 1905, box 25, Baker Collection.

21. Wentworth, *Solid Geometry*.

22. Richard P. Baker, "A List of Mathematical Models," box 14, Baker Collection. This copy of the catalog lacks a title page.

23. Virgil Snyder to Richard P. Baker, Apr. 29, 1904, box 25, Baker Collection.

24. *Bulletin of the American Mathematical Society* 11 (1905): 329; *Jahresbericht der Deutsche Mathematiker Vereinigung* 14 (1905): 210.

25. This advertisement is mentioned in W. L. Wright to R. P. Baker, Oct. 26, 1905, box 21, Baker Collection. The subject of mathematics was added to the magazine *School Science* in 1905.

26. Roy E. Dickson to R. P. Baker, June 30, 1905; and Floyd Field to R. P. Baker, July 21, [1909?], box 21, Baker Collection.

27. J. W. A. Young, *The Teaching of Mathematics in the Elementary and the Secondary School* (New York: Longmans, Green & Co., 1916), 284.

28. See Charles Carter to R. P. Baker, Mar. 18, 1907; Marie Walker to R. P. Baker, Sept. 7, 1927; and E. B. Hodges to R. P. Baker, Jan. 28, 1930, box 21, Baker Collection.

29. Arnold Emch, "Mathematical Models," University of Illinois, Urbana. This work was published in four parts, which appeared in the 1920s, the first two in the *University of Illinois Bulletin* 18, no. 12 (1920), and 20, no. 42 (1923).

30. Edwin S. Crawley to R. P. Baker, Dec. 12, 1905, box 25, Baker Collection.

31. There are 10 orders for a total of 228 models in correspondence dated 1905 in boxes 21 and 25 of the Baker Collection. In addition, there was an order for which Baker was paid $24.60. There also is later correspondence referring to models that were probably ordered and/or made at this time.

32. Wells College Bookstore to R. P. Baker, Feb. 8, 1906, box 21; and J. S. Kendall to R. P. Baker, June 21, 1905, box 25, Baker Collection.

33. R. P. Baker to O. W. Albert, July 16, 1930, box 25, Baker Collection.

34. Patterns and computations for many of these models are in the Baker Collection. On the number of models designed and built, see Frances E. Baker to Randolph Church, July 18, 1949, Accession File 211257, National Museum of American History (NMAH), Smithsonian Institution, Washington, D.C.

35. Heinrich Maschke, "Note on the Unilateral Surface of Moebius," *Transactions of the American Mathematical Society* 1 (1900): 39.

36. These models are listed in R. P. Baker, *Mathematical Models* (Iowa City: R. P. Baker, Jan., 1931), 18. See also H. W. Randall to R. P. Baker, Jan. 15, 1908; and Columbia University to R. P. Baker, Aug. 6, 1918, box 25, Baker Collection.

37. Eduard Study, *Geometrie der Dynamen: die Zusammensetzung von Kraften und verwandte Gegenstände der Geometrie* (Leipzig: B. G. Teubner, 1903).

38. Albert Einstein to Richard P. Baker, June 1, 1937, Dibner Library, NMAH, Smithsonian Institution.

39. These models are on exhibit at the University of Arizona and are also shown online at www.math.arizona.edu/~models/. See also William Mueller, "Mathematical Wunderkammern," *American Mathematical Monthly* 108 (Nov. 2001): 785–96.

40. A portion of Wheeler's models found their way to the Smithsonian's National

Museum of American History after his death, along with some of his correspondence and private papers. Much of the discussion of Wheeler in this chapter is based on this material.

41. On the early years of the Clark mathematics department, see Roger Cooke and V. Frederick Rickey, "W. E. Story of Hopkins and Clark," in *A Century of Mathematics in America,* pt. 3, ed. Peter Duren, with the assistance of Richard A. Askey, Harold M. Edwards, and Uta C. Merzbach (Providence, R.I.: American Mathematical Society, 1989), 47–76.

42. Albert Harry Wheeler, *Examples in Algebra* (Boston: Little, Brown, 1914); and Wheeler, *First Course in Algebra* (Boston: Little, Brown, 1907). The annual reports of the Worcester schools indicate that Wheeler's texts were used in the Worcester high schools in the 1910s and 1920s.

43. For the history of polyhedral investigations, see Marjorie Senechal and George Fleck, *Shaping Space: A Polyhedral Approach* (Boston: Birkhäuser, 1988), 80–92; the "historical remarks" sections in Max Brückner, *Vielecke und Vielflache* (Leipzig: B. G. Teubner, 1900); and H. S. M. Coxeter, *Regular Polytopes,* 3rd ed. (New York: Dover Publications, 1973). Wheeler invoked Kepler, Poinsot, and others as precursors in "Certain Forms of the Icosahedron and a Method for Deriving and Designating Higher Polyhedra," *Proceedings of the International Mathematical Congress Held in Toronto, Aug. 11–16, 1924* (Toronto: University of Toronto Press, 1928), 1:701.

44. A. H. Wheeler, "List of Models," A. H. Wheeler Papers, Mathematics Collections, NMAH, Smithsonian Institution. Hereafter this collection will be cited as Wheeler Papers.

45. A. H. Wheeler to C. E. Melville, June 7, 1923, Wheeler Papers.

46. C. E. Melville to Wheeler, July 12, 1923, Wheeler Papers.

47. According to the Congress's "Provisional Daily Programme of Sessions of Sections," the original title was "New Methods for Constructing Geometry Models. Illustrated by Models, Charts, and Photographs," Wheeler Papers. Wheeler also presented an exhibit at the meeting of the British Association for the Advancement of Science being held in Toronto at the same time as the Mathematical Congress, entitled "A Collection of Models and Photographs of New Forms of Higher Polyhedra," Wheeler Papers.

48. *Wellesley College Bulletin* (1927–28), 112.

49. A. A. Bennett to Wheeler, Aug. 4, 1930; and Maine Mineralogical Society to Wheeler, Feb. 1, 1939, Wheeler Papers.

50. Abstracts of Wheeler's American Mathematical Society papers are: "A general formula for polyhedra with applications to higher dodecahedra and icosahedra," *Bulletin of the American Mathematical Society* 42 (Sept. 1936): 639; "Groups of inscribed tetrahedra," *Bulletin of the American Mathematical Society* 44 (May 1938): 353; and "A one-sided polyhedron having one hundred twenty faces," *Bulletin of the American Mathematical Society* 45 (Nov. 1939): 839. As far as we are aware, the 1924 Toronto paper was the only research paper he ever published; the rest were delivered only orally at meetings. Whether he tried and failed to publish his work or whether

he simply felt more comfortable in a situation in which he could always fall back on a razzle-dazzle display of his beloved models, we do not know.

51. W. T. Martin to Wheeler, July 26, 1950, Wheeler Papers.

52. Worcester Polytechnic Institute (WPI), Office of the Registrar.

53. Advertisements for these models are found in the WPI *Annual Catalogue* from 1872 through 1894. The models seem to have included numerous cones and prisms as well as other figures not generally employed in mathematics.

54. Cooke and Rickey, "W. E. Story of Hopkins and Clark," 43.

55. G. Stanley Hall Papers, box 10, folder 13, Clark University Archives, Worcester, Mass.

56. In 1910 Schilling did publish a set of paper models of stellated polyhedra as his ser. 37. These were designed by Friedrich Schilling of the technical high school in Danzig. Whether Wheeler knew of this series is unclear. See Martin Schilling, *Catalog mathematischer Modelle . . .* (Leipzig: Martin Schilling, 1911), 100–101, 149.

57. Max Brückner, *Vielecke und Vielflache* (Leipzig: B. G. Teubner, 1900). Brückner studied at Leipzig from 1880 to 1885, receiving a doctorate in 1886. His career at Leipzig would thus have largely intersected with that of Felix Klein, although the relationship between the two is unclear. Brückner spent most of his career teaching at the Gymnasium at Bautzen. See J. C. Poggendorff, *J. C. Poggendorffs biographisch-literarisches Handwörterbuch* (Leipzig: Verlag Chemie, 1926), 5:175; and (Berlin: Verlag Chemie, 1936), vol. 6, pt. 1, 350.

58. The inspiration of Brückner's photographs for later geometry enthusiasts is testified to in Senechal and Fleck, *Shaping Space,* 88.

59. For a 1940 view of the decline of older geometric traditions in the twentieth century, see E. T. Bell, *The Development of Mathematics* (New York: McGraw-Hill, 1940), 298–301. For a more recent sociological/historical meditation on this theme, see Philip J. Davis, "The Rise, Fall, and Possible Transfiguration of Triangle Geometry: A Mini-History," *American Mathematical Monthly,* 102 (Mar. 1995), 204–14. Comments on Morley's "old-fashioned mathematics" can be found in Parshall and Rowe, *Emergence of the American Mathematical Research Community,* 433, 435.

60. In the last few decades there has been a revival of interest in classical geometry at the research level, partly fueled by computer graphics, and Coxeter is now hailed as one who kept the spark alive. See Chandler Davis et al., eds., *The Geometric Vein: The Coxeter Festschrift* (New York: Springer-Verlag, 1981). On efforts to revive geometry at more elementary levels, see Senechal and Fleck, *Shaping Space.* On Coxeter's career, see the interview in *Mathematical People,* ed. Donald J. Albers and G. L. Alexanderson (Boston: Birkhäuser, 1985), 51–64.

61. H.S.M. Coxeter to Wheeler, Oct. 23, 1934, Wheeler Papers. Italics added.

62. P. A. Caris, "The Twelfth Annual Meeting of the Philadelphia Section," *American Mathematical Monthly* 45 (Feb. 1938): 69–70.

63. J. R. Killian Jr. to Wheeler, Oct. 27, 1937; George Sarton to Wheeler, July 24, 1944, Wheeler Papers.

64. *Worcester Post,* Apr. 27, 1937.

65. The Brill models and Felix Klein's efforts for educational reform in Germany stand in sharp contrast.

66. On the history of movements in the United States to promote utilitarian secondary education, including industrial education, see Edward A. Krug, *The Shaping of the American High School, 1880–1920* (New York: Harper & Row, 1964), 217–447, esp. 308–9 and 347–53.

67. *Wellesley College News,* Oct. 21, 1926.

CHAPTER FOURTEEN Linkages

Epigraph: Advertisement, *Mathematics Teacher* 44 (Oct. 1951): back cover.

1. Brief historical sketches on linkages can be found in the following: Dickson H. Leavens, "Linkages," *American Mathematical Monthly* 22 (Dec. 1915): 330–34; Joseph Hilsenrath, "Linkages," *Mathematics Teacher* 30 (Oct. 1937): 277–84; Bruce E. Meserve, "Linkages as Visual Aids," *Mathematics Teacher* 39 (Dec. 1946): 372–79; Meserve, *Fundamental Concepts of Algebra* (Cambridge, Mass.: Addison-Wesley, 1953), 244–47; R. C. Archibald, "Outline of the History of Mathematics," *American Mathematical Monthly* 56 (Jan. 1949): 52, 99; Eugene S. Ferguson, "Kinematics of Mechanisms from the Time of Watt," *United States National Museum Bulletin* 228 (1962): 185–230; Daina Taimina, "How to Draw a Straight Line," in Cornell University's Kinematic Models for Design Digital Library, http://kmoddl.library.cornell.edu/tutorials/04/.

2. On Watt and linkages, see Leavens, "Linkages," 331–32; Hilsenrath, "Linkages," 277–78; Meserve, "Linkages," 372; Meserve, *Algebra,* 246; and Ferguson, "Kinematics," 191–98. Watt in fact produced a portion of a lemniscate, a curve defined in polar coordinates with trigonometric functions. See Hilsenrath, "Linkages," 277.

3. Ferguson, "Kinematics," 198–201. For technical details see Douglas P. Adams, "Straight-line Mechanism," in *McGraw-Hill Encyclopedia of Science and Technology* (New York: McGraw-Hill, 1977).

4. See Douglas P. Adams, "Slider-Crank Mechanism," in *McGraw-Hill Encyclopedia of Science and Technology* (New York: McGraw-Hill, 1977).

5. "Notes and News," *American Mathematical Monthly* 29 (Oct. 1922): 363. Presumably, this was written by the editor of this section of the journal, R. W. Burgess of Brown University.

6. Ferguson, "Kinematics," 199.

7. Wm. Woolsey Johnson, "The Peaucellier Machine and Other Linkages," *Analyst* 2 (Mar. 1875): 41; George Bruce Halsted, "Tchebychev," *Science,* n.s. 1, Feb. 1, 1895, 130. Ferguson regards these reports of Chebyshev trying to prove the impossibility of producing straight-line motion as mere "rumor." Ferguson, "Kinematics," 204.

8. For the technical details, see Richard Courant and Herbert Robbins, *What Is Mathematics? An Elementary Approach to Ideas and Methods,* 2nd ed., ed. Ian Stewart (New York: Oxford University Press, 1996), 140–44, 156–57. Engineers consider the

Peaucellier cell to be an eight-bar linkage and in general count one more bar than mathematicians do in describing linkages. Mathematicians, in essence, consider a linkage as a free-floating object, whereas engineers include the supporting plane, as explained in 1876 by English engineering professor Alexander Kennedy: "The plane in which the two fixed points are supposed to be, and relatively to which some other point describes some particular curve, is in every sense as much a 'bar' as any of the other links." F. Reuleaux, *The Kinematics of Machinery*, ed. and trans. Alex. B. W. Kennedy (London: Macmillan, 1876), 590. See also Ferguson, "Kinematics," 204, 207. We use the mathematicians' convention in this chapter.

9. Leavens, "Linkages," 332–33; Ferguson, "Kinematics," 204–5.

10. At least in the nineteenth century the French referred to linkages as "systèmes articulées." Leavens, "Linkages," 333. In German one finds *Gelenkwerk, Gelenksystem,* and *Gelenkmechanismen.* See Arnold Emch, *An Introduction to Projective Geometry and Its Applications* (New York: John Wiley & Sons, 1905), 243; *Catalog mathematischer Modelle fur den höheren mathematischen Unterricht* (Leipzig: Martin Schilling, 1911), 166; and Felix Klein, *Elementary Geometry from an Advanced Standpoint: Geometry* (1908; rpt., n.p.: Dover Publications, 1939), 101.

11. Karen Hunger Parshall, *James Joseph Sylvester: Life and Work in Letters* (Oxford: Clarendon Press, 1998), 142.

12. Hilsenrath, "Linkages," 277; Ferguson, "Kinematics," 206. See also A. B. Kempe, "On a General Method of Producing Exact Rectilinear Motion by Linkwork," *Proceedings of the Royal Society of London* 23 (1874–75): 566.

13. Quoted in Robert C. Yates, *Tools: A Mathematical Sketch and Model Book* (Baton Rouge, La.: n.p., 1941), 82.

14. Quoted in Hilsenrath, "Linkages," 283.

15. Kempe, "General Method," 565. For technical details, see Courant and Robbins, *What Is Mathematics,* 157.

16. Meserve, *Algebra,* 246. Kempe is best known for his fallacious proof of the famous four-color problem, a question that engaged him shortly after his initial interest in linkages. Indeed, in the paper containing his "proof" Kempe mentioned that he was working on "the general theory of linkages." See A. B. Kempe, "On the Geographical Problem of the Four Colors," *American Journal of Mathematics* 2 (Sept. 1879): 200. In retrospect Kempe was perceptively proposing that map coloring be studied by what has become known as graph theory. N. L. Biggs, E. K. Lloyd, and R. J. Wilson, *Graph Theory: 1736–1936* (Oxford: Clarendon Press, 1976), 90–108.

17. A. B. Kempe, *How to Draw a Straight Line* (London: Macmillan, 1877). See Leavens, "Linkages," 330.

18. Meserve, *Algebra,* 245.

19. Leavens, "Linkages," 333.

20. Yates, *Tools,* 82.

21. Ferguson, "Kinematics," 209–13.

22. Robert Willis, *Principles of Mechanism, Designed for the Use of Students in the Universities, and for Engineering Students Generally* (London: John W. Parker, 1841),

20, 399–402; William Fairbairn, *The Principles of Mechanism and Machinery of Transmission* (Philadelphia: H. C. Baird & Co., 1871), 31–38.

23. F. Reuleaux, *Lehrbuch der Kinematik* (Braunschweig: Friedrich Vieweg und Sohn, 1875).

24. Reuleaux, *Kinematics of Machinery*. See n. 8.

25. *Slider-crank* was Kennedy's translation of Reuleaux's *Schubkurbel*. Cf. Reuleaux, *Kinematics*, xiii; and *Kinematik*, xix. On Reuleaux, see Otto Mayr, "Reuleaux, Franz," in *Dictionary of Scientific Biography* (New York: Charles Scribner's Sons, 1975), 11:383–85.

26. See Reuleaux, *Kinematics*, 46; and *Kinematik*, 49.

27. Reuleaux, *Kinematik*, 354, 415.

28. Reuleaux, *Kinematics*, 590.

29. Kinematic Models for Design Digital Library, Cornell University, http://kmoddl.library.cornell.edu/rx_collection.php.

30. Alexander Ziwet, *Elements of Theoretical Mechanics* (New York: Macmillan, 1908), 125–26. Ziwet was a professor of mathematics at the University of Michigan.

31. Adams, "Straight-Line Mechanism."

32. J. Michael McCarthy, *Geometric Design of Linkages* (New York: Springer, 2000), 9–10.

33. Ferguson, "Kinematics," 209.

34. Henry Barnard, *Practical Illustrations of the Principles of School Architecture* (Hartford: Press of Case, Tiffany & Co., 1851), 157. On the gonigraph, see also N. A. Calkins, *Primary Object Lessons for a Graduated Course of Development* (New York: Harper & Bros., 1861), 46.

35. See Phillip McCann and Francis A. Young, *Samuel Wilderspin and the Infant School Movement* (London: Croom Helm, 1982), 168.

36. Frederick A. P. Barnard, ed., *Johnson's New Universal Cyclopedia* (New York: A. J. Johnson & Son, 1875–78), 185; John M. Ross, ed., *The Globe Encyclopedia of Universal Information* (Boston: Estes & Lauriat, 1876–79), 18.

37. Johnson, "Peaucellier Machine," 41. Johnson also wrote "Recent Results in the Study of Linkages," *Analyst* 3 (Mar. 1876): 42–46; and "Recent Results in the Study of Linkages [Continued]," *Analyst* 3 (May 1876): 70–74. Other American contributions from this period include Wm. D. Marks, "Peaucellier's Compound Compass and Other Linkages," *Journal of the Franklin Institute* 107 (June 1879): 372 (cited in Meserve, "Linkages," 378); and Frank T. Freeland, "Linkages for x^m," *American Journal of Mathematics* 3 (Dec. 1880): 316–19.

38. Personal communication from Sylvester scholar Karen Hunger Parshall, Aug. 10, 2005.

39. George Bruce Halsted, *The Elements of Geometry* (New York: John Wiley & Sons, 1885), 350–52. This book is dedicated to Sylvester.

40. Halsted, "Tchebychev," 131.

41. Percey F. Smith and Arthur Sullivan Gale, *The Elements of Analytic Geometry* (Boston: Ginn & Co., 1904), 309; Emch, *Projective Geometry*, 242–49. Emch pub-

lished several research papers on linkages; see, e.g., "Algebraic Transformations of a Complex Variable Realized by Linkages," *Transactions of the American Mathematical Society* 3 (Oct. 1902): 493–98.

42. E. H. Moore, "The Cross-Section Paper as a Mathematical Instrument," *School Review* 14 (May 1906): 317–38.

43. J. W. A. Young, *The Teaching of Mathematics in the Elementary and Secondary School* (New York: Longmans, Green & Co., 1907), 167–68.

44. In 1905 R. P. Baker began advertising linkages among his models for sale. "Notes," *American Mathematical Monthly* 12 (Apr. 1905): 99. See also "Machinery Instructions," box 14, Richard P. Baker Collection, University Archives, Department of Special Collections, University of Iowa Libraries, Iowa City.

45. Schilling, *Catalog mathematischer Modelle,* 166.

46. See the Mathematics Genealogy Project, North Dakota State University, http://genealogy.math.ndsu.nodak.edu; and the MacTutor History of Mathematics Archive, University of St. Andrews, Scotland, www-gap.dcs.st-and.ac.uk/~history/Mathematicians/Morley.html.

47. Frank V. Morley, "Linkages," *Scientific Monthly* 9 (Oct. 1919): 366–78. F. V. Morley, "The Three-Bar Curve," *American Mathematical Monthly* 31 (Feb. 1924): 71–77.

48. Frank Morley and F. V. Morley, *Inversive Geometry* (Boston: Ginn & Co., 1933), 42–43, 183–85.

49. Felix Klein, *Elementary Mathematics from an Advanced Standpoint: Geometry* (New York: Macmillan, 1939), 100–101; D. Hilbert and S. Cohn-Vossen, *Geometry and the Imagination* (New York: Chelsea Publishing Co., 1952), 272–73; Courant and Robbins, *What Is Mathematics,* 155–58. The Klein book was originally published in German in 1908 and the Hilbert and Cohn-Vossen in 1932.

50. Many such presentations are documented in the *American Mathematical Monthly.* Among the institutions represented were Harvard, University of Illinois, Cooper Union, Illinois State Normal University, University of Kentucky, Washington University of St. Louis, Milwaukee Downer College, and Nebraska State Teachers College at Chadron. See *American Mathematical Monthly* 27 (Dec. 1920): 480; 29 (Feb. 1922): 79; 32 (Oct. 1925): 430; 36 (Nov. 1929): 487; 38 (Feb. 1931): 99; 42 (June 1935): 391; 45 (Aug.–Sept. 1938): 477; 46 (June–July 1939): 360. Such presentations persisted until 1954, after which the *Monthly* ceased reporting on college mathematics clubs.

51. Hilsenrath, "Linkages," 284.

52. Robert Carl Yates, "Small Vibrations of Certain Mechanical Systems" (Ph.D. diss., Johns Hopkins University, 1930). Acquaintance between Yates and the Morleys is hypothetical but plausible. They cite Yates on p. 200 of *Inversive Geometry.* Yates's supervisor at Johns Hopkins was F. D. Murnaghan, himself a student of the elder Morley.

53. Robert C. Yates, "Mechanically Described Curves," *National Mathematics Magazine* 10 (Jan. 1936): 134–38; "An Ellipsograph," *National Mathematics Maga-*

zine 12 (Feb. 1938): 213–15; "A Trisector," *National Mathematics Magazine* 12 (Apr. 1938): 323–24; "A Linkage for Describing Curves Parallel to the Ellipse," *American Mathematical Monthly* 45 (Nov. 1938): 607–8; "Line Motion and Trisection," *National Mathematics Magazine* 13 (Nov. 1938): 63–66.

54. Yates, *Tools;* see n. 13. Robert C. Yates, *Geometrical Tools: A Mathematical Sketch and Model Book* (St. Louis: Educational Publishers, 1949).

55. Yates was the first West Point mathematics instructor to hold a Ph.D. degree in the subject. See Frederick Rickey and Amy Shell-Gellasch, "201 Years of Mathematics at West Point," in *West Point: Two Centuries and Beyond,* ed. Lance Betros (Abilene, Tex.: McWhiney Foundation, 2004), 601. He had graduated from the Virginia Military Institute before attending Johns Hopkins and was a captain in the army at the time of his West Point appointment. See Yates diss. vita. In 1946 he wrote a short pamphlet for his West Point students containing material on linkages: "Curves," prepared for use in the Department of Mathematics, U.S. Military Academy, by Lt. Col. Robert C. Yates AUS, 1946.

56. Robert C. Yates, "Linkages," 117–29; Yates, "Trisection," 146–53; Michael Goldberg, "Linkages in Three Dimensions," 160–63, all in *Multisensory Aids in the Teaching of Mathematics,* ed. W. D. Reeve, NCTM 18th Yearbook (New York: Teachers College Press, 1945).

57. Meserve wrote an undergraduate thesis at Bates College in the late 1930s on "Linkages: Their Development and Utility." He then proceeded to use them in teaching at a private secondary school in Providence, R.I. Meserve served as president of the NCTM in the 1960s. Bruce E. Meserve, Oral History Interview with David L. Roberts, Apr. 3, 2003, National Council of Teachers of Mathematics Oral History Project Records, 1992–1993, 2002–2004, Archives of American Mathematics, Center for American History, University of Texas at Austin.

58. Meserve, "Linkages," 372.

59. Advertisement, *Mathematics Teacher* 43 (Apr. 1950): 183.

60. *Mathematics Teacher* 44 (Jan. 1951): back cover.

61. *Mathematics Teacher* 45 (Nov. 1951): back matter.

62. James D. Gates, Oral History Interviewed with David L. Roberts, Mar. 12–13, 2004, National Council of Teachers of Mathematics Oral History Project Records, 1992–1993, 2002–2004, Archives of American Mathematics, Center for American History, University of Texas at Austin.

63. On the New Math, see David L. Roberts and Angela L. E. Walmsley, "The Original 'New Math': Storytelling versus History," *Mathematics Teacher* 96 (Oct. 2003): 468–73.

64. *Curriculum and Evaluation Standards for School Mathematics* (Reston, Va.: NCTM, 1989); *Principles and Standards for School Mathematics* (Reston, Va.: NCTM, 2000).

65. Courant and Robbins, "How to Use the Book," *What Is Mathematics,* n.p. (immediately before contents page).

66. Ibid., 563.

67. See, e.g., Don Shimamoto and Catherine Vanderwaart, "Spaces of Polygons in the Plane and Morse Theory," *American Mathematical Monthly* 112 (Apr. 2005): 289–310; and M. Kapovish and J. J. Millson, "Universality Theorems for Configuration Spaces of Planar Linkages," *Topology* 41 (2002): 1051–1107. The latter paper, using an array of concepts from recent topology and algebraic geometry, points out deficiencies in Kempe's 1877 proof that any algebraic curve can be constructed via a suitable linkage and reformulates and proves a rigorous version.

CHAPTER FIFTEEN Calculators

Epigraph: "Editor's Note" to Mildred D. Schaughency, "Teaching Arithmetic with Calculators," *Arithmetic Teacher* 2 (Feb. 1955): 22.

1. Frederick A. P. Barnard, "Machinery and Processes of the Industrial Arts and Apparatus of the Exact Sciences," in *Reports of the United States Commissioners to the Paris Universal Exposition,* ed. W. P. Blake (Washington, D.C.: Government Printing Office, 1869), 3:629. On Barnard and other early users of calculating machines, see P. A. Kidwell, "American Scientists and Calculating Machines—From Novelty to Commonplace," *Annals of the History of Computing* 12, no. 1 (1990): 31–40.

2. This sum may not sound extravagant. Recall, however, that in this period the total annual expenses of a frugal student at the Thayer School of Engineering at Dartmouth College were estimated as running from $350 to $400, including tuition, housing, and food. Dartmouth College, *Catalogue,* 1898–99, 181.

3. James W. Cortada, *Before the Computer: IBM, NCR, Burroughs, and Remington Rand and the Industry They Created, 1865–1956* (Princeton: Princeton University Press, 1993); Sharon Hartman Strom, *Beyond the Typewriter: Gender, Class, and the Origins of Modern American Office Work, 1900–1930* (Urbana: University of Illinois Press, 1992), esp. 178–226, 273–94.

4. George W. Myers, "The Laboratory Method in the Secondary School," *School Review* 11 (1903): 737.

5. The Monroe Educator actually was introduced in the late 1930s, but the advent of World War II apparently diverted the company from seriously attempting to market it. The machine is mentioned as the Model LA 1307 in American Office Machines Research, Inc., "Monroe," May 1940, 12. (This is one of several descriptions of office machines prepared by American Office Machines Research that the company circulated in black loose-leaf binders. It is in volume 3 of the set of binders. The machine is not called the Educator. A penciled note in the copy of this document that came to the Smithsonian Institution from Victor Comptometer Corporation indicates that in Oct. 1938 the machine sold for "170.00 Net, $148.73." But the Oct. 1938 listing of Monroe machines does not include any machine with the properties of the Educator.) Records of the U.S. Patent and Trademark Office indicate that Monroe Calculating Machine Company filed for a trademark for the term *Educator* as applied to calculating machines on Oct. 19, 1949, receiving the registration Feb. 27, 1951. The trademark application indicates that the firm first used the term in commerce in Oct. 1938, whatever the Office Machines Research Service documentation may

say. It is possible that Monroe initially did not include the machine in its list of offerings because the machine was sold only to schools. See trademark no. 71,586,540, U.S. Patent and Trademark Office. A copy of this trademark is available electronically at www.uspto.gov/. On the Hunter College experiment, see also Catherine D. Carney, "An Experiment in Teaching 'Mechanical Arithmetic' in an Elementary School," *Business Education World* 32 (Nov. 1951): 128–29.

6. Mildred D. Schaughency, "Teaching Arithmetic with Calculators," *Arithmetic Teacher* 2 (Feb. 1955): 21–22.

7. "Editor's Note" to Schaughency, "Teaching Arithmetic with Calculators," 22.

8. Howard F. Fehr, George McMeen, and Max Sobel, "Using Hand-Operated Computing Machines in Learning Arithmetic," *Arithmetic Teacher* 3 (Oct. 1956): 145–50.

9. "Machine Age Arithmetic," *American School Board Journal* (Apr. 1957). This clipping is one of several relating to the development of the Monroe Educator in a scrapbook held by Monroe Systems for Business in Levittown, Pa. We thank Ira Chinoy for information about these materials.

10. National Education Association Service, "School Finds Machines Create Interest in Arithmetic," *Pontiac Press,* May 18, 1957. A copy of this article is in the scrapbook cited in n. 9.

11. "Editor's Note" to Howard F. Fehr, George McMeen, and Max Sobel, "Using Hand-Operated Computing Machines in Learning Arithmetic," *Arithmetic Teacher* 3 (Oct. 1956): 150.

12. Milton W. Beckmann, "New Devices Elucidate Arithmetic," *Arithmetic Teacher* 7 (Oct. 1960): 296–301.

13. Lois L. Beck, "A Report on the Use of Calculators," *Arithmetic Teacher* 7 (Feb. 1960): 103. Advertisements for the Monroe Educator appeared on the back cover of *Arithmetic Teacher* in Mar. 1957, in a two-page spread in the Oct. 1957 issue of the periodical, and in *College and University Business* for Mar. 1957.

14. Henry W. Syer and Donovan A. Johnson, "Aids to Teaching," *Mathematics Teacher* 45 (Dec. 1952): 624.

15. *Arithmetic Teacher* 3 (Apr. 1956): 127.

16. Burroughs Corporation news releases gave biographical information about Schott and reported on the development of his project. See documents dated Jan. 19, 1954, and Sept. 8, 1955, Burroughs Corporation Records, Press Releases (CBI 90), Charles Babbage Institute (CBI), University of Minnesota, Minn.; "We Try Something New in Milwaukee," *Burroughs B-Line International* 3, no. 4 (Mar. 1, 1954): 4, Burroughs Corporation Records, Serial Publications (CBI 90), CBI.

17. Glenda Lappan and Jeffrey J. Wanko, "The Changing Roles and Priorities of the Federal Government in Mathematics Education in the United States," in *A History of School Mathematics,* ed. George M. A. Stanic and Jeremy Kilpatrick (Reston, Va.: NCTM, 2003), 2:897–930.

18. "How to Multiply Student Achievement," *Mathematics Teacher* 61, no. 8 (Dec. 1968): 813. This advertisement also ran in the January, Feb., and Mar. 1969 issues of

the journal. A somewhat different ad ran in December 1969 (62, no. 8: 570). Wang's interest in the education market did not end in 1969. See, e.g., the ad "Thinking Tools," in *Mathematics Teacher* 64, no. 2 (Feb. 1971): 172. Soon after this period, however, the company shifted its emphasis from calculators to computers and word processing.

19. Walter J. Koetke, "Computers," *Mathematics Teacher* 64, no. 8 (Dec. 1971): 726–27.

20. "Singer Proudly Announces the First Survival Courses for the 21st Century," *Mathematics Teacher* 62, no. 7 (Nov. 1969): 560–61.

21. "Pythagoras, He's Not," *Mathematics Teacher* 63, no. 7 (Nov. 1970): 586.

22. "Something New to Help You Turn Unrest into Interest," *Mathematics Teacher* 64, no. 8 (Dec. 1971): 736–37.

23. It is difficult to know precisely what calculator companies charged schools, particularly in this time of rapidly decreasing prices. A Nov. 18, 1969, advertisement for Sharp electronic calculators listed desktop models that cost between $395 and $1,275 wholesale. See *New York Times*, Nov. 18, 1969, 74.

24. "Introducing the Classmate 88," *Mathematics Teacher* 69, no. 1 (Jan. 1976): 52; "Most Teachers Won't Believe What a 9830 Calculator Will Do . . . ," *Mathematics Teacher* 68, no. 2 (Feb. 1975): 166; Roberta Schwartz, "Terminals Enter More Classrooms," *Electronics* 45, no. 16, July 31, 1972, 56–57.

25. Of course, the slide rule was also considerably cheaper. Introductory textbooks sold with their own paper slide rules for as little as $6.95. See "Books Received," *Science*, n.s. 169, no. 3945, Aug. 7, 1970, 619. Durable slide rules cost more than that but still far less than $300.

26. *HP-35 Operating Manual* (Cupertino, Calif.: Hewlett-Packard, n.d.), i.

27. Bruce Flamm, "The Story of the HP-35," *International Calculator Collector* 2, no. 1 (Spring 1994): 4–6. On the durability of the HP-35, see "Hewlett-Packard Introduces a Smaller Uncompromising Calculator," *Science*, n.s. 187, Feb. 28, 1975, 686–87. Several articles on the calculator were published in the *Hewlett-Packard Journal* for June 1972 (2–9). They are available online at the HP Virtual Museum, www.hp.com/hpinfo/abouthp/histnfacts/museum/index.html.

28. This advertisement ran in *Science*, n.s. 177, Aug. 4, 1972, 384–85; Nov. 3, 1972, 446–47; and Dec. 1, 1972, 930–31.

29. "Some Things Are Changing for the Better," *Science*, n.s. 177, Aug. 4, 1972, 384.

30. "The Ultimate Slide Rule?" *Mathematics Teacher* 65, no. 5 (May 1972): 436–37.

31. S. Greitzer, "The Second U.S.A. Mathematical Olympiad," *American Mathematical Monthly* 81, no. 3 (Mar. 1974): 254; "The Third U.S.A. Mathematical Olympiad," *American Mathematical Monthly* 82, no. 3 (Mar. 1975): 220; and "The Fourth U.S.A. Mathematical Olympiad," *American Mathematical Monthly* 83, no. 2 (Feb. 1976): 120.

32. Gene Smith, "Texas Instruments Puts 3 Calculators on Market," *New York*

Times, Sept. 21, 1972, 67. Texas Instruments had made a prototype handheld calculator known as the Cal Tech as early as 1967, but it did not become a commercial product.

33. Edward B. Fiske, "Educators Feel That Calculators Have Both Pluses and Minuses," *New York Times,* Jan. 5, 1975, E7.

34. "Calculators in the Classroom," *Mathematics Teacher* 68, no. 3 (Mar. 1975): 226.

35. V. Frederick Rickey and Amy Shell-Gellasch, "201 Years of Mathematics at West Point," in *West Point: Two Centuries and Beyond,* ed. Lance Betros (Abilene, Tex.: McWhiney Foundation Press, 2004), 601.

36. Lowell Leake Jr., "Editor's Comment," *Mathematics Teacher* 68, no. 5 (May 1975): 415.

37. "The Electronic Slide Rule," *Physics Today* 26, no. 1 (Jan. 1973): 88.

38. "Alas Poor Mastodon of Mathematics," *New York Times,* July 11, 1976, 97.

39. Patrick J. Boyle, "Calculator Charades," *Mathematics Teacher* 69, no. 4 (Apr. 1976): 281–82.

40. "Computational Skill Is Passé," *Mathematics Teacher* 67, no. 6 (Oct. 1974): 485.

41. Ibid., 487.

42. For an overview of NCTM activities at this time, see James T. Fey and Anna O. Graeber, "From the New Math to the *Agenda for Action,*" in *A History of School Mathematics,* ed. George M. A. Stanic and Jeremy Kilpatrick (Reston, Va.: NCTM, 2003), 1:521–55. Other references to NACOME are scattered through this work.

43. National Advisory Committee on Mathematics Education, *Overview and Analysis of School Mathematics, Grades K–12* (Washington, D.C.: Conference Board of the Mathematical Sciences, 1975), 138.

44. Ross Taylor, "NACOME: Implications for Teaching K-12," *Mathematics Teacher* 69, no. 6 (Oct. 1976): 458–63.

45. NCTM Instructional Affairs Committee, "Minicalculators in Schools," *Mathematics Teacher* 69, no. 1 (Jan. 1976): 92–95.

46. Richard S. Pieters, "Statistics in the High School Curriculum," *American Statistician* 30, no. 3 (Aug. 1976): 138–39.

47. Henry O. Pollak, "Hand-Held Calculators and Potential Redesign of the School Mathematics Curriculum," *Mathematics Teacher* 70, no. 4 (Apr. 1977): 293–96.

48. David Moursund, "Calculators and the Elementary School: An Idea and Some Implications," *Proceedings of the Annual Conference of the Association for Computing Machinery* (New York: ACM Press, 1976), 135–37.

49. Brendan Kelly, "The Emergence of Technology in Mathematics Education," in *A History of School Mathematics,* ed. George M. A. Stanic and Jeremy Kilpatrick (Reston, Va.: NCTM, 2003), 2:1051–1057. On the use of calculators on the examinations of the College Entrance Examination Board, see Chancy O. Jones, "The Use of Calculators on College Board Mathematics Examinations in the 1990s," in *Pro-*

ceedings of the Third International Conference on Technology and Collegiate Mathematics, ed. John Harvey, Franklin Demana, and Bert K. Waits (Reading, Mass.: Addison-Wesley, 1992), 157–61. A May 9, 2005, personal communication from Julie Duminiak, archivist of the Educational Testing Service, indicates that the planned use of calculators in fact took place.

50. Jerome C. Meyer and James A. Tillotson III, "Teaching Device Having Means Producing a Self-Generated Program," U.S. Patent 3,584,398, June 15, 1971.

51. William R. Hafel, "Mathematical Problem and Number Generating System," U.S. Patent 3,947,976, Apr. 6, 1976.

52. Thomas E. Rowan, [review of Digitor], *Mathematics Teacher* 69, no. 3 (Mar. 1976): 248–50.

53. David H. Ahl, "Centurion Educational Computers: Ten Grapefruit-Sized Computers Provide Drill and Practice on Everything from Math to English," *Creative Computing* 9, no. 10 (Oct. 1983): 44.

54. Eric F. Burtis to Robert Adams, Feb. 14, 1976, Accession File 1986.0507, NMAH.

55. *New York Times,* Dec. 23, 1975, 4. Novus had entered the calculator business by buying out the calculator division of National Semiconductor, and some of its devices were sold as the National Semiconductor Quiz Kid.

56. Nathaniel C. Nash, "Calculated Gadgetry," *New York Times,* May 23, 1976, F3.

57. On some of these rival toys, see the article cited in n. 56.

58. Prices for the Little Professor are given in advertisements in the *New York Times, Washington Post,* and *Wall Street Journal.* See also "Business Bulletin: A Special Background Report on Trends in Industry and Finance," *Wall Street Journal,* Nov. 4, 1976, 1.

59. On the Mathemagician, see Joseph W. Willhide and Henry L. Viarengo, "APF Mathemagician," *Creative Computing* 4, no. 2 (Mar.–Apr. 1978): 92–93; and David H. Ahl, "APF Mathemagician Games," *Creative Computing* 4, no. 2 (Mar.–Apr. 1978): 93–94.

60. James Wilson and Jeremy Kilpatrick, interview by David L. Roberts, May 24, 1999, R. L. Moore Legacy Collection, 1890–1900, 1900–2003, Archives of American Mathematics, Center for American History, University of Texas at Austin, Austin.

CHAPTER SIXTEEN Minicomputers

Epigraph: Anthony G. Oettinger, with the collaboration of Sema Marks, *Run, Computer, Run: The Mythology of Educational Innovation* (Cambridge: Harvard University Press, 1969), 4.

1. For overviews of the history of computing, see Martin Campbell-Kelly and William Aspray, *Computer: A History of the Information Machine* (New York: Basic Books, 1996); and Paul E. Ceruzzi, *A History of Modern Computing* (Cambridge, Mass.: MIT Press, 1999).

2. For an overview of the subject of distance learning, see Marina Stock McIsaac and Charlotte Nirmalani Gunawardena, "Distance Education," in *Handbook of Re-*

search for Educational Communications and Technology, ed. David H. Jonassen (New York: Macmillan, 1996), 403–37; as well as related articles in that volume.

3. Simon Ramo, "A New Technique of Education," *IRE Transactions on Education*, E-1, no. 2 (June 1958): 37–42. The paper was reprinted from the Oct. 1957 issue of *Engineering and Science Monthly*.

4. Paul K. Weimer, "A Proposed 'Automatic' Teaching Device," *IRE Transactions on Education*, E-1, no. 2 (June 1958): 53.

5. PLATO staff described their programming logic as a form of Crowder's branched style of programmed learning. An incorrect response did not lead to a different series of frames, however, as in other branched programs. Donald Bitzer, Peter Braunfeld, and Wayne W. Lichtenberger, "PLATO: An Automatic Teaching Device," *IRE Transactions on Education*, E-4, no. 4 (Dec. 1961): 157–61.

6. Peter G. Braunfeld and Lloyd D. Fosdick, "The Use of an Automatic Computer System in Teaching," *IRE Transactions on Education*, E-5, nos. 3–4 (Sept.–Dec. 1962): 156–67.

7. Elisabeth Van Meer, "PLATO: From Computer-Based Education to Corporate Social Responsibility," *Iterations: An Interdisciplinary Journal of Software History*, Nov. 5, 2003. The journal is available at the Web site of the Charles Babbage Institute of the University of Minnesota. On the use of electronic mail and flat-screen consoles in PLATO, see Paul Ceruzzi, "Internet Dreams," *Invention and Technology* 19, no. 3 (Winter 2004): 34–37.

8. Douglas D. Noble, *The Classroom Arsenal: Military Research, Information Technology, and Public Education* (London: Falmer, 1991), 110–20; and George Buck and Steve Hunka, "Development of the IBM 1500 Computer-Assisted Instructional System," *IEEE Annals of the History of Computing* 17 (1995): 19–31. We thank Steve Hunka for explanations relating to the IBM 1500 system.

9. Patrick Suppes et al., *Sets and Numbers* (Stanford, Calif.: Stanford University Press, 1961); Patrick Suppes and Newton Hawley, *Geometry for Primary Grades* (San Francisco: Holden-Day, 1960).

10. This account is based largely on Patrick Suppes, "Intellectual Autobiography (Written 1978)," MS, available online at www.stanford.edu/~psuppes/autobio1.html.

11. Patrick Suppes et al., *Computer-Assisted Instruction: Stanford's 1965–66 Arithmetic Program* (New York: Academic Press, 1968), 10.

12. Patrick Suppes and Mona Morningstar, *Computer-Assisted Instruction at Stanford, 1966–1968* (New York: Academic Press, 1972).

13. Nan Robertson, "$2-Billion Set Aside by U.S. to Help the Poor Learn Better," *New York Times*, Jan. 12, 1966, 45, 51.

14. Buck and Hunka, "Development of the IBM 1500 Computer-Assisted Instructional System."

15. Suppes, "Intellectual Autobiography."

16. In addition to Suppes's autobiography and Noble's *Classroom Arsenal*, see Roger Cohen, "Simon & Schuster Buys School Software Maker," *New York Times*,

Mar. 12, 1990, D6; and Doreen Carvajal, "Pearson Receives U.S. Approval to Buy Simon & Schuster Unit," *New York Times,* Nov. 24, 1998, C10.

17. George H. Litman, "CAI in Chicago," paper presented at the Association for Educational Data Systems Annual Convention (New Orleans, La., Apr. 16–19, 1973), ERIC identifier ED087423; Max Weiner et al., "An Evaluation of the 1968–1969 New York City Computer Assisted Instruction Project in Elementary Arithmetic," paper presented at Annual Meeting of the American Educational Research Association, Minneapolis, Feb., 1970), ERIC identifier ED040576; Roberta Schwartz, "Terminals Enter More Classrooms," *Electronics* 45, no. 16, July 31, 1972, 56–57. In addition to projects at Stanford and Illinois, the National Science Foundation sponsored "Time-Shared, Interactive, Computer-Controlled Information Television" (TICCIT) at the MITRE Corporation. This project, begun in 1969, initially delivered lessons over four terminals to elementary school students. By the early 1970s the focus had shifted to providing mathematics and English instruction to community colleges; mathematics programs focused on algebra. See Paul Saettler, *The Evolution of American Educational Technology* (Englewood, Colo.: Libraries Unlimited, 1990), 309–11; and Wanda Rappaport and Elizabeth Olenbush, "Tailor-Made Teaching through TICCIT," *Mitre Matrix* 8, no. 4 (Fall 1975): 2–17.

18. Stewart A. Denenberg, "A Personal Evaluation of the PLATO System," *ACM SIGCUE Bulletin* 12, no. 2 (Apr. 1978): 6.

19. Dean Donald Morrison of Dartmouth had been a graduate student at Princeton from 1936 until 1941 and consulted with Albert W. Tucker of the Princeton mathematics department in selecting new faculty for the mathematics department of the college. See Walter Meyer, "The Origins of Finite Mathematics: The Social Science Connection," *College Mathematics Journal* 39, no. 2 (Mar. 2007): 106–18. Here Meyer also discusses the writing and influence of Kemeny, Snell, and Thompson's textbook *Introduction to Finite Mathematics.*

20. John G. Kemeny, J. Laurie Small, and Gerald L. Thompson, *Introduction to Finite Mathematics* (Englewood Cliffs, N.J.: Prentice-Hall, 1957).

21. John G. Kemeny, Hazelton Mirkil, J. Laurie Snell, and Gerald L. Thompson, *Finite Mathematical Structures* (Englewood Cliffs, N.J.: Prentice-Hall, 1959).

22. Dartmouth College Writing Group, *Modern Mathematical Methods and Models: A Book of Experimental Text Materials* ([Ann Arbor? Mich., 1958] distributed by the Mathematical Association of America).

23. John G. Kemeny and J. Laurie Snell, *Mathematical Models in the Social Sciences* (Boston: Ginn, 1962).

24. Thomas E. Kurtz, *Basic Statistics* (Englewood Cliffs, N.J.: Prentice-Hall, 1963).

25. The Dartmouth Mathematics Project, funded by the National Science Foundation and the Office of Naval Research, was active from at least 1955 through 1960. Copies of many of its reports are in the Dartmouth College Library. See Dartmouth College, Dartmouth Mathematics Project, *Project Report,* Hanover, N.H. Dartmouth also sponsored a conference on new directions in mathematics education in 1961,

which was attended by both secondary and college teachers. On this conference, see "Mathematics Conference—New Directions," Vertical Files, Rauner Special Collections, Dartmouth College. We are grateful to Sarah Hartsell of the Rauner Special Collections for this information.

26. Eric A. Weiss, "John George Kemeny," *Annals of the History of Computing* 15, no. 2 (1993): 58–59.

27. On the history of BASIC, see Thomas E. Kurtz, "BASIC," in *History of Programming Languages,* ed. Richard L. Wexelblat (New York: Academic Press, 1981), 515–49; and John G. Kemeny and Thomas E. Kurtz, *Back to BASIC: The History, Corruption, and Future of the Language* (Reading, Mass.: Addison-Wesley, 1985). On the General Electric computers, see J. A. N. Lee, "The Rise and Fall of the General Electric Corporation Computer Department," *Annals of the History of Computing* 17, no. 4 (1995): 24–45.

28. John G. Kemeny, *Man and the Computer* (New York: Charles Scribners, 1972), esp. 32–37; Kurtz, "BASIC," 519.

29. Dartmouth College, *BASIC: A Manual for BASIC, the Elementary Algebraic Language Designed for Use with the Dartmouth Time Sharing System* ([Hanover, N.H.]: Trustees of Dartmouth College, 1965). A copy of this manual is in the Computer Documentation Collection, Computers and Mathematics, National Museum of American History.

30. J. A. N. Lee, "The Rise and Fall of the General Electric Corporation Computer Department," *Annals of the History of Computing* 17, no. 4 (1995): 33.

31. John G. Kemeny, *Man and the Computer* (New York: Charles Scribners, 1972), esp. 33–37. Some of these games would be published for microcomputer users in paperback books such as David H. Ahl, ed., *BASIC Computer Games, TRS-80 Edition* (Morristown, N.J.: Creative Computing Press, 1979).

32. James W. Cortada, "BASIC," *Historical Dictionary of Data Processing Technology* (New York: Greenwood Press, 1987), 39–43.

33. For biographical information about Licklider, see M. Mitchell Waldrop, *The Dream Machine: J.C.R. Licklider and the Revolution That Made Computing Personal* (New York: Viking, 2001).

34. This account of early educational efforts at BBN is based largely on Wallace Feurzeig, "Educational Technology at BBN," *Annals of the History of Computing,* 28, no. 1 (Jan.–Mar. 2006): 18–31. We thank the author for allowing us to see this MS.

35. John McCarthy, "Paper: History of LISP," in *History of Programming Languages,* ed. Richard L. Wexelblat (New York: Academic Press, 1981), 173–85. For an early statement of McCarthy's views on time sharing, see John McCarthy, "Time-Sharing Computer Systems," in *Management and the Computer of the Future,* ed. Martin Greenberger (Cambridge, Mass.: MIT Press, 1962), 220–48.

36. On the history of time sharing, see Arthur L. Norberg and Judy O'Neill, *Transforming Computer Technology: Information Processing for the Pentagon, 1962–1986* (Baltimore: Johns Hopkins University Press, 1996), esp. 68–118.

37. We use the form *LOGO,* except in quotations in which the word is written *Logo.*

38. For basic biographical information about Papert, see Richard Hull, "Seymour Papert," in *Encyclopedia of Computers and Computer History,* ed. Raúl Rojas (Chicago: Fitzroy Dearborn Publishers, 2001), 2:608–9.

39. Wallace Feurzeig, Seymour Papert, Marjorie Bloom, Richard Grant, and Cynthia Solomon, "Programming-Languages as a Conceptual Framework for Teaching Mathematics," *Interface* 4, no. 2 (Apr. 1970): 13–17. This article was abstracted from a much longer report of the same title sent to NSF in Nov. 1969.

40. Seymour Papert, [draft chaps.], 1970, Computer Documentation Collection, Computers and Mathematics, NMAH.

41. Ibid.

42. Allen L. Hammond, "Computer-Assisted Instruction: Many Efforts, Mixed Results," *Science* 176, June 2, 1972, 1005–6.

43. Seymour Papert, *Mindstorms: Children, Computers, and Powerful Ideas* (New York: Basic Books, 1980). The idea that mathematics education might benefit if children were able to explore the patterns they could generate by programming computers was not confined to MIT. David Kibbey, who worked on the fractions curriculum at the University of Illinois, also took this view. In the early 1970s he developed programs that allowed children to move about an icon representing an "airplane" or "spider," either in straight lines or fractions of a circle, and to save the resulting patterns. On Kibbey's Skywriting and Spider Web programs, see Sharon Dugdale, "Research in a Pioneer Constructive Network-Based Curriculum Project for Children's Learning of Fractions," in *Research on Technology in the Learning and Teaching of Mathematics: Syntheses and Perspectives,* ed. Heid and G. Blume, in press. We thank Sharon Dugdale for allowing us to see a draft of this chapter.

44. Harvard University, *Final Report SD-265 (PROJECT TACT),* ESD-TR 69-121, Oct. 1, 1968. A copy of this report is in the Harvard University Archives, collection HUF 300.885.

45. For brief discussion of Project MAC, see Arthur Norberg and Judy E. O'Neill, *Transforming Computer Technology: Information Processing for the Pentagon, 1962–1986* (Baltimore: Johns Hopkins University Press, 1996), 94–102, 221–22. See also Albert R. Meyer et al., eds., *Research Directions in Computer Science: An MIT Perspective* (Cambridge, Mass.: MIT Press, 1991), 321–23, 449–81; and Paulo Ney de Souza, Richard J. Fateman, Joel Moses, and Cliff Yapp, *The Maxima Book* (Sept. 19, 2004), which is available online at http://maxima.sourceforge.net/docs/maximabook/.

46. D. Alpert and D. L. Bitzer, "Advances in Computer-Based Education," *Science* 167, no. 3925, Mar. 20, 1970, 1582–90.

47. Dugdale, "Research in a Pioneer Constructive Network-Based Curriculum Project."

48. Feurzeig, "Educational Technology at BBN."

49. Glenn D. Allinger et al., "Reviewing and Viewing," *Arithmetic Teacher* 36, no. 1 (Sept. 1988): 50.

50. Sharon Dugdale, "Computers: Applications Unlimited," in *Computers in Mathematics Education, 1984 Yearbook,* ed. Viggo P. Hansen and Marilyn J. Zweng (Reston, Va.: NCTM, 1984), 82–88.

51. Dugdale, "Research in a Pioneer Constructive Network-Based Curriculum Project."

52. Patrick H. McCann, "Mathematics Instruction with Games," *Journal of Experimental Education* 45, no. 3 (Spring 1977): 61–68.

CHAPTER SEVENTEEN Early Microcomputers

Epigraph: Janet F. Asteroff, "Computing in Higher Education," *SIGCUE Outlook* 19, nos. 1–2 (Spring–Summer 1986): 3.

1. Peggy Schmidt, "The Computer as Tutor," *New York Times,* Apr. 14, 1985, ES50.

2. For an overview of the history of software, see Martin Campbell-Kelly, *From Airline Reservations to Sonic the Hedgehog: A History of the Software Industry* (Cambridge, Mass.: MIT Press, 2003). On general policies relating to early school use of computers, see Leonard J. Waks, "The New World of Technology in U.S. Education: A Case Study in Policy Formation and Succession," *Technology in Society* 13, no. 3 (1991): 233–53. Much recent material on the history of word processing and of spreadsheets has appeared in the *IEEE Annals of the History of Computing.*

3. For an account of the development of educational microcomputers in Sweden, see Thomas Kaiserfeld, "Computerizing the Swedish Welfare State," *Technology and Culture* 37, no. 2 (Apr. 1996): 249–79.

4. Thomas E. Kurtz, "BASIC," in *History of Programming Languages,* ed. Richard L. Wexelblat (New York: Academic Press, 1981), 534.

5. On the early history of Microsoft, see Paul Ceruzzi, *A History of Modern Computing* (Cambridge, Mass.: MIT Press, 1998), 232–36; and Paul Freiberger and Michael Swaine, *Fire in the Valley: The Making of the Personal Computer* (Berkeley: Osborne/McGraw Hill, 1984), 140–43.

6. During the same period the People's Computer Company of Menlo Park, Calif., developed several versions of the language under the general name of Tiny BASIC and distributed them in print and later on paper tape. See *Dr. Dobb's Computer Calisthenics and Orthodontia: Running Light without Overbyte* 1 (1976) (articles about Tiny BASIC appeared in several issues of this volume); and Richard Rosner, "A Review of Tom Pittman's Tiny BASIC," *Byte* 2, no. 4 (Apr. 1977): 34–38. Gordon Eubanks, a student at the Naval Post Graduate School in Pacific Grove, Calif., developed BASIC for the IMSAI microcomputer as well as a version sold slightly later as CBASIC. See Freiberger and Swaine, *Fire in the Valley,* 144–46.

7. Gary McGath, "A Look at LISP," *Byte* 2, no. 12 (Dec. 1977): 156.

8. Christopher Lett to Peggy Kidwell, personal communication, Aug. 27, 2005.

9. Ibid.

10. Christopher Lett, "A High School Computer System," *Byte* 1, no. 10 (June 1976): 28–30.

11. See an advertisement for the TRS-80 in *ROM: Computer Applications for Liv-*

ing 1, no. 4 (Oct. 1977): back cover; Jules H. Gilder on the PET in "Home Computers: A Look at What's Coming," *ROM: Computer Applications for Living* 1, no. 7 (Jan. 1978): 58–63; and Stephen Wozniak, "System Description: The Apple II," *Byte* 2, no. 5 (May 1977): 34–43, esp. 41–43. Hobbyists also were encouraged to add a special circuit board to ease their use of BASIC. See Jim Kreitner, "Pick Up BASIC by PROM Bootstraps," *Byte* 2, no. 1 (Jan. 1977): 50–51. On the TRS-80 machines at John F. Kennedy, we are again indebted to the personal communication cited previously.

12. Advertisement in *80-U.S.: The TRS-80 Users Journal* 2, no. 3 (May–June 1979): 3.

13. Apple Computer, Inc., *Apple in Depth* (Cupertino, Calif.: Apple Computer, 1981), 29.

14. Apple Computer, Inc., *Apple Software in Depth* (Cupertino, Calif.: Apple Computer, 1982), 47, 56.

15. See, e.g., Robert S. Siegler and Douglas R. Thompson, "'Hey, Would You Like a Nice Cold Cup of Lemonade on This Hot Day?' Children's Understanding of Economic Causation," *Developmental Psychology* 34, no. 1 (Jan. 1998): 146–60.

16. For descriptions of this game, see Thomas W. Malone, "Microcomputers in Education: Cognitive and Social Design Principles," *SIGCUE Outlook* 17, no. 2 (Spring 1983): 3–11; Sharon Dugdale, "Computers: Applications Unlimited," in *Computers in Mathematics Education, 1984 Yearbook,* ed. Viggo P. Hansen and Marilyn J. Zweng (Reston, Va.: NCTM, 1984), 82–88.

17. Jim Perry and Chris Brown, *80 Programs for the TRS-80* (Peterborough, N.H.: 1001001, Inc., 1979).

18. Bertram Gader and Manuael V. Nodar, *Apple Software for Pennies* (New York: Warner Books, 1985).

19. *80-U.S.: The TRS-80 Users Journal* 2, no. 6 (Nov.–Dec. 1979): 50.

20. EduWare Services obtained a trademark for the term EDUWARE in 1981. At that time the Maynard, Mass., software company took the name EDU-WARE-EAST. See the Web site of the U.S. Patent and Trademark Office as well as advertisements in volume 4 of *80-U.S.: The TRS-80 Users Journal.*

21. See advertisement, *Creative Computing* 6, no. 9 (Sept. 1980): 67.

22. David Lubar, "Educational Software," *Creative Computing* 6, no. 9 (Sept. 1980): 64.

23. See advertisement, *Creative Computing* 6, no. 9 (Sept. 1980): 67.

24. See advertisement, *Byte* 7, no. 2 (Feb. 1982): 319. According to the catalog of the Los Angeles Public Library, it holds a copy of this disk, copyrighted 1981.

25. Andrew Pollack, "Slugging It Out on the Software Front," *New York Times,* Oct. 16, 1983, F1.

26. Spinnaker Software Corp., "In Search of the Most Amazing Thing," 1983 (object 1987.3128.195), Computer Collections, National Museum of American History (NMAH), Smithsonian Institution, Washington, D.C.

27. Richard A. Shaffer, "Software Firm Taps Market for Education," *Wall Street Journal,* Nov. 15, 1983, 37.

28. See the advertisement for a "Graphics Package" in *80-U.S.: The TRS-80 Users Journal* 3, no. 1 (Jan.–Feb. 1980): 34. An example of the programs in the Computer Collections, NMAH, gives the name of the authors of the programs and uses the title "Graphing Package." The cassette was the property of Jon Eklund, a TRS-80 owner, *Creative Computing* subscriber, and Smithsonian curator of the history of chemistry and computing.

29. Dugdale, "Computers," 82–88; Sharon Dugdale, "Green Globs: A Microcomputer Application for Graphing of Equations," *Mathematics Teacher* 75, no. 3 (Mar. 1982): 208. For a review, see *Mathematics Teacher* 76, no. 9 (Dec. 1983): 693. This review indicated that the program sold for the Apple II for $45. On the early history of the program, see the report of a Mar. 12–14, 1981, conference at the University of California, San Diego, where the program was first described. Thomas W. Malone and James Levin, "Microcomputers in Education: Cognitive and Social Design Principles," *SIGCUE Outlook* 17, no. 1 (Winter 1983): 13–14; and Thomas W. Malone, "Microcomputers in Education: Cognitive and Social Design Principles," *SIGCUE Outlook* 17, no. 2 (Spring 1983): 3–11. See also Sharon Dugdale, "From Network to Microcomputers and Fractions to Functions: Continuity in Software Research and Design," in *Research on Technology in the Learning and Teaching of Mathematics: Syntheses and Perspectives*, ed. K. Heid and G. Blume, in press. Thanks to Sharon Dugdale for allowing me to see a draft of this chapter.

30. John G. Kemeny and Thomas E. Kurtz, *Back to BASIC: The History, Corruption, and Future of the Language* (Reading, Mass.: Addison-Wesley, 1985).

31. GPL was used by John C. Plaster in preparing a series of drill and practice programs known as the Milliken Math Sequence. See Gary M. Kaplan, "An Interview with John C. Plaster: Designer and Programmer of 'Tombstone City: 21st Century'" *99'er Magazine* 1, no. 5 (Feb. 1982): 34–35.

32. Mark Stevens, "This Robot Can Act but Can't Think," *Christian Science Monitor*, May 24, 1978, 4; "What's New?" *Byte*, 3, no. 8 (Aug. 1978): 200.

33. *Byte* 3, no. 10 (Oct. 1978): 138. By June 1979 the price was listed as $400 for the kit and $600 fully assembled. See *Byte* 4, no. 6 (June 1979): 235. The staff of Terrapin, Inc., was not the only group to consider building inexpensive turtles. Michael Folk, then of the Mathematics Department of Drake University, had used the MIT turtle to teach mathematics and computer science to elementary schoolchildren and built a much less expensive robotic tortoise, which was to be controlled by a Motorola microprocessor. Partway through this project, Folk and his colleagues decided that it might be possible to implement a more powerful version of LOGO on their system and set out to do so. See Michael Folk, "Building an Inexpensive Turtle," *ACM SIGCUE Bulletin* 12, no. 2 (Apr. 1978): 11–12. According to Folk, this work did not lead to a commercial product. For this information we are indebted to a personal communication dated Oct. 14, 2005.

34. James A. Guptan Jr., "Talk to a Turtle: Build a Computer Controlled Robot," *Byte* 4, no. 6 (June 1979): 74–76, 78–80, 82, 84.

35. *Byte* 6, no. 11 (Nov. 1981): 54.

36. Seymour Papert, *Mindstorms: Children, Computers, and Powerful Ideas* (New York: Basic Books, 1980); Robert Taylor, ed., *The Computer in the School: Tutor, Tool, Tutee* (New York: Teachers College Press, 1980), 161–212.

37. Richard Roth, "A Comparison of Logos: Today's Turtle Is No Slowpoke," *Creative Computing* 10, no. 12 (Dec. 1984): 94.

38. *Byte* 6, no. 6 (June 1981): 195.

39. *Byte* 6, no. 10 (Oct. 1981): 365.

40. Apple Logo was favorably reviewed in *Mathematics Teacher* 75, no. 10 (Dec., 1982): 778. Radio Shack's Color Logo was examined in *Mathematics Teacher* 76, no. 8 (Nov. 1983), 627–28. Several articles about LOGO appeared in *Arithmetic Teacher* over the course of the 1980s and 1990s.

41. Andrew Pollack, "Texas Instruments' Pullout," *New York Times,* Oct. 31, 1983, D1.

42. For biographical information about Rich we have relied on a personal communication from him of Nov. 15, 1999.

43. For an introduction to early computer algebra systems, see David R. Stoutemyer, "LISP Based Symbolic Math Systems," *Byte* 4, no. 8 (Aug. 1979): 176–92. Limited biographical information about him is available in his publications.

44. Albert D. Rich and David R. Stoutemyer, "Capabilities of the MUMATH-78 Computer Algebra System for the INTEL-8080 Microprocessor," *Lecture Notes in Computer Science,* vol. 72: *Proceedings of the International Symposium on Symbolic and Algebraic Computation* (London: Springer, 1979): 241. A brief historical account of Rich and Stoutemyer's work is on p. 248.

45. Albert D. Rich and David R. Stoutemyer, "An Exciting New Era in Mathematics muMath-78 [*sic*]: A Symbolic Math System," *Creative Computing* 5, no. 7 (July 1979): 82–86; and 5, no. 8 (Aug. 1979): 74–79. The second article recounts the early history of Rich and Stoutemyer's collaboration.

46. David R. Stoutemyer, "LISP Based Symbolic Math Systems," *Byte,* 4, no. 8 (Aug. 1979): 176–92.

47. *Byte* 4, no. 11 (Nov. 1979): 43.

48. *Byte* 5, no. 1 (Jan. 1980): 42.

49. David R. Stoutemyer, "LISP Based Symbolic Math Systems," *Byte* 4, no. 8 (Aug. 1979): 192.

50. *Byte* 5, no. 3 (Mar. 1980): 109.

51. *Byte* 5, no. 4 (Apr. 1980): 193. The advertisement also ran in *80-U.S.: The TRS-80 Users Journal* 4, no. 2 (Mar./Apr. 1981): 1. This issue of *80-U.S.* also included two introductory articles on muMATH: Cameron C. Brown, "MμMath [*sic*]," 62–63; and Terry Dettmann, "MuMath [*sic*] a Second Look," 64–65.

52. Greg Williams, "The muSIMP/muMATH-79 Symbolic Math System," *Byte* 5, no. 11 (Nov. 1980): 324–39.

53. Ibid., 338.

54. Ibid.

55. Richard Pavelle, Michael Rothstein, and John Fitch, "Computer Algebra,"

Scientific American 245, no. 6 (Dec. 1981): 146. For other public notice, see Lynn A. Steen, "Computer Calculus," *Science News* 119, Apr. 18, 1981, 250–51.

56. Herbert S. Wilf, "The Disk with the College Education," *American Mathematical Monthly* 89, no. 1 (Jan. 1982): 4–5.

57. Ibid., 8.

58. Beverly Mugrage, "muMath/muSimp-80 [*sic*]," *Mathematics Teacher* 76, no. 6 (Sept. 1983): 441. This version of the program, distributed by Microsoft, was for the Apple II microcomputer.

59. Sharon Burrowes, "muMath/muSimp-80 [*sic*]," *Mathematics Teacher* 76, no. 6 (Sept. 1983): 442.

60. David R. Stoutemyer, "Nonnumeric Computer Applications to Algebra, Trigonometry and Calculus," *The Two-Year College Mathematics Journal* 14, no. 3 (June 1983): 233–39.

CHAPTER EIGHTEEN Graphing Calculators and Software Systems

Epigraphs: M. Kathleen Heid, "The Technological Revolution and the Reform of School Mathematics," *American Journal of Education* 106 (Nov. 1997): 5; Alvin White, quoted in Barry Cipra, "Calculus: Crisis Looms in Mathematics' Future," *Science* 239, no. 4847, Mar. 25, 1988, 1492.

1. On the history of distance learning, see Marina Stock McIsaac and Charlotte Nirmalani Gunawardena, "Distance Education," in *Handbook of Research for Educational Communications and Technology: A Project of the Association for Educational Communications and Technology*, ed. David. H. Jonassen (New York: Macmillan, 1996), 403–37. For views from around 2000, see Robin Mason, *Globalising Education* (London: Routledge, 1998); and Paul S. Goodman, ed., *Technology Enhanced Learning: Opportunities for Change* (Mahwah, N.J.: Lawrence Erlbaum Associates, 2002).

2. Here, as previously, we focus on the implications of technologies for mathematical learning. For a more general approach, an overview of developing psychological theories, and extensive references, see David H. Jonassen, ed., *Handbook of Research for Educational Communications and Technology* (New York: Macmillan, 1996); and Ann Kovalchick and Kara Dawson, eds., *Education and Technology: An Encyclopedia* (Santa Barbara: ABC-CLIO, 2004). For an overview of the implications of electronic technologies on mathematics instruction, with extensive citations, see Heid, "Technological Revolution," 5–61.

3. Keith Devlin, "Computers and Mathematics," *Notices of the American Mathematical Society* 38, no. 3 (Mar. 1991): 190.

4. Keith Devlin, "Computers and Mathematics," *Notices of the American Mathematical Society* 41, no. 9 (Nov.–Dec. 1994): 1155.

5. Mike Beilby, "The Graphics Calculator," *Math & Stats* 1, no. 2 (May 1990): 1.

6. These dates are taken from the Web site of Casio at http://world.casio.com/corporate/history/chronology.html.

7. The Casio fx-8000G was noted as a new product in *Electronics* 59, no. 5 (Feb.

3, 1986): 61. According to this notice, the calculator sold for $69.95. The first reference we have found to the fx-8000G in newspaper databases is part of a general advertising supplement in the Oct. 2, 1988, *Washington Post*. There it is listed as being available for $59.97, marked down from $79.95. An advertisement in the Sept. 13, 1988, issue of the *Ubyssey*, a student paper of the University of British Columbia, gives a price of $149.95 (Canadian), or about $120 American (based on the exchange rate given in the *Wall Street Journal*, Sept. 7, 1988, 38). This advertisement also lists a Casio graphing calculator with slightly larger memory for $179.95 (issues of the *Ubyssey* are available online at the Web site of the archives of the University of British Columbia). A price of $75 is given in R. W. W. Taylor, "Affordable Technology: What Should We Do with What We've Got?" in *Proceedings of the Second Annual Conference on Technology in Collegiate Mathematics*, ed. Franklin Demana, Bert K. Waits, and John Harvey (Reading, Mass.: Addison-Wesley, 1989), 328. The fx-8000G does not seem to have been advertised in such journals as *Science*, *Mathematics Teacher*, *Notices of the American Mathematical Society*, and the *College Mathematics Journal*.

8. This account is based largely on Bert K. Waits and Franklin Demana, "The Calculator and Computer Precalculus Project (C^2PC): What Have We Learned in Ten Years?" in *Impact of Calculators on Mathematics Instruction*, ed. George Bright, H. C. Waxman, and S. E. Williams (Lanham, Md.: University Press of America, 1994), 91–110.

9. Paul Eckert, George Kitchen, Cameron Nichols, and Charles Von der Embse, "Graphing Calculators in the Secondary Mathematics Classroom," *Michigan Council of Teachers of Mathematics Monograph*, no. 21 (Sept. 1989).

10. For high school students there was Franklin Demana and Bert K. Waits, *Precalculus Mathematics: A Graphing Approach* (Reading, Mass.: Addison-Wesley, 1989). Other books in the series concerned college algebra and trigonometry.

11. Franklin Demana and Bert K. Waits, eds., *The Twilight of the Pencil and Paper: 1st Annual Conference on Technology in Collegiate Mathematics* (Reading, Mass.: Addison-Wesley, [1989]).

12. Robert H. White, "Calculus of Reality," in *Calculus for a New Century: A Pump, Not a Filter*, ed. Lynn Arthur Steen (Washington, D.C.: Mathematical Association of America, 1988), 9.

13. For an early evaluation of these programs, see Alan C. Tucker and James R. C. Leitzel, *Assessing Calculus Reform Efforts: A Report to the Community* (Washington, D.C.: Mathematical Association of America, 1994).

14. William C. Wickes, *Synthetic Programming on the HP-41C* (College Park, Md.: Larkin Publications, 1980). For information about Wickes, see Jay Coleman, "It All Adds Up," *Measure* (July–Aug. 1991): 18–20.

15. Barry Cipra, "Recent Innovations in Calculus Instruction," in Steen, *Calculus for a New Century*, 97–98.

16. Thomas Tucker, "Calculators with a College Education?" *FOCUS: The Newsletter of the Mathematical Association of America* 7, no. 1 (Jan. 1987): 1, 5. This article was republished in Steen, *Calculus for a New Century*, 229–31.

17. Yves Nievergelt, "The Chip with the College Education: The HP-28C," *American Mathematical Monthly* 94, no. 9 (Nov. 1987): 895–902.

18. "Hewlett [*sic*] Unveils Calculator for Conceptual Algebra," *Wall Street Journal*, Jan. 12, 1987, 8.

19. See the advertisement "Hewlett-Packard Re-invents the Calculator," *Science*, n.s. 235, no. 4787, Jan. 23, 1987, 417–18. This ad also appeared in *American Mathematical Monthly* 94, no. 6 (June 1987): n.p.; and in *Notices of the American Mathematical Society* 34, no. 3 (Apr. 1987): n.p.

20. On the introduction of the HP-28S, see Ivars Peterson, "Calculus in the Palm of Your Hand," *Science News* 133, Jan. 23, 1988, 62; *Notices of the American Mathematical Society* 34, no. 6 (Oct. 1987): 959; and *Notices* 34, no. 7 (Nov. 1987): 1203.

21. "Products and Materials," *Science* 235, no. 4793, Mar. 6, 1987, 1267; Barry Cipra, "Recent Innovations in Calculus Instruction," in Steen, *Calculus for a New Century*, 97–98.

22. National Council of Teachers of Mathematics (NCTM), *Curriculum and Evaluation Standards for School Mathematics* (Reston, Va.: NCTM, 1989), 124.

23. "Introducing the TI-81 Graphics Calculator: An Educated Solution Tailored to Educational Needs," *Mathematics Teacher* 83, no. 4 (Apr. 1990): 325.

24. For an overview of the development of the NCTM standards, see Douglas B. McLeod, "From Consensus to Controversy: The Story of the NCTM *Standards*," in *A History of School Mathematics*, ed. George M. A. Stanic and Jeremy Kilpatrick (Reston, Va.: NCTM, 2003), 1:753–818 and additional articles in the same volume.

25. Staff of Texas Instruments Instructional Communications, *TI-81 Guidebook* (1990; rpt., n.p.: Texas Instruments, 1992). Waits and Demana are acknowledged on the title page. They went on to spearhead a program that came to be known as T^3 (Teachers Teaching with Technology), which provided instruction in graphing calculator use for mathematics teachers.

26. On the development of electronic calculators for mathematics education, particularly products of Texas Instruments, see Brendon Kelly, "The Emergence of Technology in Mathematics Education," in *A History of School Mathematics*, ed. George M. A. Stanic and Jeremy Kilpatrick (Reston, Va.: NCTM, 2003), 1:1037–81, esp. 1071–73. See also Kathy B. Hamrick, "The History of the Hand-Held Electronic Calculator," *American Mathematical Monthly* 103, no. 8 (Oct. 1996): 633–39.

27. Chancy O. Jones, "The Use of Calculators on College Board Mathematics Examinations in the 1990s," in *Proceedings of the Third International Conference on Technology and Collegiate Mathematics*, ed. John Harvey, Franklin Demana, and Bert K. Waits (Reading, Mass.: Addison-Wesley, 1992), 157–61. The paper was delivered in Nov. 1990. The TI-83 calculator was released in 1996, between the time that the first AP Statistics examination was written and the time it was given. Reportedly, the new capabilities of this calculator caused considerable headaches for those giving the examination. See Rosemary Roberts, Richards Schaeffer, and Ann Watkins, "Advanced Placement Statistics—Past, Present, and Future," *American Statistician* 53, no. 4 (Nov. 1999): 307–20, esp. 316–17.

28. "Calculators in Standardized Testing in Mathematics," in Steen, *Calculus for a New Century,* 219–21.

29. Jones, "Use of Calculators on College Board Mathematics Examinations," 157–61. Brent Bridgeman, Anne Harvey, and James Braswell, "Effects of Calculator Use on Scores on a Test of Mathematical Reasoning," *Journal of Educational Measurement* 32, no. 4 (Winter 1995): 323–40; Jennifer S. Lee, "Calculators Throw Teachers a New Curve," *New York Times,* September 2, 1999, G1.

30. See, e.g., Deirdre Fernandes, "Student's Persistence Plus Calculator Flaw Equals a Big Recall," *Hampton Roads Pilot,* June 7, 2005; available online at http://home.hamptonroads.com/stories/story.cfm?story=87425&ran=225030. An article based on Fernandes's story ran on the Associated Press wire and was published, e.g., in the *Annapolis Capital* on June 8, 2005.

31. John C. Nash and Chris Olsen, "The Texas Instruments TI-83 Graphing Calculator," *American Statistician* 52, no. 3 (Aug. 1998): 285.

32. T. A. Ryan Jr. and Brian L. Joiner, "Minitab: A Statistical Computing System for Students and Researchers," *American Statistician* 27, no. 5 (Dec. 1973): 225. On the history of Minitab, see R. A. Ryan, B. L. Joiner, and B. F. Ryan, "Minitab," in *Encyclopedia of Statistical Science,* ed. Samuel Kotz and Norman L. Johnson (New York: Wiley, 1982), 5:539–42.

33. Eric L. Grinberg, "The Menu with the College Education," *Notices of the American Mathematical Society* 36, no. 7 (Sept. 1989): 842. For Herman's comments, see Eugene Herman, "Derive [*sic*], A Mathematical Assistant," *American Mathematical Monthly* 96, no. 10 (Dec. 1989): 948–58.

34. David Harper, Chris Wooff, and David Hodgkinson, *Computer Algebra Support Project: A Guide to Computer Algebra Systems* (Liverpool: University, 1990), 26; available online at www.inf.ed.ac.uk/teaching/courses/ca/caguide.pdf.

35. Barry Cipra, "How Number Theory Got the Best of the Pentium Chip," *Science* 267, no. 5195, Jan. 13 1995, 175. See also the advertisement for DERIVE version 3 published in *Mathematics Magazine* 68, no. 2 (Apr. 1995): n.p.

36. For an overall view of the history of MAPLE and two of its competitors, see Norman Chonacky and David Winch, "Maple, Mathematica, and Matlab: The 3M's without the Tape," *Computing in Science and Engineering* 7, no. 1 (Jan.–Feb. 2005): 8–16 as well as electronic references given there. See also Keith O. Geddes, Gastón H. Gonnet, and Bruce W. Char, *MAPLE User's Manual* ([Waterloo, Ont.]: University of Waterloo, Computer Science Dept., 1982); Bruce W. Char and Keith O. Geddes, *Maple User's Guide: First Leaves: A Tutorial Introduction to Maple and Maple Reference Manual,* 4th ed. (Waterloo, Ont.: WATCOM Publications, ca. 1985); and Bruce W. Char et al., *First Leaves: A Tutorial Introduction to Maple* (Waterloo, Ont.: WATCOM Publications, 1988). For advertisements, see *American Mathematical Monthly* 96, no. 2 (Feb. 1989): U4, U13; and 99, no. 2 (Feb. 1992): U1. On MAPLE V, see the review in "Software Reviews," *College Mathematics Journal* 25, no. 1 (Jan. 1994): 56–63.

37. Gina Kolata, "Caltech Torn by Dispute over Software," *Science,* n.s. 220, no.

4600, May 27, 1983, 932–34. See also a related letter to the editor by John D. Roberts of Caltech, published in *Science,* n.s. 221, no. 4606, July 8, 1983, 110, 112.

38. John M. Hosack, "A Guide to Computer Algebra Systems," *College Mathematics Journal* 17, no. 5 (Nov. 1986): 434–41.

39. Stephen Wolfram, *Mathematica: A System for Doing Mathematics by Computer* (Reading, Mass.: Addison-Wesley, 1988).

40. "New Products," *Physics Today* 41, no. 8 (Aug. 1988): 91–92; "Telegraphic Reviews," *American Mathematical Monthly* 95, no. 8 (Oct. 1988): 791–92; Eugene A. Herman, "Mathematica—A Review," *Notices of the American Mathematical Society* 35, no. 9 (Nov. 1988): 1333–44; Alan Hoenig, "Mathematics by Machine with Mathematica," *College Mathematics Journal* 21, no. 2 (Mar. 1990): 146–49. G. A. Taubes, "Physics Whiz Goes into Biz," *Fortune* 117, Apr. 11, 1988, 90–93; Andrew Pollack, "A Top Scientist's Latest: Math Software," *New York Times,* June 24, 1988, D1; Peter H. Lewis, "Liberating the 'Prose' of Math from Its Grammar," *New York Times,* July 19, 1988, C9. Although the first version of Mathematica was bundled with all NeXT computers, the second edition went only with NeXT machines purchased for higher education. Commercial users paid $1,495. See Bruce Berkoff, "Math for the Masses," *Nextworld* 1, no. 4 (Winter 1991): 49–50.

41. John A. Wass, "A Mathematical Swiss Army Knife," *Science* 286, no. 5548, Dec. 17, 1999, 2291–92.

42. For further comments on the history of Mathematica, see Chonacky and Winch, "Maple, Mathematica, and Matlab," 8–16.

43. On software for plane geometry, see "Software Reviews," *College Mathematics Journal* 24, no. 4 (Sept. 1993): 370–76.

44. For general discussions of the use of computers in the schools, see Todd Oppenheimer, *The Flickering Mind: Saving Education from the False Promise of Technology* (New York: Random House, 2003); and William D. Pflaum, *The Technology Fix: The Promise and Reality of Computers in Our Schools* (Alexandria, Va.: Association for Supervision and Curriculum Development, 2004).

Index

Page numbers followed by an *f* indicate figures.